70

Fortschritte der Chemie organischer Naturstoffe

Progress in the Chemistry of Organic Natural Products

Founded by
L. Zechmeister

Edited by
W. Herz, G. W. Kirby,
R. E. Moore, W. Steglich,
and Ch. Tamm

Authors:
A. Cavé, D. Cortes,
B. Figadère, A. Laurens,
G. R. Pettit

Springer-Verlag Wien GmbH

Prof. W. Herz, Department of Chemistry,
The Florida State University, Tallahassee, Florida, U.S.A.

Prof. G. W. Kirby, Chemistry Department,
The University of Glasgow, Glasgow, Scotland

Prof. R. E. Moore, Department of Chemistry,
University of Hawaii at Manoa, Honolulu, Hawaii, U.S.A.

Prof. Dr. W. Steglich, Institut für Organische Chemie der Universität
München, München, Federal Republic of Germany

Prof. Dr. Ch. Tamm, Institut für Organische Chemie der Universität Basel,
Basel, Switzerland

Originally published by Springer-Verlag/Wien in 1997
Softcover reprint of the hardcover 1st edition 1997
Library of Congress Catalog Card Number AC 39-1015

Typesetting: Thomson Press (India) Ltd., New Delhi

Graphic design: Ecke Bonk
Printed on acid-free and chlorine-free bleached paper

With 86 partly coloured Figures

ISSN 0071-7886
ISBN 978-3-7091-7349-7 ISBN 978-3-7091-6551-5 (eBook)
DOI 10.1007/978-3-7091-6551-5

Contents

List of Contributors VII

The Dolastatins.
By G. R. PETTIT 1

1. Introduction 2

2. *Dolabella auricularia* 3

3. Isolation and Structure Determination of the Dolastatins 4

4. Synthesis of the Dolastatins 12
 4.1 Dolastatin 3 12
 4.2 Dolastatin 10 14
  4.2.1 Dolaphenine (Doe) 14
  4.2.2 Dolaproine (Dap) 18
  4.2.3 Dolaisoleuine (Dil) 24
  4.2.4 Dolaproinyl-Dolaphenine (Dap-Doe) 32
  4.2.5 Conversion of Dap-Doe to Dolastatin 10 33
  4.2.6 Chromatography of Dolastatin 10 36
  4.2.7 High-Field NMR Analysis of Dolastatin 10 39
 4.3 Syntheses of (6*R*)-Isodolastatin 10 and Other Chiral modifications of Dolastatin 10 39
  4.3.1 Crystal Structure of (6*R*)-Isodolastatin 10 40
  4.3.2 Molecular Modeling of (6*R*)-Isodolastatin 10 41
 4.4 Structural Modifications of Dolastatin 10 43
 4.5 Synthesis of Dolastatin 15 45
 4.6 Synthesis of Dolastatin C 47

5. Spectral and Analytical Characterization 48
 5.1 Dolastatin 10 48
 5.2 Dolastatins 11 and 12 51
 5.3 Dolastatin 13 52
 5.4 Dolastatin 14 55
 5.5 Dolastatin 15 55

6. Cytostatic and Antineoplastic Activities 57
 6.1 Dolastatins 10–15 57
 6.2 Dolastatin 10 Structural Modifications 63
  6.2.1 Chiral Isomers 63
  6.2.2 Structural Substitutions 63

7. Biochemical Mechanisms of Action 68
7.1 Inhibition of Tubulin Assembly and Mitosis 68
7.2 Effects on Actin 69

8. Pharmacology and Toxicology 69

9. Conclusion 70

Acknowledgements 70

References 70

Acetogenins from Annonaceae.
By A. Cavé, B. Figadère, A. Laurens, and D. Cortes 81

I. Introduction 81

II. Classification 82
Synonyms of Acetogenins 84
Listing (Structures and Physical Data) 89
Distribution 180

III. Extraction and Isolation 188

IV. Structure Elucidation 189

V. Biogenetic Hypotheses 226

VI. Syntheses 231

VII. Biological Activities 264

Acknowledgements 273

References 273

Author Index 289

Subject Index 297

List of Contributors

CAVÉ, Prof. A., Laboratoire de Pharmacognosie, Faculté de Pharmacie, F-92296 Châtenay-Malabry, France.

CORTES, Prof. D., Departemento de Farmacologia, Farmacognosia y Farmacodinamica, Faculdad de Farmacia, Avenida Andres Estelles, E-46100 Burjasot (Valencia), Spain.

FIGADÈRE, Dr. B., Laboratoire de Pharmacognosie, Faculté de Pharmacie, F-92296 Châtenay-Malabry, France.

LAURENS, Dr. A., Laboratoire de Pharmacognosie, Faculté de Pharmacie, F-92296 Châtenay-Malabry, France.

PETTIT, Prof. G. R., Cancer Research Institute, Arizona State University, Tempe, AZ 85287-1604, U.S.A.

The Dolastatins

G. R. PETTIT, Cancer Research Institute and
Department of Chemistry, Arizona State University,
Tempe, Arizona, USA

Contents

1. Introduction 2

2. *Dolabella auricularia* 3

3. Isolation and Structure Determination of the Dolastatins 4

4. Synthesis of the Dolastatins 12
   4.1 Dolastatin 3 12
   4.2 Dolastatin 10 14
      4.2.1 Dolaphenine (Doe) 14
      4.2.2 Dolaproine (Dap) 18
      4.2.3 Dolaisoleuine (Dil) 24
      4.2.4 Dolaproinyl-Dolaphenine (Dap-Doe) 32
      4.2.5 Conversion of Dap-Doe to Dolastatin 10 33
      4.2.6 Chromatography of Dolastatin 10 36
      4.2.7 High-Field NMR Analysis of Dolastatin 10 39
   4.3 Syntheses of (6*R*)-Isodolastatin 10 and Other Chiral Modifications of Dolastatin 10 39
      4.3.1 Crystal Structure of (6*R*)-Isodolastatin 10 40
      4.3.2 Molecular Modeling of (6*R*)-Isodolastatin 10 41
   4.4 Structural Modifications of Dolastatin 10 43
   4.5 Synthesis of Dolastatin 15 45
   4.6 Synthesis of Dolastatin C 47

5. Spectral and Analytical Characterizations 48
   5.1 Dolastatin 10 48
   5.2 Dolastatins 11 and 12 51
   5.3 Dolastatin 13 52
   5.4 Dolastatin 14 55
   5.5 Dolastatin 15 55

6. Cytostatic and Antineoplastic Activities 57
   6.1 Dolastatins 10–15 57

6.2 Dolastatin 10 Structural Modifications 63
6.2.1 Chiral Isomers 63
6.2.2 Structural Substitutions 63

7. Biochemical Mechanisms of Action 68
7.1 Inhibition of Tubulin Assembly and Mitosis 68
7.2 Effects on Actin 69

8. Pharmacology and Toxicology 69

9. Conclusion 70

Acknowledgements 70

References 70

1. Introduction

Certain marine animals were known to the ancients for their potent biological constituents and presumed use in primitive medicine. The early periods of recorded history contain references to support these assumptions (*1*). Illustrative are hieroglyphics on the Egyptian Pharaoh Ti's tomb (approximately 2700 BC) that describe the poisonous puffer fish *Tetraodon stellatus*. One of the earliest recorded uses of a marine organism in primitive medical practice has been attributed to the Roman, Gaius Plinius Secundus (AD 29–79, Pliny the Elder), who recommended that the sting system of the stingray be ground up and used for treatment of toothache and in obstetrics. One of the first modern pharmacological and chemical studies of potent marine animal constituents involved tetrodotoxin from the poisonous puffer fish (*2, 3*). About 40 years ago some especially important observations began to be recorded. Illustrative was the fact that extracts from certain sponges and coelenterates were shown to have antibiotic properties (*4–8*) and that marine invertebrates produce various other potentially medically useful components (*9–11*). Very importantly, from the viewpoint of anticancer drug discovery, were reports that starfish meal (*12–14*) and fractions from the peanut worm *Bonellia fulginosa* (*15, 16*), certain sea cucumbers, and molluscs [clams (*16*) and oysters (*9, 11*)] exhibited antitumor activity against sarcoma-180 and Krebs-2 ascites tumor (*10, 17–19*).

In 1965–66, we began the first systematic study of marine invertebrates, vertebrates, and plants as a vast untapped resource for discovery of promising new anticancer drugs with the presumed unprecedented structures necessary to improve human cancer treatment. During the next four years, we evaluated components from many such marine organisms

References, pp. 70–79

from a broad geographic area that included the Atlantic and Pacific coasts of North and South America and the coasts of Asia. Antineoplastic activity was assessed by use of the Walker 256 carcinoma (intramuscular) and both a lymphoid (L1210) and lymphocytic leukemia (PS) as developed at the U.S. National Cancer Institute. By 1968 our original expectations concerning certain marine animals as potential sources of new anticancer drugs were amply confirmed and reported in 1970 (*20*). In the same period, we began the first such investigation of terrestrial arthropods for antineoplastic constituents, and initial results were reported in 1968 (*21*). Subsequently, we have isolated and characterized a substantial number of new cytotoxic, cytostatic and/or antineoplastic marine animal constituents (*22, 23*). We continue to devote considerable efforts to isolating such very active, albeit trace, constituents of certain exceptionally promising marine animal extracts. Two of the earliest leads we uncovered (1968 and 1972), which provided unprecedented structural types and led to clinical candidates that meet the rigorous criteria of the U.S. National Cancer Institute, were the bryostatin (*22*) and dolastatin series (*23*). Discovery of the bryostatins in the phylum Ectoprocta (Bryozoa) species *Bugula neritina* has been reviewed in a preceding chapter (*22*). The present review will be focused on discovery of the dolastatins.

2. *Dolabella auricularia*

The phylum Mollusca contains a great variety of terrestrial and marine organisms that have evolved an even more astonishing spectrum of reproductive and defensive strategies. Among the marine Mollusca, the Aplysiomorpha, Nudibranchia, and Sacoglossa constitute the three largest orders of opisthobranch sea slugs [shell-less molluscs (*24*)]. The aplysiomorphs feed on marine algae. The herbivorous Sacoglossa even have the ability to cultivate in their digestive glands ingested chloroplasts and then survive for weeks at a time on solar radiation and dissolved atmospheric gases. In contrast, the nudibranchs are carnivorous and free swimming. The dorid nudibranchs are even capable of consuming bryozoans, ascidians, acorn barnacles, sponges, and tunicates. Such varied dietary selections in turn serve as useful sources of potent compounds for devising powerful chemical defenses for these soft-bodied and slow-moving animals. In the aplysiomorpha class, Gastropoda species in the family Aplysiidae are commonly known as sea hares (*24*). The Romans first designated Mollusca of the family Aplysiidae in this fashion owing to a similarity between the ears of a hare and the auriculate tentacles of these gastropods (*25*).

The great Roman natural scientist Pliny the Elder in his comprehensive study (*26*) of about 60 A.D. first described a most potent Indian Ocean sea hare of the genus *Dolabella*. Extracts from this animal and two related *Aplysia* species from the Mediterranean were well known for their toxic properties during the reign of Nero. Such toxic mixtures are believed to have been used by Agrippina, mother of Nero (A.D. 37–68), to dispatch relatives in the way of his ascent to being Roman emperor. Indeed, Locusta, servant to Agrippina Minor, is believed to have murdered Caesar Augustus and Claudius Britannicus, among others, with potions from a *Dolabella* sp. believed to be *auricularia* (*27*). That species of sea hare was probably the one first described by Pliny, and the minor variations recorded in subsequent literature as, *e.g.*, *D. andersoni*, *D. californica*, *D. ecaudata*, and *D. scapula* are actually one species, namely, *D. auricularia* (*28*). By 150 A.D. Nicander (*25*) recognized the possibility of using such extracts for treatment of certain diseases. In 1568 the French scholar Grevin described in vivid details the potency of extracts prepared from a sea hare presumed to be *Dolabella auricularia* (*29*). In addition to the zoological studies already noted (*24*, *28*) and one published in 1974 (*31*), early scientific studies of *Dolabella auricularia* were focused on various biological properties ranging from cardiac behavior (*32–36*) to calcium transport (*37*), wall muscle physiology (*38*), and hemocyanin content (*39*). Evaluations of toxic fractions were first conducted in the 1969–73 period (*40–42*). However, the potential of the Indian Ocean *Dolabella* with respect to modern medical problems was not recognized until we uncovered evidence in 1972 for extremely active anticancer constituents in the Indian Ocean *Dolabella auricularia* (*30*). Presumably the dolastatins are important representatives of the potent *D. auricularia* components recognized from ancient (*26*, *27*) to modern (*40–42*) times.

3. Isolation and Structure Determination of the Dolastatins

By October, 1972 our broad geographic exploratory survey of marine organisms for antineoplastic constituents had been extended to the Western Indian Ocean and concentrated in the region from Mauritius to South Africa. With the capable assistance of my marine zoologist colleague, Claude Michel, we were able to evaluate Mauritius specimens of the olive green (and pear-shaped) *Dolabella auricularia*. Against the U. S. National Cancer Institute's (NCI) P388 lymphocytic leukemia (PS system), ethanol extracts of *D. auricularia* gave 67 to 135% life extension at doses of 176 to 600 mg/kg. In short, it was a very high priority lead and was pursued on that basis.

References, pp. 70–79

Recollections of *D. auricularia* from Mauritius were either extracted with denatured ethanol on site or shipped as the whole animal preserved in ethanol. Of the ethanol extracts, the first (2 kg) was received in October, 1975 and subjected to isolation studies. After a series of solvent partition separations (9:1 → 4:1 → 3:2 methanol-water with ligroin → carbon tetrachloride → chloroform), the PS antineoplastic activity was found to reside in the carbon tetrachloride and chloroform extracts. Many attempts at isolation of the active constituent(s) failed and additional recollections were required. A fall 1975 recollection of *D. auricularia* in ethanol was extracted to yield a greenish-black oil (1.6 kg). Again, only small quantities of unproductive and very complex fractions were obtained. In November, 1976, an ethanol extract (1.6 kg) prepared in Mauritius was received and used to provide PS-active carbon tetrachloride (3.3 g) and chloroform fractions (3.3 g). Separation of the carbon tetrachloride fraction on silica gel columns by use of hexane → ethyl acetate → methanol solvent systems led to dolatriol (**1a**, 5 mg) and dolatriol 6-acetate (**1b**, 15 mg) (*30*). Loliolide (**2**, 10 mg) (*43*) and the dolatriols were again isolated from the chloroform fraction.

The next supply (12 kg wet weight in ethanol) of *D. auricularia* was collected in the fall of 1977 and gave 585 g of ethanol extract. Solvent partitioning again gave PS-active carbon tetrachloride (7.9 g) and chloroform (16 g) fractions. Active fractions from the earlier recollections were combined and added to the comparable fraction prepared from this new extract. Further separation was performed by use of a series of gel permeation and partition chromatographic procedures employing Sephadex LH-20. After a final separation on an LH-20 partition column with 5:5:1 hexane-chloroform-methanol as mobile phase, successive column chromatographic steps that employed silica gel with 85:15:1.6 chloroform-methanol-water, 95:5 ethyl acetate-methanol, 90:10:0.8 chloroform-methanol-water, and finally 97:3 chloroform-ethanol, provided dolastatin 1 (3.76 mg) and dolastatin 2 (6.02 mg) (*44*). Both exhibited very potent inhibition of the PS leukemia. Although the P388 results for dolastatins 1 and 2 proved to be very promising, consumption of material by the biological screening evaluations, combined with unexpected decomposition at ambient temperatures, resulted in insufficient product to complete the structure determinations.

Larger scale recollections under way in early 1979 provided a 100-kg (wet weight) amount of the sea hare in ethanol. In this case the concentrated aqueous residue was partitioned with methylene chloride in place of chloroform (*45*), and the solvent partition scheme was modified by use of 9:1 → 1:1 methanol-water with ligroin and methylene chloride. The combined active methylene chloride fraction (129 g) was separated in portions on Sephadex LH-20 in methanol, followed by separation of the PS-active fractions in 4:1 methanol-methylene chloride. The active fractions were further separated by methods analogous to those utilized for isolation of dolastatins 1 and 2. Dolastatins 3–9 were separated in submilligram amounts with the exception of dolastatin 3 (**3**, 3.5 mg, PS ED_{50} 0.14 µg/mL) (*46, 47*). After preliminary biological and chemical studies the sample required repurification. As a result, about 1 mg remained (*47*). Such trace quantities of dolastatins 4–9 were not sufficient,

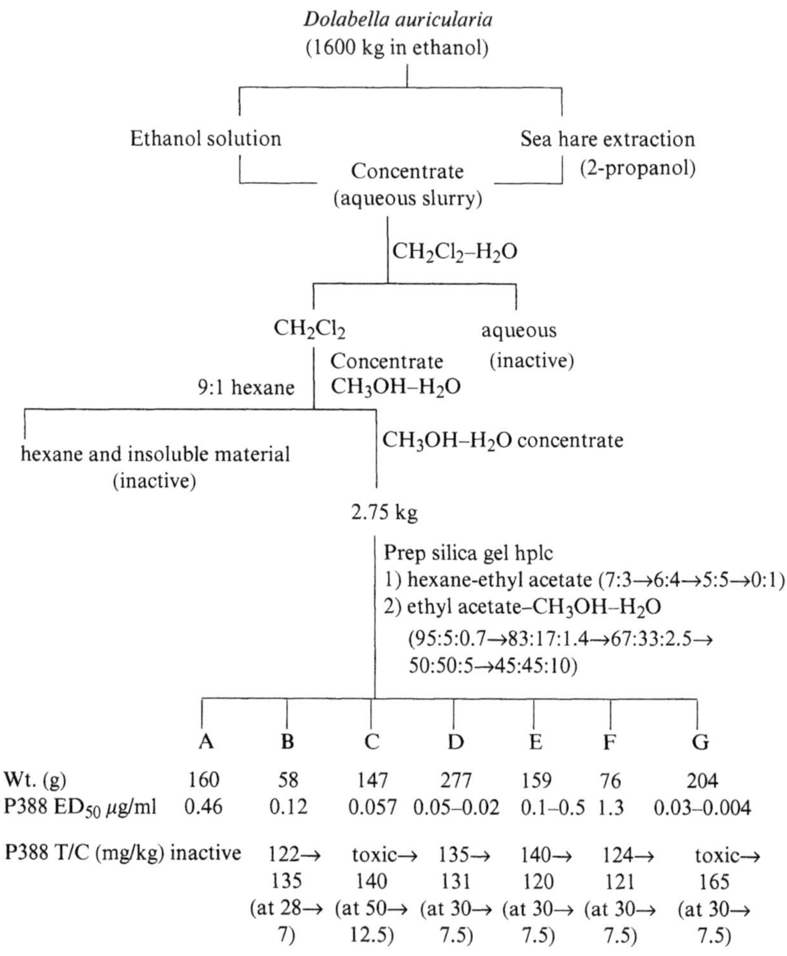

Scheme 1 Part 1

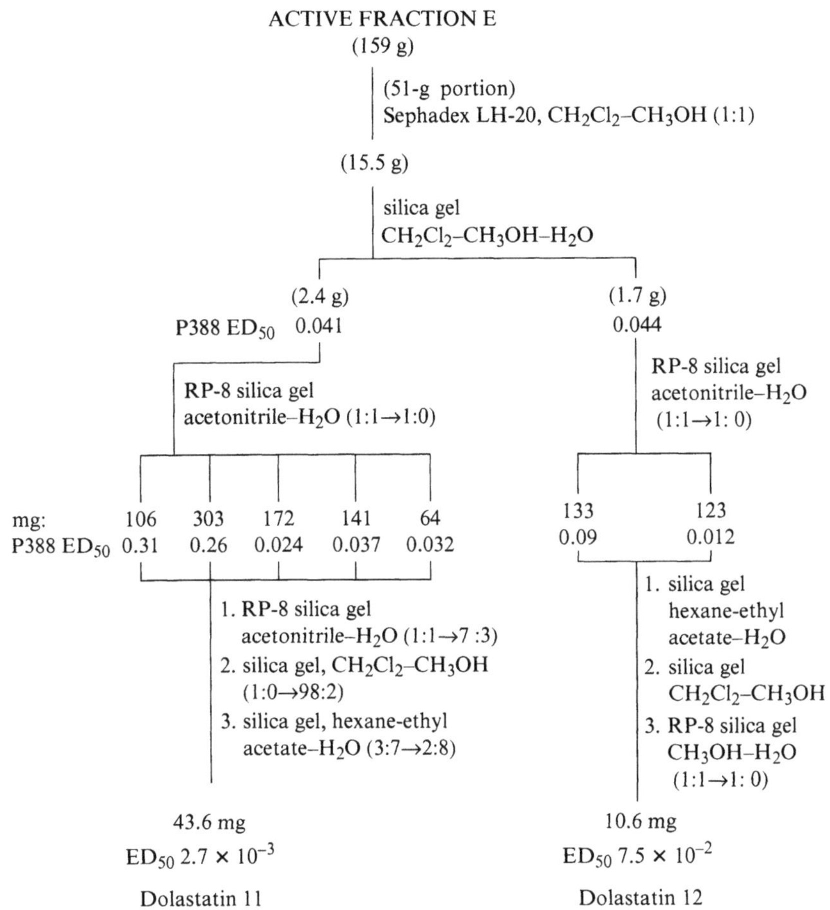

Scheme 1 Part 2

with the NMR facilities available at that period, for structure elucidation. By this time it was clear that the principal antineoplastic constituent(s) had still eluded us and a large-scale recollection (1600 kg wet weight in ethanol for a total volume of 2800 liters) completed in 1982 was required to discover the most promising constituents, namely dolastatins 10–15 (*48, 49*). Because of the complexity of the bioassay-guided (PS *in vivo* cell line) isolation of dolastatins 10–15 (*48*), a summary has been outlined in Scheme 1. From the original 1600 kg of *D. auricularia* (1982 recollection)

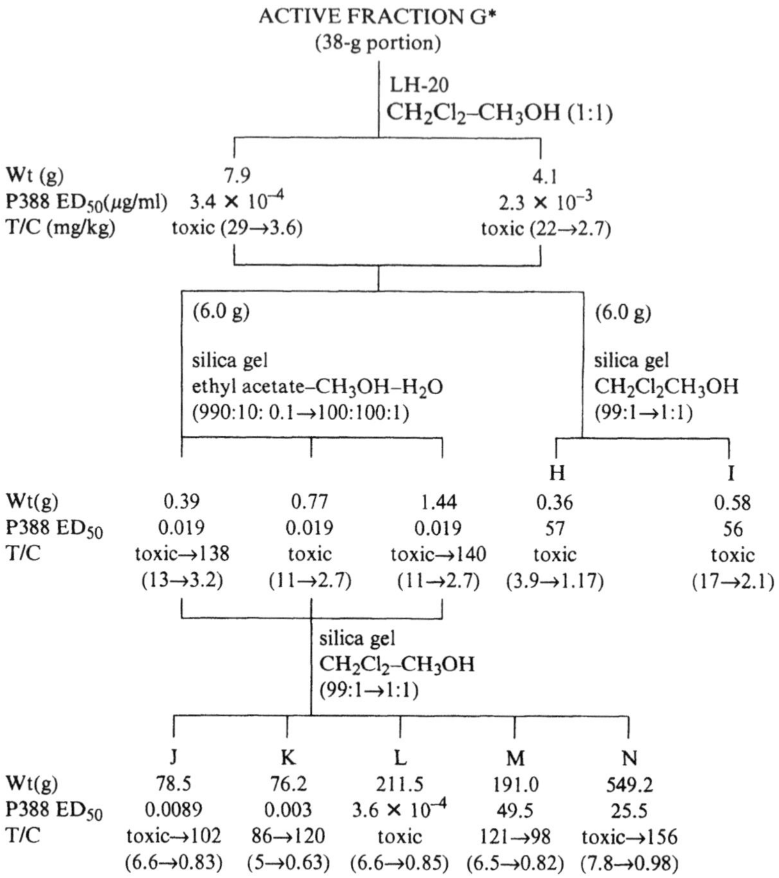

*Dolastatin 15 was isolated from extensive separation of the major portion of this fraction

Scheme 1 Part 3

the following total yields of dolastatins 10–15 respectively were obtained: 28.7 mg (1.8×10^{-6}%), 43.6 (2.7×10^{-6}%), 10.6 mg (6.6×10^{-7}%), 25.2 mg (1.6×10^{-6}%), 12.0 mg (7.5×10^{-7}%) and 6.2 mg (3.9×10^{-7}%). Dolastatins 10–15 were chromatographically pure (by TLC and HPLC) and the purity was further confirmed by high-field (400-MHz) NMR and high-resolution mass spectral studies.

References, pp. 70–79

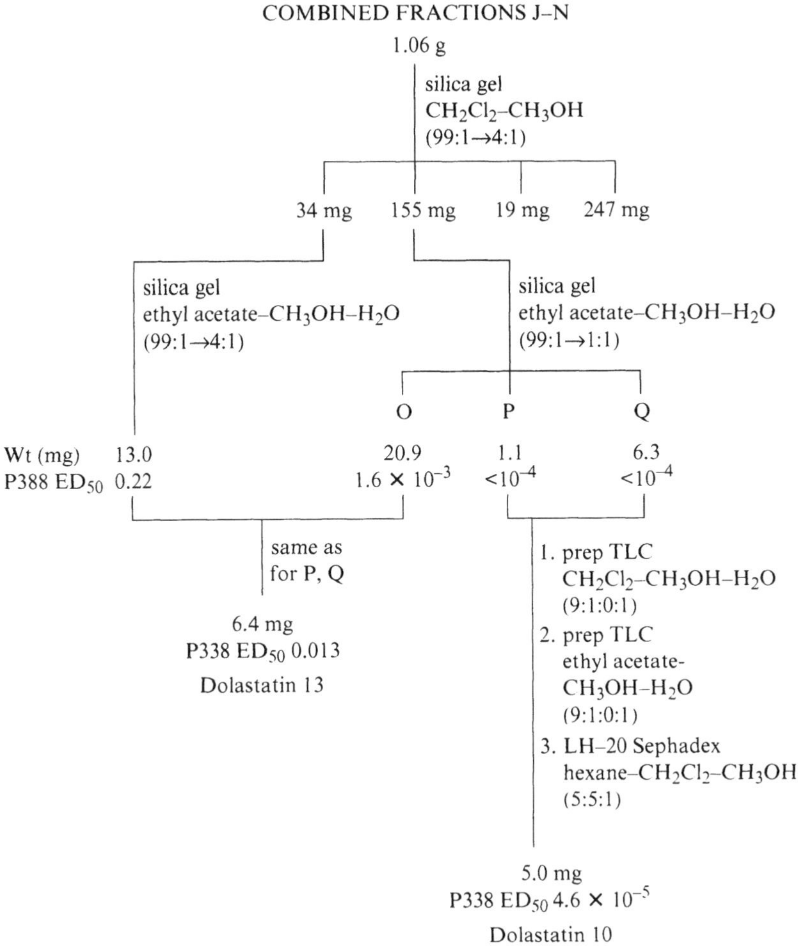

Scheme 1 Part 4

Each of the dolastatin structure determinations presented a new series of challenges that were compounded by the very small amounts (*e.g.*, of dolastatins 3 and 10 about 1 mg each was isolated at the time of structural analysis) available following initial biological studies. By employing primarily high-field (400-MHz) 2D-NMR techniques, combined with high-resolution mass spectral analyses for molecular formulae determina-

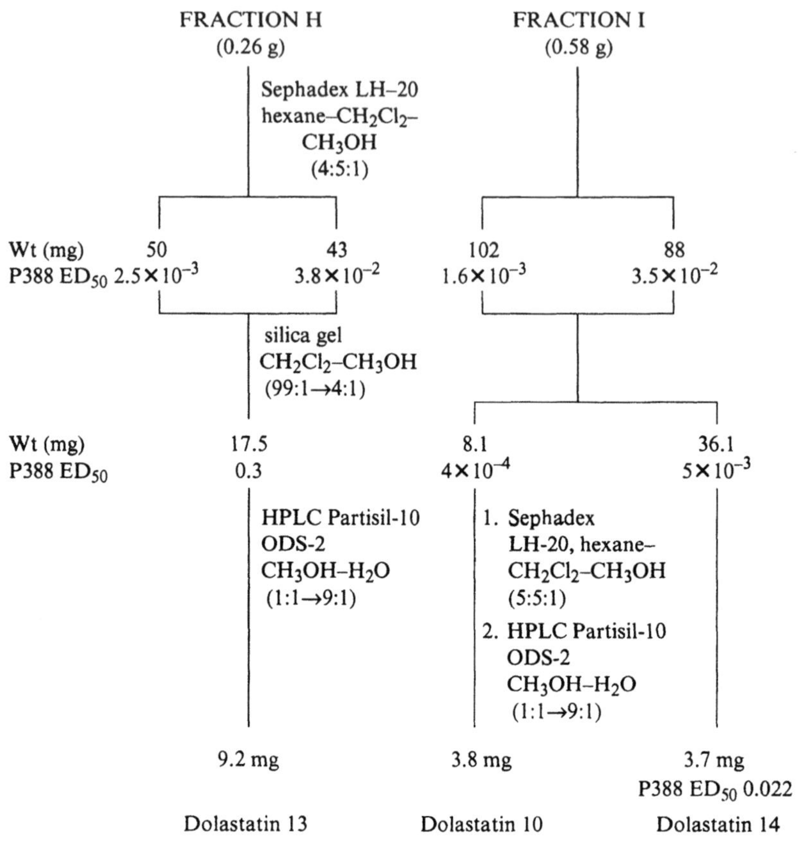

Scheme 1 Part 5

tions, the overall structures of dolastatins 3 (**3**) and 10–15 (**4–9**) were elucidated (*47, 49–53*). We were unable to grow crystals suitable for X-ray crystal structure determination and thereby unequivocally obtain relative and absolute configurations. However, total synthetic methods were successively employed to reach these chirality objectives for dolastatins 3 (**3**), 10 (**4**), and 15 (**9**). The total synthesis of dolastatin 3 also allowed a minor correction (*47*) in the original structure assignment (*46*) that arose from a malfunction in the mass spectrometer used for sequencing this cyclic peptide. Detailed discussions of the dolastatin syntheses and NMR data have been reserved for Sections 4 and 5 of this review.

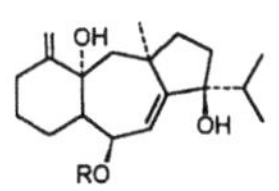

1a, R = H, *Dolatriol*
1b, R = COCH$_3$

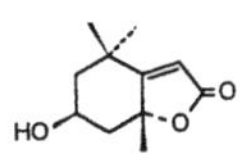

2, *Loliolide*

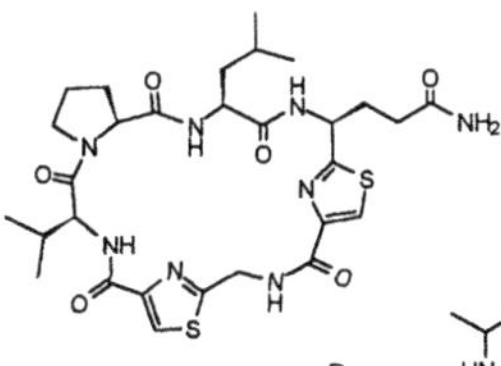

3, Dolastatin 3

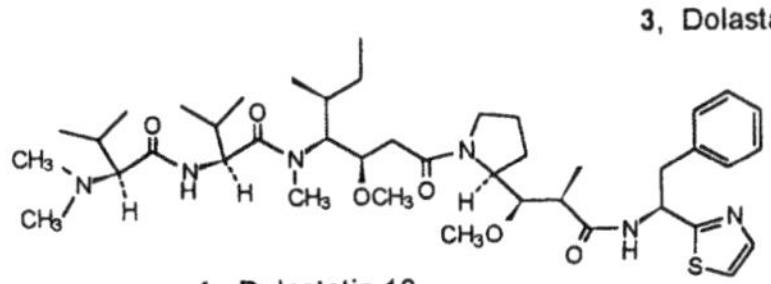

4, Dolastatin 10

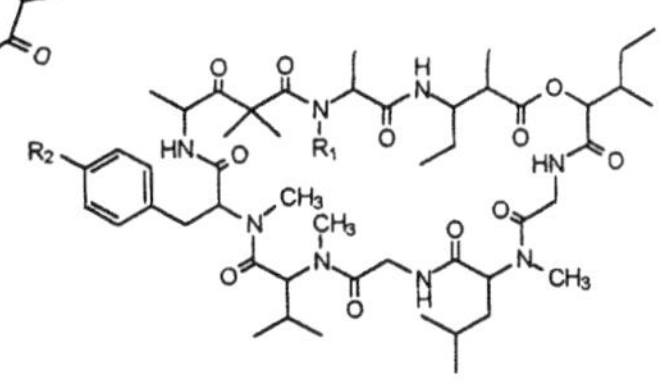

5, R$_1$ = H, R$_2$ = OCH$_3$, Dolastatin 11
6, R$_1$ = CH$_3$, R$_2$ = H, Dolastatin 12

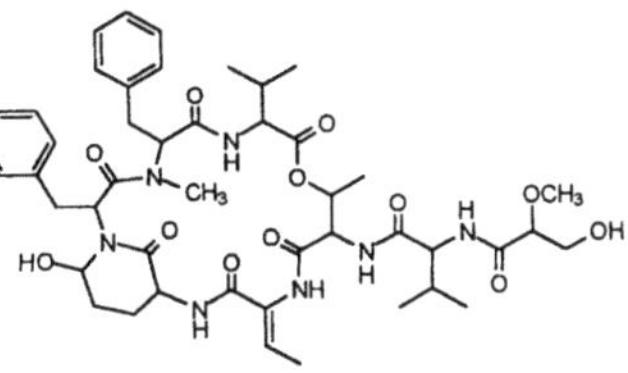

7, Dolastatin 13

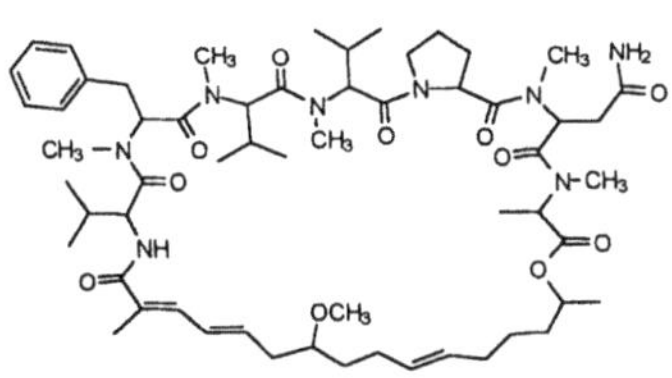

8, Dolastatin 14

Recently, YAMADA and colleagues have investigated the cell growth inhibitory constituents of *D. auricularia* collected on the coast of the Shima Peninsula, Japan. So far those studies have led to isolation and structural determination of dolastatins C (**10**) (*54*) and D (**11**) (*55*) through H (*56*) as well as doliculols A and B (**12**) (*57*) and doliculide (structurally similar to jasplakinolide) (*58*). The structures of dolastatins C and D were confirmed by total syntheses. Interestingly, YAMAZAKI (*59*) has also isolated two antineoplastic glycoproteins, namely dolabellanins A and P, from this species.

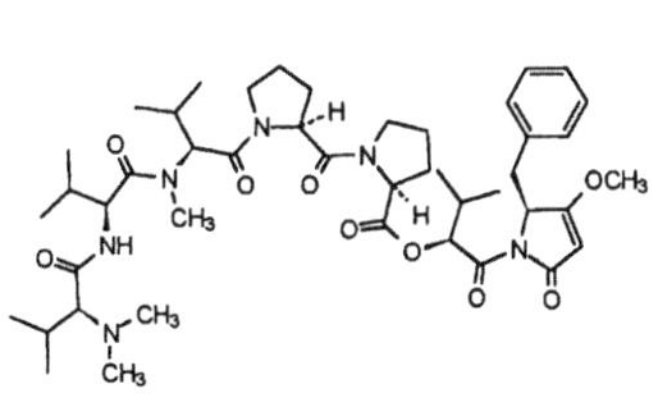

9, Dolastatin 15

N,N-DiMe-L-Ile-L-isoleucic acid-*N*-Me-L-Ile-L-Pro-L-Phe-NH$_2$

10, Dolastatin C

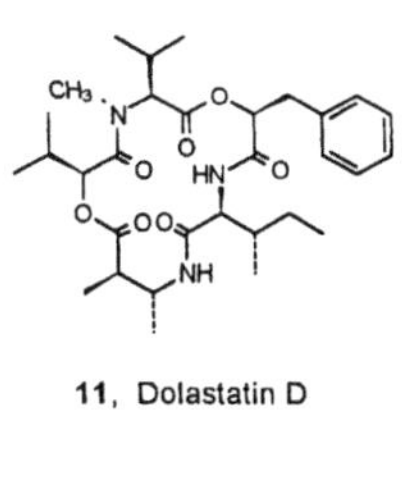

11, Dolastatin D

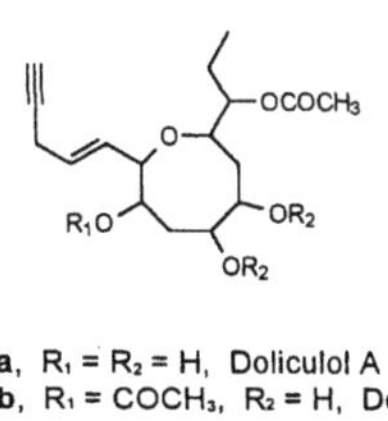

12a, R$_1$ = R$_2$ = H, Doliculol A
12b, R$_1$ = COCH$_3$, R$_2$ = H, Doliculol B

A brief investigation of the green pigments characteristic of *D. auricularia* resulted in isolation of the nickel-chlorin chelate tunichlorin from a collection of this animal in Papua New Guinea (*60*).

4. Synthesis of the Dolastatins

4.1 Dolastatin 3

After the extensive series of isolation studies guided by bioassay (PS system) as outlined in Section 3, we were able to obtain the first nine dolastatins in approximately 1-mg amounts. One of these, dolastatin 3, was subjected to detailed spectral and hydrolytic studies. As a consequence of the limited supply of dolastatin 3 and lack of crystallinity, two structural ambiguities remained unsettled. One question concerned the possibility of a reverse order of peptide bonding and the second the absolute configuration of the four chiral amino acids. Both possibilities became especially important when we synthesized the all-L isomer [with (*R*)- and (*S*)-(gln)Thz] corresponding to *cyclo*-[Pro-Leu-Val-(gln)Thz-(gly)Thz] and found it to be close to, but not identical with, dolastatin 3. In order to eliminate or confirm the reverse order of bonding for dolastatin 3, the synthesis of *cyclo*-[(gly)Thz-(*R* and *S*)-(gln)Thz-L-Val-L-Leu-L-Pro] was also undertaken. Synthesis of all 16 diastereoisomers would have required considerable efforts, so we prepared the diastereoisomer that possesses an all-L configuration in the Val-Leu-Pro segment (*61*); however that structural sequence did not correspond to dolastatin 3.

With the aid of a new computer technique for analyzing peptide mass spectral data, one of the next most likely structural sequences for dolastatin 3 was deduced to be *cyclo*-[L-Val-L-Leu-L-Pro-(*R*)- and (*S*)-(gln)Thz-(gly)Thz], and synthesis of this isomer in the all-L configuration was undertaken and again led to an isomer of dolastatin 3 (*62*). Meanwhile, we also synthesized the dolastatin 3 isomer *cyclo*-[L-Pro-L-Leu-L-Val-(*R*,*S*)-(gln)Thz-(gly)Thz] (*63*). The most significant challenge in these dolastatin isomer syntheses was to develop a practical route to the new thiazole amino acid component (gln)Thz. This was achieved as presented in Scheme 2 (*64*). Experience gained from the various dolastatin 3 isomer syntheses was then deployed in the successful synthesis of dolastatin 3 (**3**).

Once we and others (*65–69*) had eliminated *cyclo*-[Pro-Leu-Val-(gln)Thz-(gly)Thz], its chiral isomers (*63*, *65–69*), the reverse order of bonding (*61*, *65–69*), and modified (*61*, *70*) amino acid sequences by total syntheses, and since the remaining few micrograms of dolastatin 3 had decomposed in storage, efforts were directed at reisolation of this peptide.

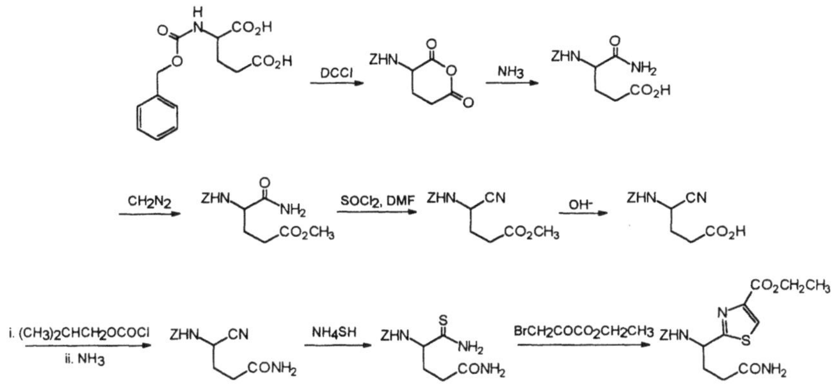

Scheme 2. Synthesis of (gln)Thz

Reisolation proved to be even more challenging and afforded only 1.8 mg [1.8×10^{-6}% yield, mp 155–159°C, $[\alpha]_D^{29}$ −48.5 (*c* 0.01, CH_3OH)] of this elusive peptide (*47*). However, by in-depth utilization of then current (1986) advances in high-field (400- and 500-MHz) ^{1}H-NMR and other necessary techniques, we deduced unequivocal structure **3** for dolastatin 3. Structure **3** was confirmed and the chirality of the (gln)Thz unit was established by total synthesis conducted by single L-amino acid unit additions from L-Pro-OMe, with employment of diethyl phosphorocyanidate (DEPC)-triethylamine for peptide bond formation and with *N*-Boc protection (trifluoroacetic acid cleavage). By this means, Boc-L-Leu-L-(gln)Thz-(gly)Thz-L-Val-L-Pro-OMe was obtained in 71% overall yield. After successive hydrolysis (1 N NaOH, dioxane, 3 N HCl), conversion (DCCI, DME, pentafluorophenol) to the OPfp active ester, Boc cleavage, and cyclization (in dioxane containing 4% *tert*-butyl alcohol and 4-pyrrolidinopyridine at 95°C, 76% yield), synthetic (−)-dolastatin 3 [colorless amorphous solid from ethanol-ethyl acetate, mp 170–173°C, $[\alpha]_D^{25}$ −53° (*c* 0.94 in $CHCl_3$)] was realized in 41% overall yield. Analogous synthesis of an isomeric dolastatin 3 containing D-(gln)Thz in place of the L-epimer gave a product that was found to differ significantly from the natural product. The ^{1}H- and ^{13}C-NMR spectra of the D-(gln)Thz isomer suggested that considerable conformational changes resulted from this otherwise simple substitution.

The synthetic (−)-dolastatin 3 was identical with the natural product. Comparison ^{1}H-NMR (400-MHz) spectra observed in methylene chloride-d_2 were superimposable, as were SP-HRSIMS (*71*) spectra and thin-layer chromatographic comparisons (on silica gel, normal and reverse phase) in four different (*e.g.*, 90:10:0.8 methylene chloride-methanol-water) solvent systems. Both specimens of dolastatin 3 inhibited growth of the PS leukemia (*46*) to the same extent (ED_{50} 0.16 *vs.* 0.17 μg/mL) and displayed an identical tendency to undergo decomposition in solution, especially in chloroform. The difficulties experienced in uncovering appropriate experimental conditions for cyclizing the linear pentapeptide precursor of (−)-dolastatin 3, combined with its sensitivity and biological activity, suggest that its overall conformational preferences are very important. Other isomers of dolastatin 3 so far examined are quite stable and *in*active against the PS system. The detailed structural elucidation and synthesis of dolastatin 3 reported herein now provides a pathway to further biological and chemical investigations of this interesting marine organism biosynthetic product.

4.2 Dolastatin 10

Dolastatin 10 (**4**) initially presented a challenging target for synthesis owing to its nine asymmetric centers (512 possible isomers) of then unknown chirality and the urgent need for obtaining a clinical supply. For the latter purpose some 700 tons of the sea hare would have been required and this was clearly untenable for ecological and other obvious reasons. A practical total synthesis of dolastatin 10 was also expected to provide the segments and synthetic analogues needed for elucidating structure/activity relationships along with providing insights into the mechanism of action. We therefore developed an efficient synthetic route to dolastatin 10 that confirmed the structure and absolute configuration (*72*) of this unique peptide. Subsequently a number of partial (*73–76*) or total (*77–80*) syntheses have been reported. The coupling of tripeptide unit A and dipeptide unit B was planned, to avoid some of the problems anticipated with a sequential approach. The absence of a chiral center next to the carbonyl group in the Dil unit made the 3 + 2 approach strategically attractive.

4.2.1 Dolaphenine (Doe)

Marine animal constituent investigations have resulted in discovery of a number of thiazole-containing cyclic peptides (*43*, *49*, *81–85*) that proved to be cytotoxic. Some of these have been synthesized and include dolastatin 3 (**3**) (*47*, *63*, *66*, *67*), ascidiacyclamide, (*86*) patellamides A (*87*), B (*88–91*) and C (*88*, *89*), ulithiacyclamide (*68*, *92*) and ulicyclamide (*93*). Each contains a 2-(1-aminoalkyl)thiazole-4-carboxylic acid unit whose biosynthesis probably involves dehydrative cyclization of an amino acid-

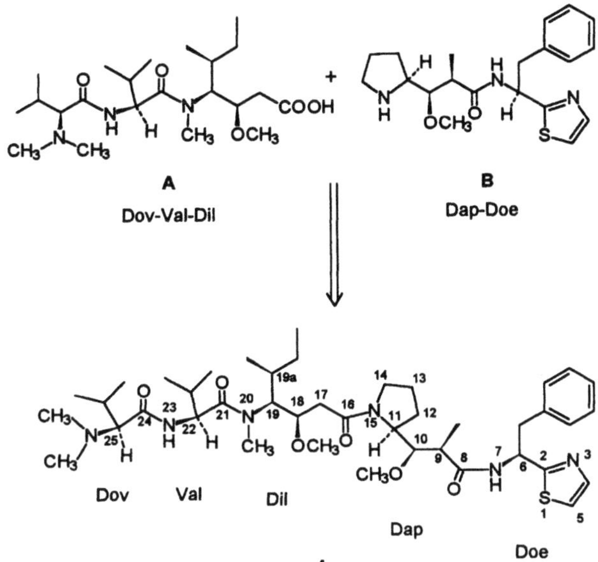

4

Dolastatin 10

cysteinyl dipeptide. As noted above, dolastatin 10 (**4**) was found to be an exception in that the thiazole ring lacks substitution at both the 4- and 5-positions.

Thiazole-4-carboxylic acids have generally been prepared by the Hantzsch condensation, but with chiral α-amino acid precursors racemization usually occurs owing to an acid-catalyzed imine-enamine type tautomerization of the intermediate thiazoline (*64, 66, 76, 94, 95*). However, syntheses of thiazole-4-carboxylic acid esters without racemization by condensation of *S*-cysteine methyl ester (**13a**) with an N-protected α-amino aldehyde have been reported (*67*). Such a route to the synthesis of Doe (**14**) was studied in detail (Scheme 3) (*96*). Reduction of Boc-phenylalanine (**15**) with borane-tetrahydrofuran gave alcohol **16** in 90–96% yield. Oxidation of Boc-phenylalaninol (**16**) with the sulfur trioxide-pyridine complex (Parikh-Doering method) (*97*) in dimethyl sulfoxide containing triethylamine produced aldehyde **17** in about 94% yield. The aldehyde could be stored at 0° for days at a time without loss of optical activity. Generally it was used immediately in the next reaction where Boc-phenylalaninal (**17**) was stirred in benzene with 2-aminoethanethiol (cysteamine, **13b**) to give a diastereomeric mixture of the thiazolidines (**18**) in quantitative yield. With the hydrochloride salt **13c** in the presence of triethylamine the product (**18**) was obtained in somewhat lower yield.

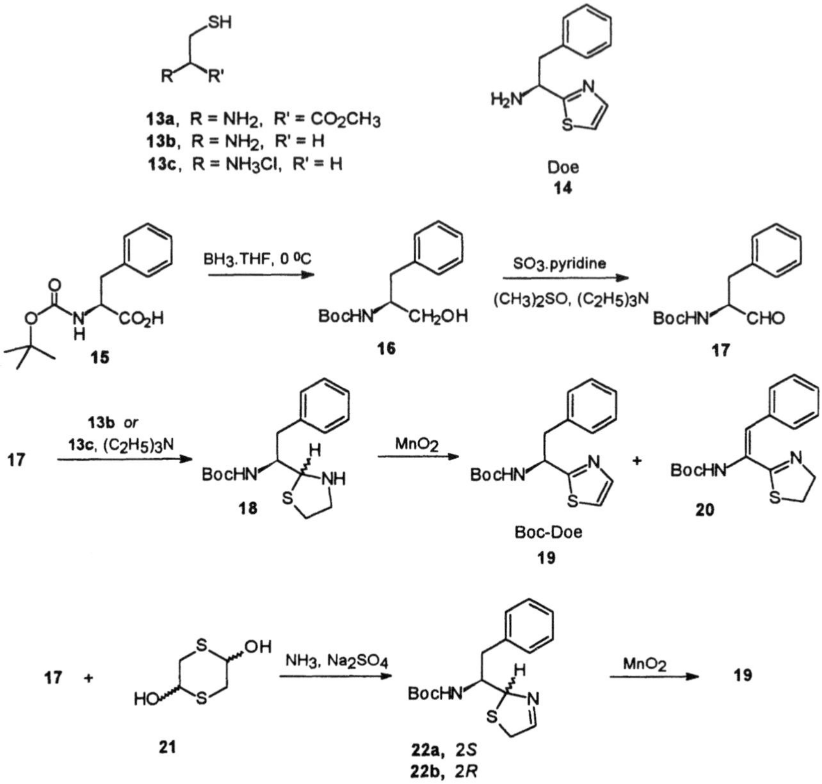

Scheme 3. Synthesis of Boc-Doe (**19**)

Various manganese dioxide preparations and other oxidizing and/or dehydrogenation reagents were explored for conversion of the thiazolidines (**18**) to thiazole **19**. We evaluated active manganese dioxide prepared by a number of methods (*98*) along with products of various activities available from commercial sources. Efforts were then concentrated on the effectiveness of types produced by Chemetals, Inc. In Table 1 are presented the yields of thiazole **19** produced when these oxides were used on a small scale (*ca.* 50 mg substrate with 2 g manganese dioxide). Interestingly, use of the battery grade manganese dioxide gave rise to racemization, from 10–100% depending on the batch of oxidant. A side product produced in even lower yield (3%) was shown to be vinyl amine **20**. Yields of Boc-Doe (**19**) varied, depending on the type of oxidant and procedure used. In general, yields were higher than those previously achieved, rising to about 70% in one case. Results were not consistently

Table 1. *Dehydrogenation of Thiazolidine* **18** *to Thiazole* **19** *by Employment of Manganese Dioxide*

Type of MnO_2[a]	% yield Flask Method	Column Method
FarM™	11	17
		16.5
177-1	36	38
		38
		43.5
177-2	45	42
		42
		51
		41
HP (batch a)	52	30
	53	48
	66	49
(batch b)	0.0	0.0
(batch c)	10	
CIR	48	29
	52.5	41
	48	47

[a] MnO_2 Type:	FarM™	171-1	177-2	HP	CIR
%Mn	60.0	59.5	61.5	62.9	61.5
$\%MnO_2$	90.0	89.9	92	99	92
BETSA (M^2/g)	110	54	40	1.4	35
Por. Vol. (cc/g)	0.29	0.12	0.06	0.002	0.05
Avg. Por. Diam (Å)	125	100	<20	200	<10
E° MnO_2/H^+ (Mv)	1275	1280	1325	1235	1320

reproducible, and it was found that yields dropped dramatically when the reaction was carried out on a larger scale (5–15 g substrate), but overall the better procedures of Table 1 gave rise to useful amounts of thiazole. We have also explored the route presented by structures **21** → **22**, but it did not offer better yields.

Recent reports by other groups (*73, 77, 78, 99, 100*) concerned with the synthesis of dolastatin 10 have also described the preparation of Doe (**14**) *via* oxidation of thiazolidine **18** by manganese dioxide (*73, 78, 99, 100*). The yields were similar to ours. They have also prepared the thiazole in 56% enantiomeric excess from Boc-phenylalanine (*73, 75, 99, 100*) according to SCHMIDT's (*101*) modification of the Hantzsch method. Furthermore, one of these reports by the SHIOIRI group (*100*) describes a more efficient synthesis utilizing asymmetric reduction of the benzyl 2-thiazolyl ketone (formed almost quantitatively from phenylacetic acid) carbonyl center, followed by a modified Mitsunobu reaction to yield Boc-Doe (**19**) in

52.5% overall yield. Another more efficient route to Doe (**14**) has been reported by Holzapfel and co-workers (*76, 94*), who synthesized optically pure thiazole amino acid derivatives by a modified Hantzsch reaction. Use of potassium or sodium bicarbonate in an aprotic solvent neutralizes the internally produced hydrobromic acid and treatment with trifluoroacetic acid anhydride and pyridine results in immediate aromatization of the intermediate thiazoline. Optically pure Boc-Doe (**19**) was thus obtained in 96% yield from the appropriate thioamide. Presently we are using a modification of this method (*76, 94*) to obtain optically pure Boc-Doe (**19**). Thiazole **14** was stored as the Boc-derivative (**19**) until coupled with dolaproine (Dap). Coupling of the resulting amide with the required tripeptide as previously outlined (*72*) (to be discussed in Section 4.2.5) gave a peptide identical with natural dolastatin 10 (**4**), thus proving that the phenylalanine-derived thiazole unit of the natural product has the *S* configuration.

4.2.2 Dolaproine (Dap)

The novel β-methoxy-γ-amino acid dolaproine (Dap, **23**) constitutes the most complex unit of dolastatin 10 and appears to be produced by a biosynthetic aldol-type condensation of a proline-derived aldehyde and a propionate equivalent. *D. auricularia* has been found (*47, 49*) to preferentially utilize such aldol-derived amino acids for synthesizing powerful antineoplastic peptides. The β-methoxy-γ-amino acids Dap (**23**) and Dil (see Section 4.2.3) are structurally related to (3*S*, 4*S*, 5*S*)-isostatine (*102*), a component of the antineoplastic didemnins (*103–105*), and to the leucine-type amino acid statine found in pepstatin (*106*), a substance isolated from a lower plant that inhibits proteases such as pepsin, renin and cathepsin D. Of the more than 500 naturally occurring amino acids currently known (*107*), apparently detoxinine (*108*), a component of the detoxin class of *Streptomyces* depsipeptides, is the only aldol-type derivative of proline similar to Dap (**23**).

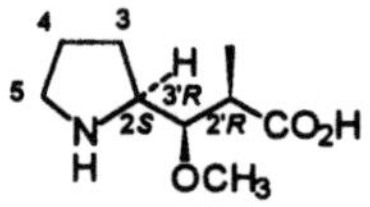

Dolaproine (Dap)
23

A common approach to the syntheses of β-hydroxy-γ-amino acids involves an aldol condensation between an aldehyde derivative of the appropriate amino acid and a suitable enolate. Current advances in improving the aldol reaction offer a wide range of techniques (*109–113*).

Because the proline moiety of dolaproine (**23**) was thought likely to have the *S*-configuration, we selected a propionate equivalent that, on condensation with N-protected (*S*)-prolinal, would produce all four diastereoisomers (Scheme 4) (*114*). Unless dolastatin 10 was derived from (*R*)-proline, one of the four diastereoisomers was expected to correspond to natural Dap (**23**). Reduction of Boc-(*S*)-proline (**24**) to alcohol **25** was performed with borane-tetrahydrofuran. Oxidation of alcohol **25** to *N*-(*tert*-butoxycarbonyl)-(*S*)-prolinal (**26**) was achieved without detectable racemization (in 92% yield) using dimethyl sulfoxide and sulfur trioxide-pyridine complex. Although this oxidation was also effected (*72*) without obvious racemization by use of the modified Swern (*115*) method, employing trifluoroacetic acid and dimethyl sulfoxide, yields were higher under the Parikh-Doering (*97*) conditions. The ^{1}H-NMR spectrum of aldehyde (**26**) in deuterochloroform indicated the presence of two rotamers. Signals corresponding to the aldehyde proton appeared at δ 9.54, the methine proton at δ 4.18 and δ 4.03, and the *tert*-butoxycarbonyl methyl groups at δ 1.46 and δ 1.41. When the temperature was increased to 55°C the signals coalesced. Change of the solvent to deuteroacetonitrile caused the room temperature spectrum to show one signal each for the aldehyde and methine protons, but the *tert*-butyoxycarbonyl methyl groups still appeared as two peaks.

The (*S*)-propionate chiral auxiliary **27** was prepared in good yield by employment of (*S*)-methyl mandelate and phenylmagnesium bromide (*108*, *116*), followed by esterification. Aldol condensation between aldehyde **26** and propionate **27** in tetrahydrofuran (at −95°C) containing magnesium bromide afforded a mixture of isomers in a ratio of 64:6:15:15 (**28a**:**28b**:**28c**:**28d**). At −78°C the ratio changed to 45:15:20:20. With increasing scale (up to 40 g of reaction products) it became necessary to develop an optimization chromatographic method (*117*) for separation of the diastereoisomers. In order to define the stereochemistry of isomer **28a** its methyl ester (**29**) was subjected to deprotection and cyclization with a trifluoroacetic acid → potassium carbonate sequence to yield lactam **30a** and the minor (presumably racemized) product **30b**. The absolute configuration of **30a** was deduced by X-ray analysis (*114*) which established the absolute configuration at the relevant chiral centers in aldol product **28a** as 2*S*, 2′*S*, 3′*R*. *Anti*-isomer **28a** was obtained with high selectivity and the likely transition state appears in Fig. 1. At this point the absolute configuration of natural Dap was still unknown and the major aldol product (**28a**) was employed for subsequent reactions leading to an isodolastatin 10 (*118*).

Trimethyloxonium tetrafluoroborate (*119*) in combination with proton sponge was found very suitable for both small and multigram scale

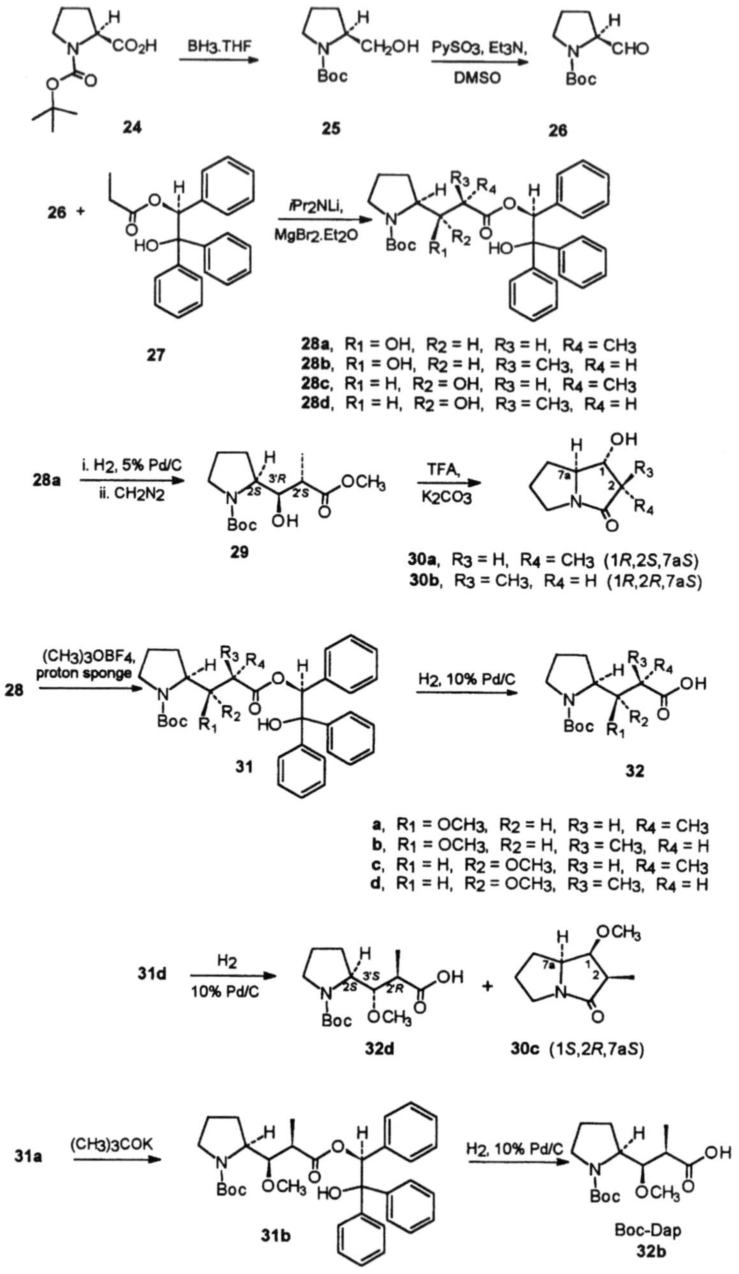

Scheme 4. Synthesis of Boc-Dap (**32b**)

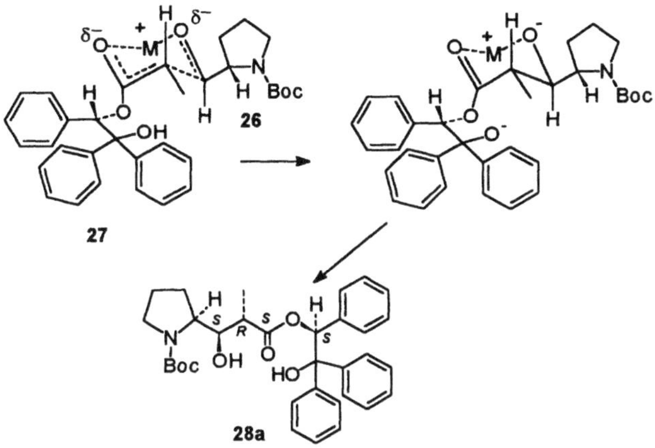

Fig. 1. Probable transition state in the synthesis of **28a**

methylation of alcohol **28a** to give methyl ether **31a**. Deprotection of the C-terminal carboxyl group was achieved by hydrogenolysis to yield Boc-Dap isomer **32a**. X-ray analysis confirmed the 2*S*, 2′*S*, 3′*R* configuration which in turn confirmed that the stereochemistry of 2′-(*S*)-iso-Dap **29** had been retained in the major product of cyclization (**30a**). Since the other products of the aldol condensation (**28b–d**) were produced in small quantities relative to isomer **28a**, the aldol mixture was subjected to detailed chromatographic separation and two of the diastereoisomers (**28c** and **28d**) were resolved. Both were methylated to afford Dap isomers **31c** and **31d**, respectively. After deesterification the resultant *N-tert*-butoxycarbonyl protected acids (**32c** and **32d**, respectively) were used to synthesize (*118*), as it turned out, peptides isomeric with dolastatin 10. Lactam **30c** was formed as a side-product during the hydrogenolysis of **31d** when ethyl acetate was included in the solvent mixture. The ^{1}H-NMR spectrum of **30c** was unequivocally assigned on the basis of its two-dimensional COSY spectrum.

If the original choice of (*S*)-proline was correct, the required aldol product was the diastereomer of lowest yield, namely, **28b**, which had either the 2′*S*,3′*S* or the 2′*R*,3′*R* configuration. The latter differed from the most abundant aldol product (**28a**) only at the, presumably, epimerizable α-carbon atom. Epimerization of the α-methyl group of isomer **31a** with potassium *tert*-butoxide under carefully controlled (−20°C) conditions gave methyl ether **31b** in 57% yield. Dap ester **31b** was subjected to hydrogenolysis to yield (91%) carboxylic acid **32b** and this γ-amino acid was converted (*72*) to natural dolastatin 10. In turn, this confirmed isomer

32b as *N-tert*-butoxycarbonyl-dolaproine and established that dolaproine possessed the 2*S*,2′*R*,3′*R* stereochemistry. With the chirality of Dap established, we focused on a stereoselective synthesis which was accomplished as follows.

We decided that Dap (**23**), being a "non-Evans-type" *syn* aldol, could be synthesized *via* a boron-mediated aldol condensation employing the aldehyde derivative of Boc-*S*-proline and a chiral oxazolidinone enolate that could direct reaction to the desired face (Scheme 5) (*120*). Chiral auxiliary **33** was synthesized by a modification of EVANS' method (*121*) using condensation of (1*S*,2*R*)-norephedrine and diethyl carbonate to form the corresponding oxayolidinone followed by acylation with propionyl chloride in the presence of butyllithium. Aldol condensation between aldehyde **26** and propionyl amide **33** using dibutylboron triflate (*120*) and triethylamine was eventually found to afford diastereoisomer **34** with the correct Dap stereochemistry (Scheme 6). Chromatographic analyses of the product indicated stereoselectivity in the 90% range with isolated yields of isomer **34** from 60–80%. Furthermore, the reaction was

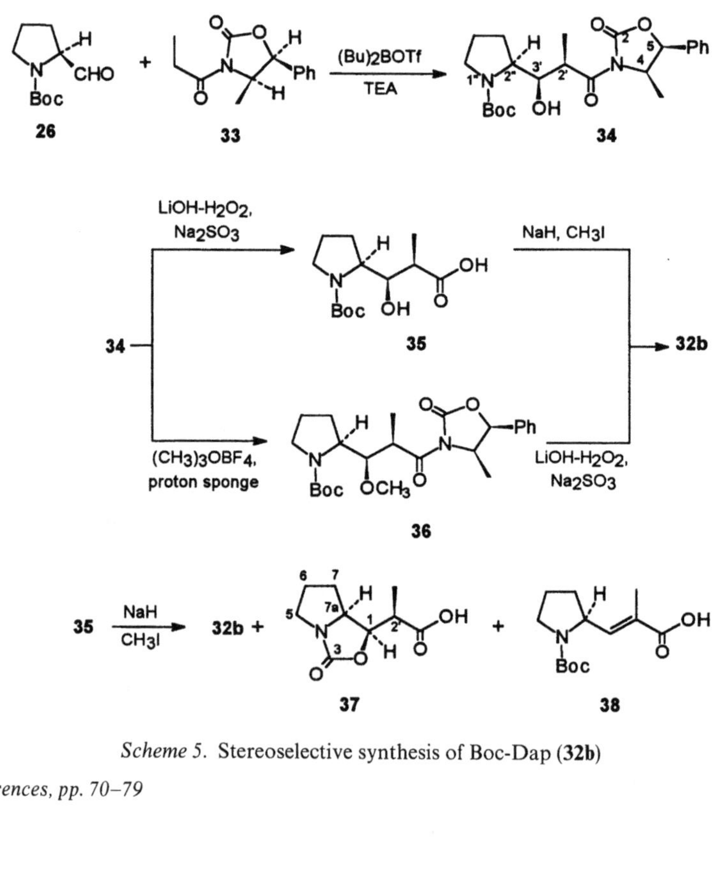

Scheme 5. Stereoselective synthesis of Boc-Dap (**32b**)

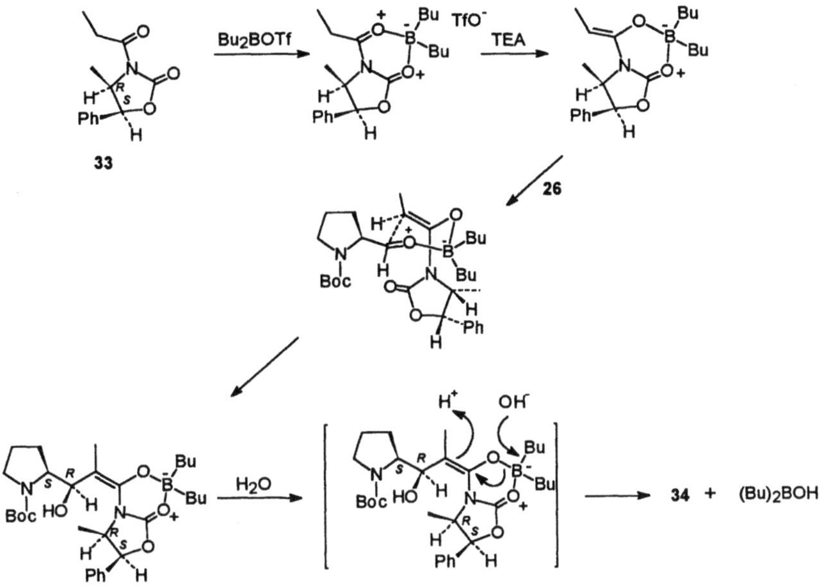

Scheme 6. Intermediates in the stereoselective synthesis of Boc-Dap

easily performed at the ⩾20-g level. The highly stereoselective dibutylboron triflate mediated type of condensation we employed is well known (*121–124*). Indeed, following completion of the study a useful synthesis of dolastatin 10 incorporating this approach was reported by Shioiri and colleagues (*77, 79*).

Once compound **34** with the desired 2′*R*,3′*R* stereochemistry was in hand the route to Dap was continued by two converging routes (**34**→**35**→**32b** and **34**→**36**→**32b**). Removal of the chiral auxiliary with hydrogen peroxide and lithium hydroxide followed by *O*-methylation with sodium hydride and iodomethane, or (preferred) initial *O*-methylation using trimethyloxonium tetrafluoroborate (*119*) and subsequent cleavage of the oxazolidinone unit gave rise to the required *N*-Boc-dolaproine (**32b**) in good yield. With the sodium hydride-iodomethane procedure, two minor side products (**37** and **38**) were produced. Tetrahydropyrrolo[1,2-*c*]oxazole derivative **37** was subjected to X-ray crystal structure determination and shown to have the 1*R*,7*aS*,2′*R* configuration. In turn this confirmed that Dap derivative **32b** had the required stereochemistry. The other sideproduct was shown by spectroscopic analysis to be olefin **38**. The stereospecific synthesis of *N*-Boc-Dap described here greatly simplified the overall synthesis of dolastatin 10 and allowed more ready access to this promising clinical candidate.

In 1994 PONCET and co-workers (*80*) described a new synthesis of Boc-Dap (**32b**), outlined by structures **24**, **39**, **26**, **40**, and **41** (Scheme 7).

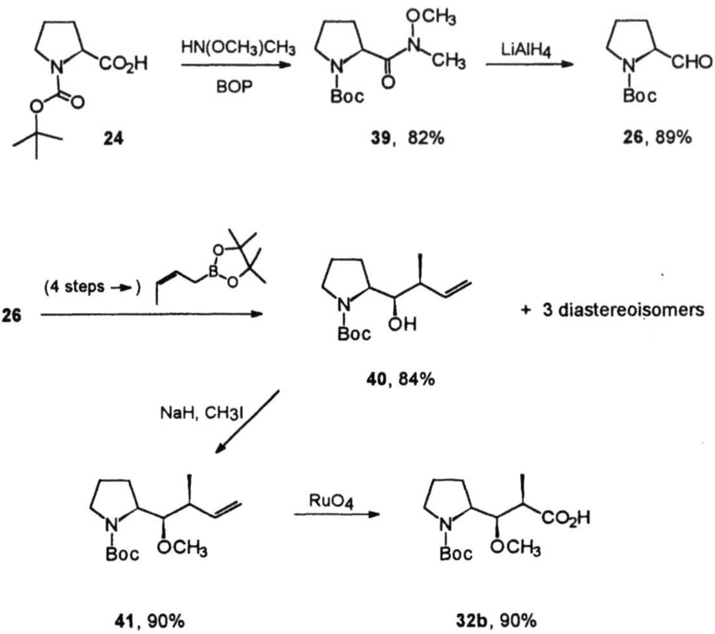

Scheme 7. Alternative synthesis of Boc-Dap (*80*)

4.2.3 *Dolaisoleuine (Dil)*

The three chiral centers of Dil (**42**), with unknown configuration, presented a difficult synthetic problem (*125*). Initial clues to solving part of the stereochemical problem posed by the dolaisoleuine unit arose from

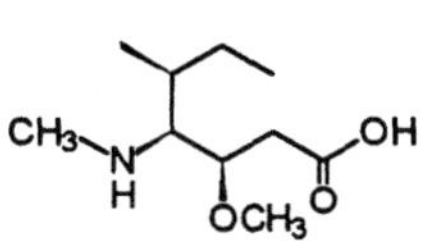

Dolaisoleuine (Dil)
42

a comparison of the relative NMR chemical shifts of the alkyl side-chain methyl signals in several *N*-methylisoleucine derivatives (*126*) and from biosynthetic consideration. Thus, it was decided to synthesize both the (3*R*,4*S*,5*S*)- and (3*S*,4*S*,5*S*)-isomers. An aldol condensation between an N-protected *N*-methyl-(*S*,*S*)-isoleucinal and a suitable acetate equivalent that would give rise to easily separable isomers suitable for stereochemical analysis was undertaken. Preparation of *N*-methyl-*N*-(benzyloxycarbonyl)-(*S*,*S*)-isoleucinal (**43**) from *N*-(benzyloxycarbonyl)-L-isoleucine was accomplished according to Scheme 8. The *N*-methylation of Z-Ile was performed without any detectable C-2 epimerization (*127*, *128*) and in almost quantitative yield by the Benoiton (*129*) procedure. The resulting *N*-methyl derivative (**44**) was reduced to alcohol **45** (90% yield) with borane-tetrahydrofuran. Oxidation of alcohol **45** to aldehyde **43** was realized in good yield (and without any epimerization) by treatment with dimethyl sulfoxide and sulfur trioxide-pyridine complex.

The aldol condensation of aldehyde **43** with *tert*-butyl acetate was carried out by use of lithium diisopropylamide in tetrahydrofuran at $-78°C$ to give a mixture of two diastereoisomers (**46a** and **46b**) in an approximate ratio of 4:3. Alcohols **46a** and **46b** were methylated with trimethyloxonium tetrafluoroborate (*119*) to provide methyl ethers **47a** and **47b** in 87 and 71% yield, respectively. A ^{1}H-NMR analysis of these ethers revealed very valuable stereochemical information about dolaisoleuine (**42**). The signal for H-4 appeared as a broad hump at δ 4.10 in the ^{1}H-NMR spectrum of ester **47a**, very similar to the signal assigned to H-19 (δ 4.70) of dolastatin 10 (**4**). In contrast, H-4 of the two stable conformers of **47b** gives rise to a doublet of doublets at δ 3.69 and δ 3.85 ($J = 10.8$ and 3.0 Hz). Such a difference in the spectra of the methyl ethers indicated that isomer **47a** was most likely to be the Dil derivative and that indication proved correct.

Selective deprotection of isomer **47a** was performed by careful (to avoid extensive lactam formation) hydrogenolysis using 5% palladium-on-carbon in ethyl acetate-methanol (3:1) followed by treatment with hydrogen chloride in ether to furnish hydrochloride **48a** (63% yield). During the hydrogenolysis lactam **49a** was obtained in 29% yield. Analogous reactions led to isomer **48b** from *N*-*Z* derivative **47b** (93% yield) with only a minor yield of lactam **49b**. When the reaction time was increased or when the *tert*-butoxy group was replaced by an ethoxy group, the γ-lactam was a major product. Lactam formation was reduced by use of a minimum amount of catalyst under anhydrous conditions. Hydrochloride salt **48a** was also derived from methyl ether **47a** by hydrogenolysis using internal hydrogen transfer with 5% palladium-on-carbon in methanol and cyclohexene (2:1) followed by treatment with ethereal

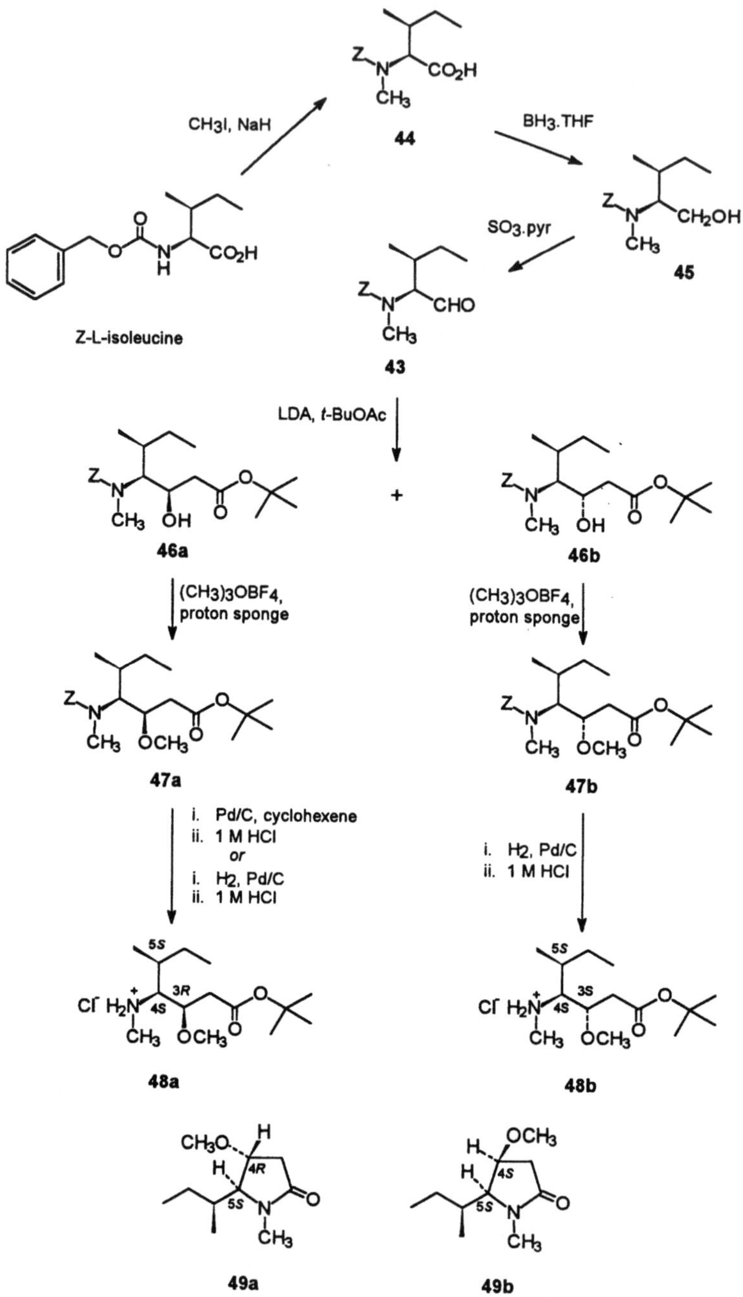

Scheme 8. Synthesis of Dil-O-*t*Bu. HCl (**48a**) and its (3*S*)-isomer

hydrogen chloride; yields were higher and pyrrolidinone **49a** formation was avoided.

Lactams **49a** and **49b** were very useful in establishing the stereochemistry of the aldol products by NMR spectroscopy. The ^{1}H-NMR spectrum of isomer **49a** exhibited a doublet of doublets at δ 3.68 ($J = 7.0$ and 1.7 Hz) for H-4 and δ 3.46 ($J = 3.4$ and 1.7 Hz) for H-5 whereas that of **49b** displayed a quartet at δ 4.04 ($J = 7.0$ Hz) for H-4 and a doublet of doublets at δ 3.50 ($J = 7.0$ and 3.6 Hz) for H-5. The spectrum of lactam **49b** also showed NOE enhancements of 6% (H-4 $\rightarrow$ H-5) and 7% (H-5 $\rightarrow$ H4), but no NOE between those ring protons was found in the spectrum of isomer **49a**. The coupling constants and NOE enhancements revealed the *cis* arrangement of H-4 and H-5 in lactam **49b**. Thus, these protons were assigned as *trans* in isomer **49a**, and the absolute configuration of lactam **49a** was assigned as 4*R*,5*S*,1′*S*. The configuration of **45a**, **46a**, and **47a** is therefore 3*R*,4*S*,5*S* and that of the **45b** series is 3*S*,4*S*,5*S*.

With Dil derivative **48a** of known stereochemistry in hand we proceeded, using the convergent synthesis previously outlined (*72*), to prepare natural dolastatin 10. That result established the chirality of the Dil unit of dolastatin 10 as 18*R*,19*S*,19*a*S and experiments directed at developing a stereoselective synthesis of Dil were pursued.

The stereospecific synthesis of Dil represented a more difficult exercise in stereocontrol than that presented by Dap. With no substituent in the 2-position of Dil (**42**), choice of the acetyl oxazolidinones **50a** or **51a** was attractive to guide an aldol condensation route. However, with two hydrogen atoms in the 2-position of Dil, no enolate *E* or *Z* character would be preset to assist stereocontrol of the aldol reaction. Those predictions were confirmed when reaction of oxazolidinone **50a** with *N*-carbobenzyloxy-*N*-methyl-isoleucinal (**43**) was implemented. The 3′*S*,4′*S*,5′*S* and 3′*R*,4′*S*,5′*S*-Dil diastereoisomers were routinely obtained in nearly equal amounts. To circumvent this problem, we utilized oxazolidinones **50b** and **51b** known (*130*) to form exclusively *Z* enolates, thereby unsuring *syn* stereocontrol.

The aldol reaction was conducted *via* enolate formation of oxazolidinone **50b**, with use of dibutylboron triflate in the presence of a slight excess of diisopropylethylamine in methylene chloride. Condensation with *N*-carbobenzyloxy-*N*-methyl-isoleucinal (**43**) at −70°C afforded only one product (**52a**). Analogous condensation with oxazolidinone **51b** again gave only one apparent product (**53a**). The *N*-methylthioacetyl oxazolidinones **50b** and **51b** both required much longer (about five-fold) reaction times for enolization and condensation compared with their somewhat less hindered *N*-propionyl oxazolidinone (**50c** and **51c**) counterparts (*131*). Tributyltin hydride in benzene, together

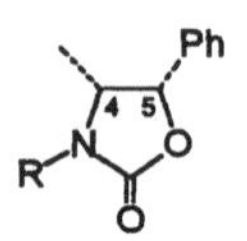

50a, R = $COCH_3$
50b, R = $COCH_2SCH_3$
50c, R = $COCH_2CH_3$

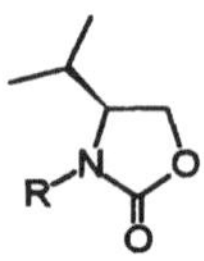

51a, R = $COCH_3$
51b, R = $COCH_2SCH_3$
51c, R = $COCH_2CH_3$

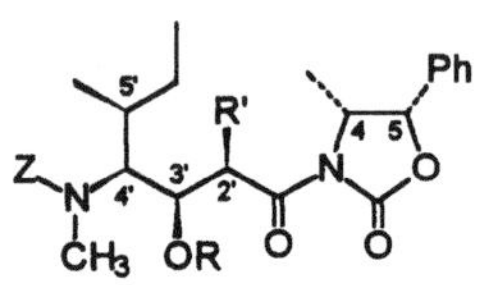

52a, R = H, R' = SCH_3
52b, R = H, R' = H
52c, R = CH_3, R' = H

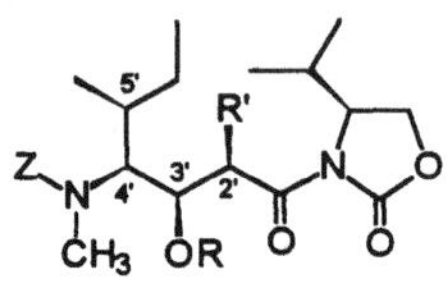

53a, R = H, R' = SCH_3
53b, R = H, R' = H
53c, R = CH_3, R' = H

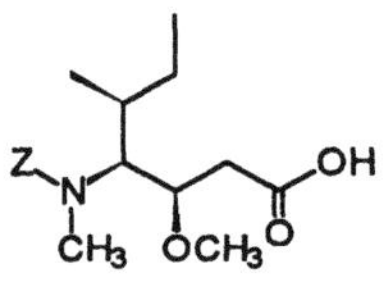

54

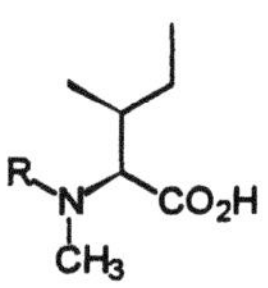

44, R = Z
55, R = $COCH_3$

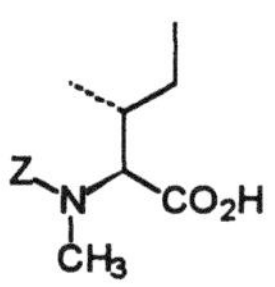

56

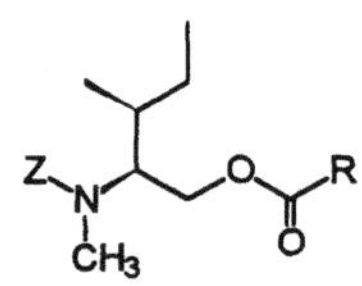

57a, R = CH_3
57b, R = $C(CH_3)_3$

with the free radical initiator 2,2′-azobis-(2-methylpropionitrile), was found to provide an efficient method for the thioether cleavage (*121*, *132*), and good yields of the Dil precursor were realized. Methylation of alcohols **52b** and **53b** with trimethyloxonium tetrafluoroborate and proton sponge proceeded well. Analysis by ^{1}H-NMR spectroscopy at 55°C of the resulting methyl ethers (**52c** and **53c**) showed coalescence of some signals including those of the 4′-H hydrogen, the methoxy hydrogens, and the *N*-methyl hydrogens, which indicates the presence of conformers rather than diastereoisomers. Cleavage of the oxazolidinone amide by

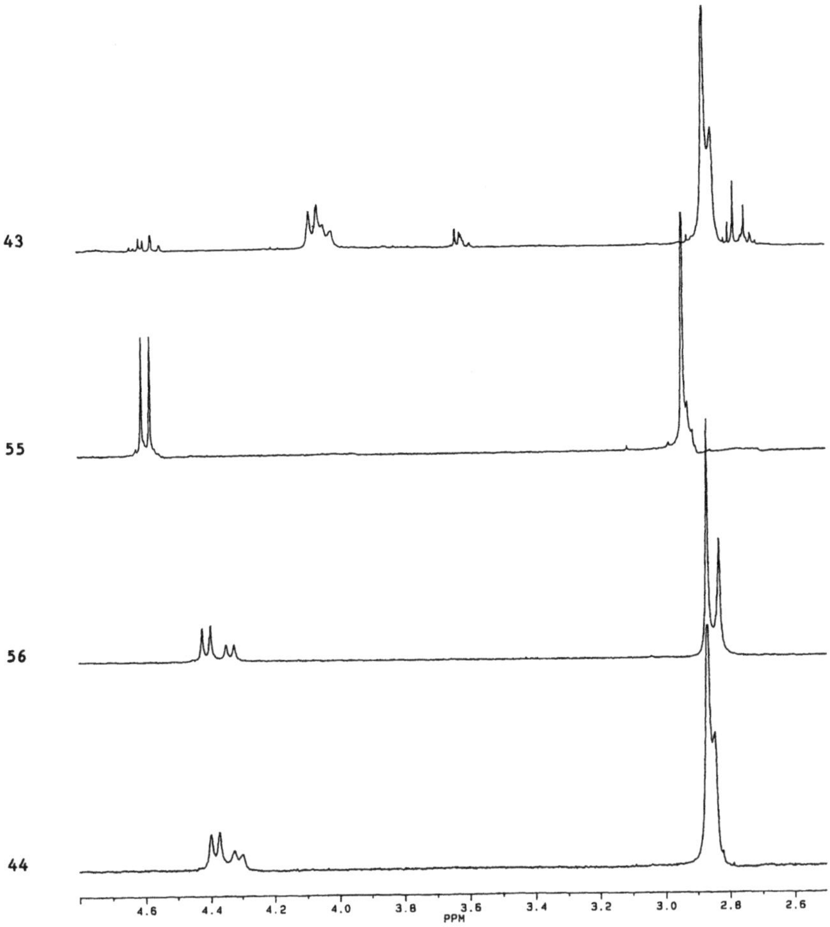

Fig. 2. ^{1}H-NMR spectra of amides **43**, **44**, **55**, and **56** at 25°C

employment of hydrogen peroxide and lithium hydroxide followed by sodium sulfite (*121*) furnished *N*-*Z*-(3*R*,4*S*,5*S*)-Dil (**54**) which was converted to ester **47a** by reaction with 2-methylpropene and a catalytic amount of sulfuric acid (*133*).

In order to study the effects of rotational or conformational isomerism in our Dil intermediates (**43–45**) and to confirm the cause of the ^{1}H-NMR signal doubling (*134*), we prepared isoleucine derivatives **55–57a, b** for spectroscopic analysis (*126*). Compounds **43**, **44**, and **56** each showed a pair of doublets for H^α and a pair of singlets representing the *N*-methyl

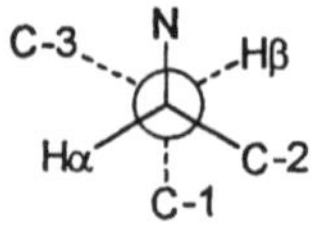

Fig. 3. Projection along the C_α–C_β axis in L-isoleucine

group at room temperature (Fig. 2). Examination of H^α–H^β NMR coupling values (**43**, $J = 9.8, 10.2$ Hz; **44**, $J = 10.6, 10.7$ Hz; **56**, $J = 10.2, 10.3$ Hz) suggested the dominance of rotamer A (Fig. 3); these values were in good agreement with ABRAHAM's (*135*) coupling constant of 10.4 Hz for *trans* H^α, H^β protons in L-leucine as well as with X-ray data for L-isoleucine (*136*). Carbamates **57a** and **57b** exhibited analogous ^{1}H-NMR results, but the H^α signal was more complex owing to interactions with the neighboring methylene group.

Interestingly, while compounds **43**, **44** and **56** with the bulky benzyloxy substituent on the carbonyl showed the doubling of signals, the non-urethane acetamide **55**, with the less bulky methyl substituent, did not demonstrate this trend. Thus we concluded that signal doubling in the spectra of amides **43**, **44** and **56** was due to restricted rotation about the carbamate amide bond. In this case, the major conformer in solution was confirmed as being *trans* by the utlization of the 1-D cross-sectional spectra of the 2-D NOESY spectrum of compound **56**. Only the *N*-methyl signal at δ 2.87 showed an NOE effect on the methylene signal at δ 5.12, demonstrating the *trans* nature of the amide bond. The ability of the *trans* conformer of amides **44** and **56** to hydrogen bond intramolecularly may have some bearing on this effect.

To evaluate the energy involved in such conformational interconversions, we studied the effect of temperature on the proton NMR spectra of amides **43, 44, 55–57a, b**. The result with amide **56** is shown in Fig. 4 along with a plot of temperature *vs.* chemical shift. From the results of this study it seems clear that *N-Z*, *N*-Me-Ile-type amino acid derivatives exist in the form of conformational isomers at ambient temperature owing to steric hindrance, and it also confirmed the lack of racemization during *N*-methylation of *Z*-Ile using the BENOITON procedure (*129*, *137*, *138*).

Of two partial stereoselective syntheses of Dil, the SHIOIRI group (*79*) employed an approach analogous to their syntheses of Boc-isostatine. The key steps involved reaction between Boc-(*S*)-isoleucinol and the ethyl

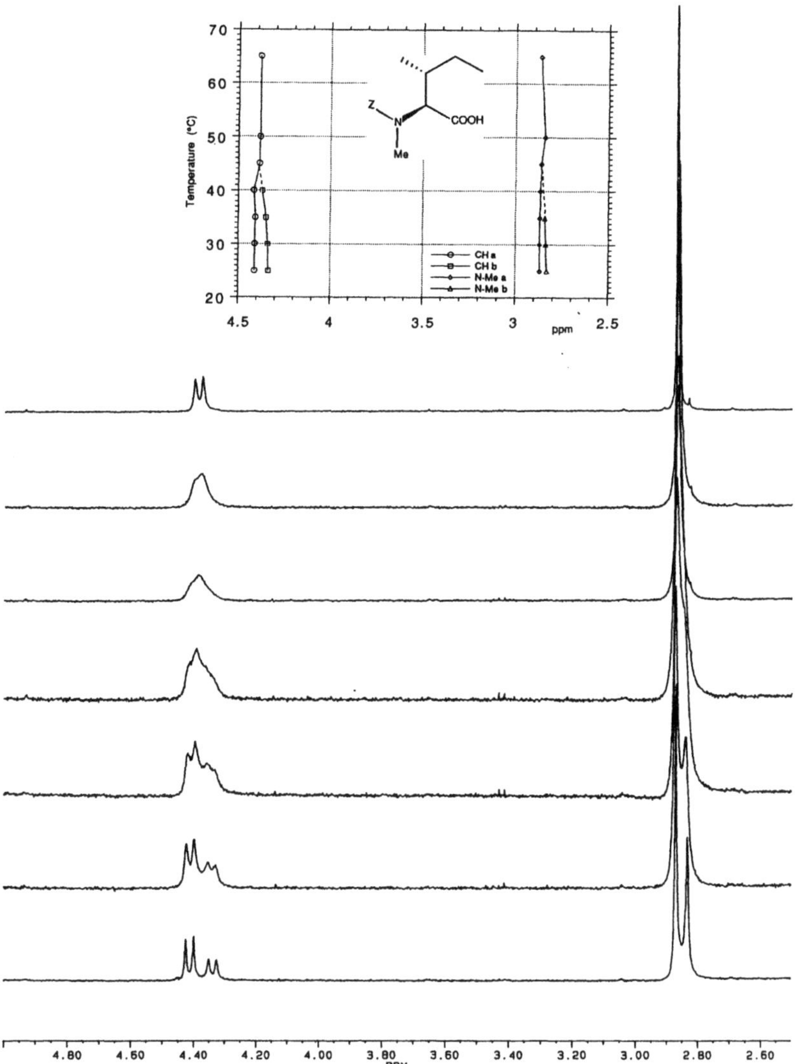

Fig. 4. Temperature-dependent ^{1}H-NMR spectra of *N*-Z-*N*-Me-*allo*-Ile (**56**)

acetate lithium enolate (yields a 19:31 isomeric mixture) followed by hydrolysis and methylation. PONCET and co-workers (*75*, *80*) extended the *Z*-crotylboronate approach, applied to obtain Boc-Dap as outlined previously (Section 4.2.2), to synthesizing Boc-Dil in an analogous multistep sequence.

4.2.4 *Dolaproinyl-Dolaphenine (Dap-Doe)*

Our approach to dolastatin 10 involved formation of dolaproinyl-dolaphenine (Dap-Doe) and the tripeptide dolavalyl-valyl-dolaisoleuine (Dov-Val-Dil) and their subsequent coupling to give dolastatin 10 (**4**). Syntheses of *N*-Boc-dolaproinyl-dolaphenine (**58b**) and three chiral isomers (**58a, c, d**) were conducted as follows (*139*). At the onset, the chirality of the side chain in dolaproine was unknown, so all four diastereoisomeric derivatives (**32a**, 2*S*,2′*S*,3′*R*; **32b**, 2*S*,2′*R*,3′*R*; **32c**, 2*S*,2′*S*,3′*S*; **32d**, 2*S*,2′*R*,3′*S*) were synthesized *via* the previously described

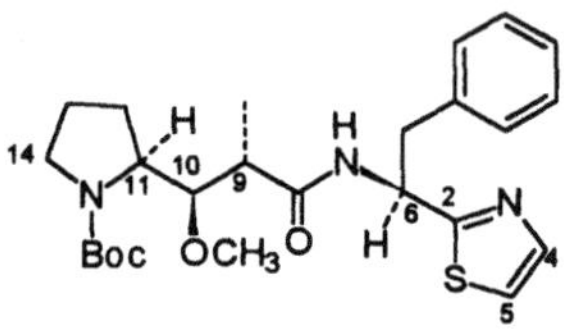

Boc-(9*S*,10*R*,11*S*)-*iso*-Dap-Doe

58a

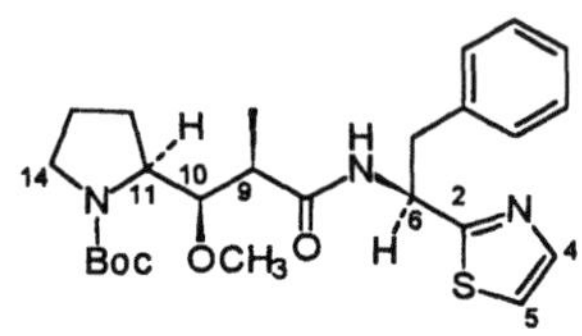

Boc-Dap-Doe

58b

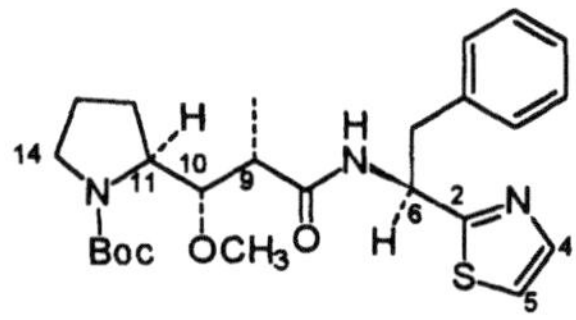

Boc-(9*S*,10*S*,11*S*)-*iso*-Dap-Doe

58c

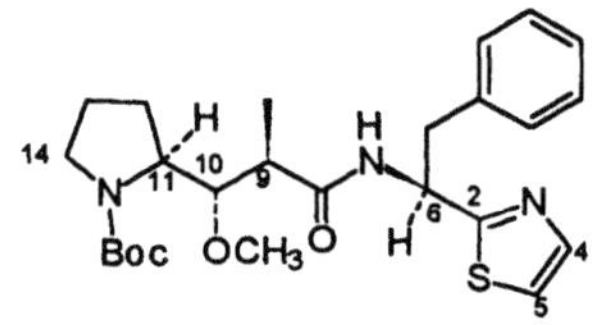

Boc-(9*R*,10*S*,11*S*)-*iso*-Dap-Doe

58d

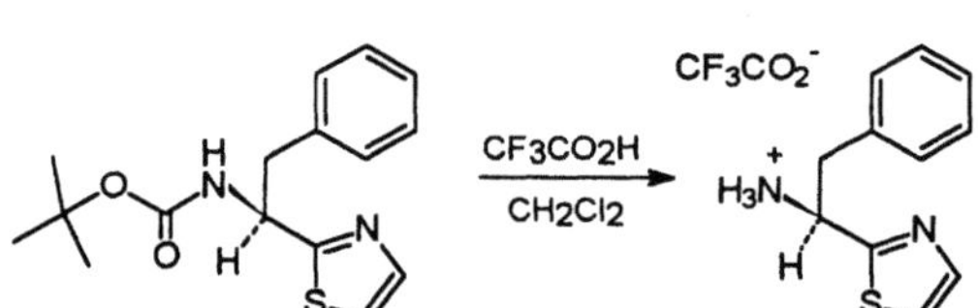

19 **59**

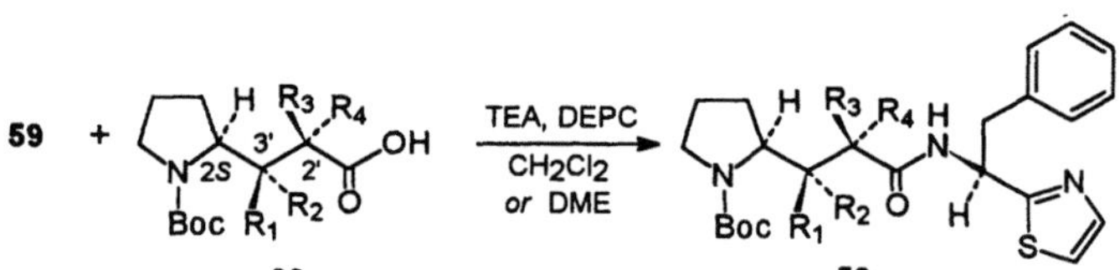

32 **58**

a, R1 = OCH3, R2 = H, R3 = H, R4 = CH3 (2′*S*, 3′*R*)
b, R1 = OCH3, R2 = H, R3 = CH3, R4 = H (2′*R*, 3′*R*)
c, R1 = H, R2 = OCH3, R3 = H, R4 = CH3 (2′*S*, 3′*S*)
d, R1 = H, R2 = OCH3, R3 = CH3, R4 = H (2′*R*, 3′*S*)

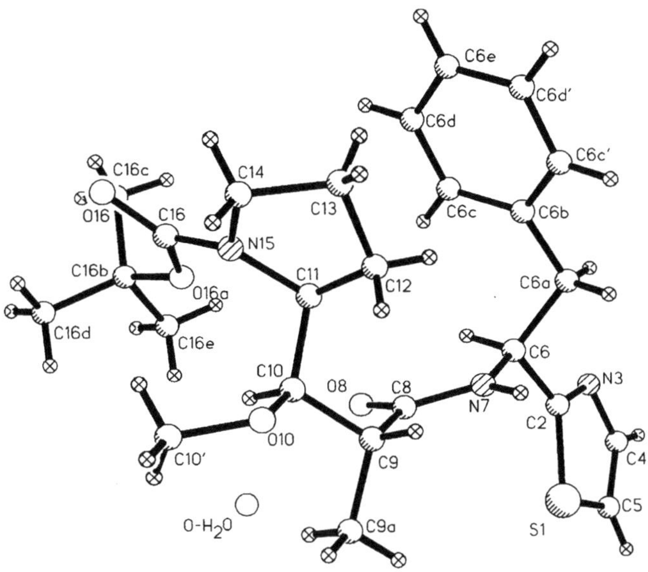

Fig. 5. Computer-generated drawing of amide **58b**

(see Dap, Section 4.2.2) aldol condensation route followed by *O*-methylation. The absolute configuration of each methyl ether was established by NMR and/or X-ray analyses (*72*, *114*).

The dipeptide derivatives **58a–d** were readily formed, diethyl phosphorocyanidate (DEPC) being used as the coupling reagent. Thiazole **19** was deprotected by treatment with trifluoroacetic acid. A mixture of the trifluoroacetate salt **59** and each methyl ether (**32a–d**) in turn was treated with DEPC and triethylamine at 0°C to yield dipeptides **58a–d**. No racemization was observed, and an X-ray analysis (Fig. 5) of amide **58b** established that the chirality was preserved in the dipeptide (6*S*,9*R*,10*R*,11*S*). The four amides (**58a–d**) of known stereochemistry were each coupled with Dov-Val-Dil (Section 4.2.5). The pentapeptide derived from dipeptide **58b** proved to be identical with natural dolastatin 10. Thus, the stereochemistry of the chiral centers in the dolaphenine and dolaproine units of dolastatin 10 was found to be 6*S*,9*R*,10*R*,11*S*.

4.2.5 Conversion of Dap-Doe to Dolastatin 10

To begin with (Scheme 9), dolaisoleuine *tert*-butyl ester hydrochloride (**48a**) was coupled with Z-L-valine by use of either pivaloyl chloride (*140*) or bromotris(dimethylamino)phosphonium hexafluorophosphate (BrOP) (*141*), the latter reagent being found preferable (82% yields and no

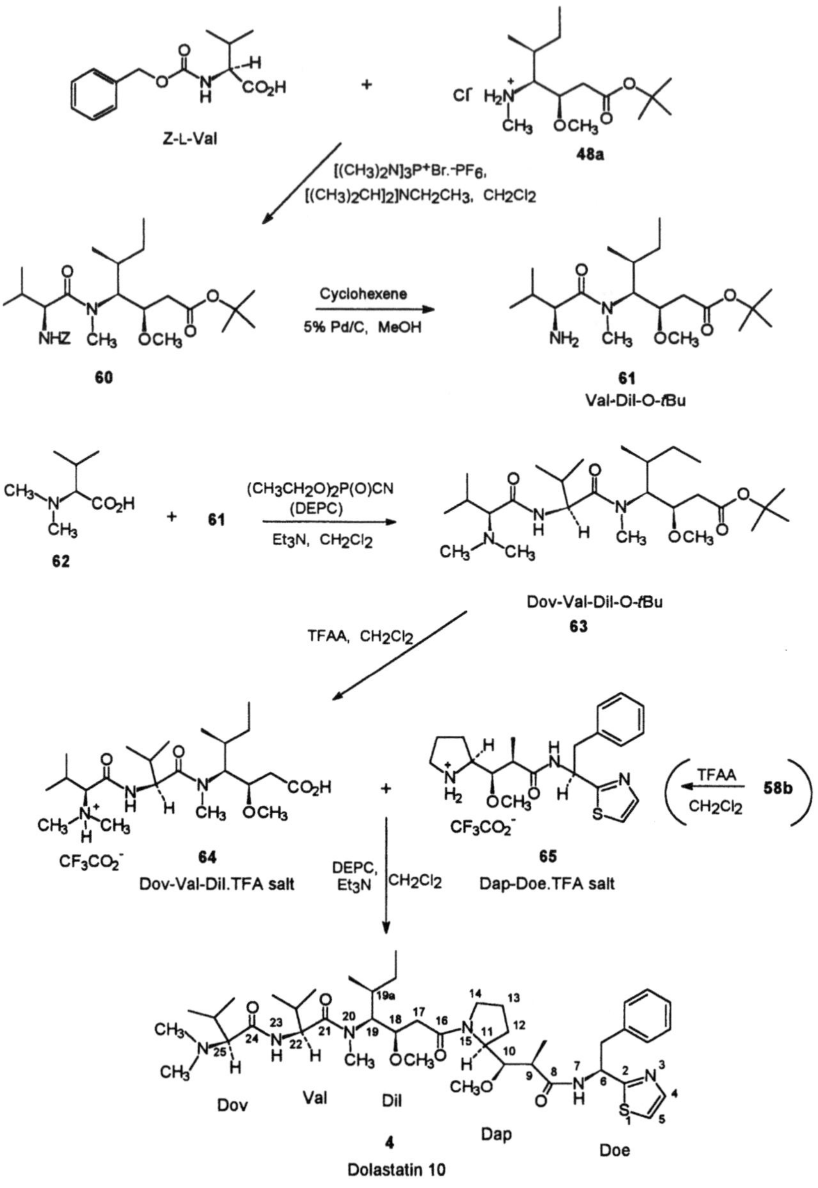

Scheme 9. Synthesis of dolastatin 10

detectable racemization). Conversion of the dipeptide product (**60**) to amine **61** was realized by use of hydrogen transfer catalyzed by 5% palladium-on-carbon suspended in cyclohexene-methanol. Diethyl phosphorocyanidate (DEPC) (*142*) was used to condense Dov [**62**, prepared (*143*) from *S*-Val] with dipeptide **61** to obtain tripeptide **63** (84% yield), and the absolute configuration was confirmed by X-ray analysis (Fig. 6). Treatment of tripeptide **63** and of Boc-Dap-Doe (**58b**) with trifluoroacetic acid furnished the trifluoroacetate derivatives, **64** and **65**, respectively, which were coupled by use of DEPC to afford dolastatin 10 (**4**) in 97% yield (*144*). The physical and spectroscopic data exhibited by the synthetic product were identical with those of the natural product, and the cell growth inhibition exhibited against the murine P388 lymphocytic leukemia was comparable (ED_{50} 4.8 × 10^{-6} μg/mL).

Recently, two new syntheses of dolastatin 10 were reported. SHIOIRI (*79*, *145*) has described single amino acid addition to Doe using diethyl phosphorocyanidate for each peptide bond-forming step except for the Boc-(*S*)-Val step where BrOP (*141*) was found superior even to bis(2-oxo-3-oxazolidinyl)phosphinic chloride (BOP-Cl). Trifluoroacetic acid was used to remove the Boc-protection. The PONCET group (*80*) has employed an analogous stepwise approach to dolastatin 10 using the BOP and BrOP reagents for coupling except for application of COMMOD for the N-terminal step. In this series both Boc- and Z-protection were employed and trifluoroacetic acid or hydrogen bromide in acetic acid were used for the cleavage steps.

Owing to increasing preclinical requirements for synthetic dolastatin 10, we recently undertook further studies (*146*) aimed at finding an improved route to the dolaphenine unit of dolastatin 10. Attempted preparation of either Boc-Doe or Z-Doe by way of the Hantzsch route at −15°C (*147*) was unsuccessful. Instead, formation of the hydroxythiazoline intermediates was favored at room temperature. Although the modified procedure was found best in the preparation of Z-Doe, partial racemization was encountered. When this modification was applied to preparation of Boc-Doe, with employment of the HOLZAPFEL (*76*) variation, a single product with properties different to those of Boc-Doe was furnished. The reaction intermediates suggested that rearrangement to thiazole **66** had occurred during the dehydration step. Apparently the

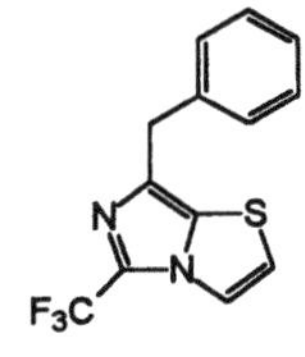

66

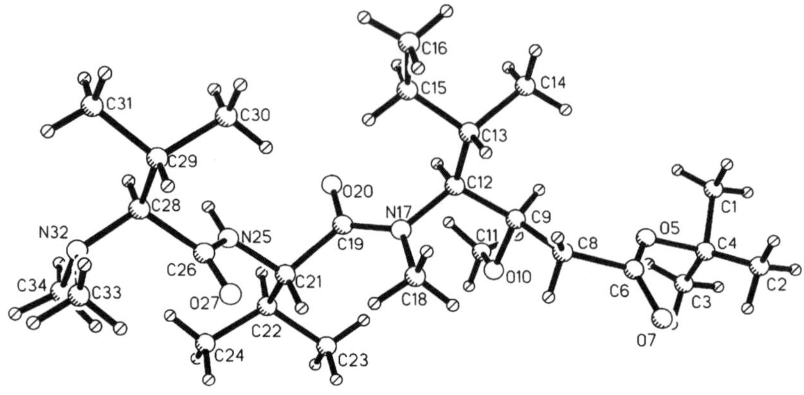

Fig. 6. Computer-generated drawing of the tripeptide derivative, Dov-Val-Dil-O-*t*Bu (**63**)

rearrangement involves anhydride-assisted cleavage of the Boc group, followed by displacement with trifluoroacetate. Excess base then allows nucleophilic attack on the carbonyl carbon by the thiazolo nitrogen, followed by dehydration to form a thiazole.

4.2.6 Chromatography of Dolastatin 10

When dolastatin 10 was first isolated (*49*) from *Dolabella auricularia*, considerable difficulty was experienced in determining whether or not it was pure; the problem was compounded by our having only a few milligrams in hand. Eventually, dolastatin 10 was found to undergo facile conformational changes with variations in experimental conditions. To assess the purity of synthetic and natural dolastatin 10, very specific chromatographic and NMR techniques were found necessary. Development of useful and reliable chromatographic procedures for analysis of dolastatin 10 (and the dolastatins in general) became an important objective (*148*). Initial HPLC analyses of dolastatin 10 were performed by use of a reverse phase (C8) column with 1% acetic acid in 3:1 acetonitrile-water. Although this system gave good resolution of dolastatin 10, results were not always reproducible and it failed to resolve a mixture of authentic dolastatin 10 and its diastereoisomer (6*R*)-isodolastatin 10 (*149*) (the *S*-Doe unit replaced by the unnatural *R*-Doe). Resolution of these two diastereoisomers was considered a stringent requirement for a useful dolastatin 10 HPLC procedure.

The use of buffered mobile phase systems proved more effective (*148*). Addition of phosphate salts (50 mM as KH_2PO_4) to a 3:1 methanol-water mobile phase (C8 column) provided well-defined peaks. However, only poor resolution of the dolastatin 10 and its (6*R*)-isodolastatin 10 isomer was realized. On the other hand, this system did result in the first firm evidence that dolastatin 10 could exist as two different conformers at room temperature. Depending upon the sample's physical history, dolastatin 10 exhibited a retention time of 3.8 min, 4.6 min, or both when the phosphate-buffered solvent system was used with a reverse phase C8 column.

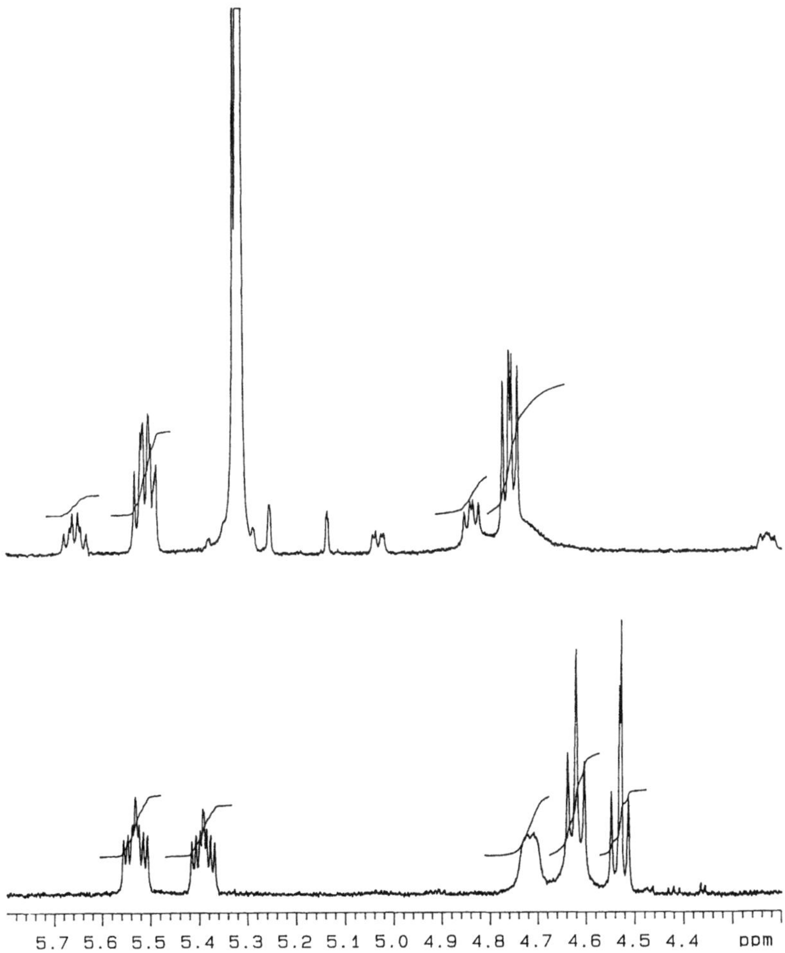

Fig. 7. Effect of solvent on conformer populations: C^{α}–H region of the ^{1}H-NMR spectrum (upper acquired in CD_2Cl_2, lower in DMSO-d_6) of dolastatin 10 (500 MHz)

In the course of evaluating other mobile phase additives we examined sodium 1-hexanesulfonate (5–10 mM $HexSO_3Na$) in a mixture of acetonitrile-2-propanol-water. Although that system could resolve dolastatin 10 and (6*R*)-isodolastatin 10 with a small difference in retention times (ΔR_f of 0.19 min), it did not give good resolution when an equimolar mixture of the isomers was chromatographed. Overall, this technique proved best for HPLC examination of other diastereomeric dolastatins, their synthetic intermediates and impurities. With this HPLC solvent, rapid equilibration to a single stable conformer occurs, and only one dolastatin 10 conformational isomer was generally observed. Although

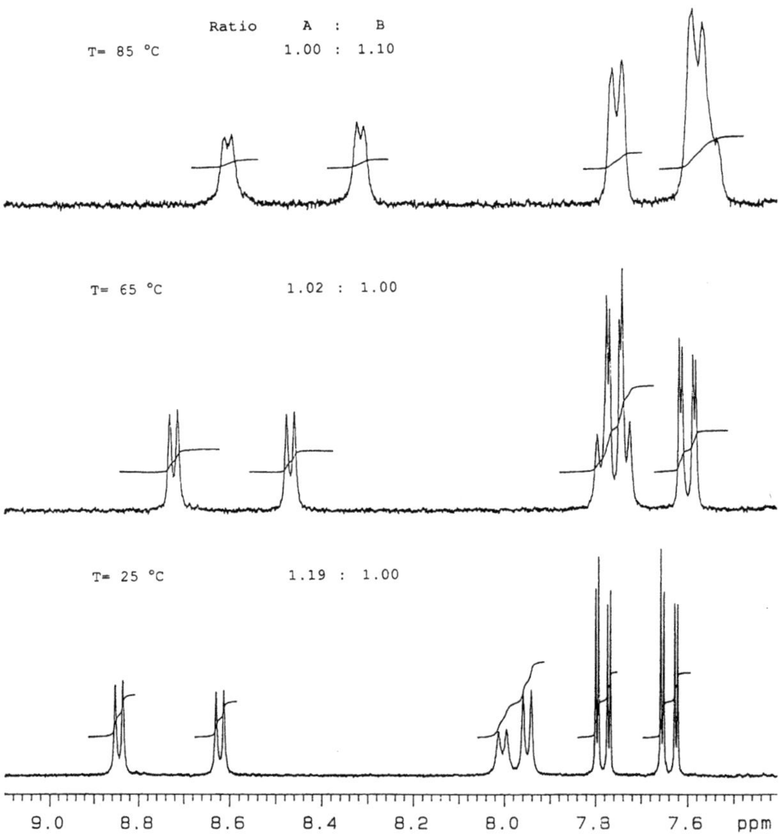

Fig. 8. Temperature dependence of dolastatin 10 conformer populations. Comparison of the amide signals of conformers A and B (500 MHz, DMSO-d_6)

References, pp. 70–79

this HPLC system proved useful for analysis of dolastatin 10 samples, it requires concomitant use of other methods (optical rotation and careful interpretation of high-field ^{1}H- and ^{13}C-NMR spectra) to assess purity.

4.2.7 High-Field NMR Analysis of Dolastatin 10

From high-field NMR studies of dolastatin 10, it is clear that this unusual peptide is a mixture of two conformers (doubling of nearly every signal in the spectra). The ratio of doubled signals varied with solvent and temperature. Spectra obtained in solvents with high dielectric constants (*e.g.*, dimethyl sulfoxide, acetone or acetonitrile) showed nearly equal conformer populations, whereas use of solvents with low dielectric constants (*e.g.*, methylene chloride or chloroform) resulted in the greatest difference in conformer populations (Fig. 7). The spectra obtained in dimethyl sulfoxide resulted in a good separation of the conformer signals, and the ratio of signals varied with temperature (Fig. 8). Later we were able to show that the conformational isomerism in dolastatin 10 arose from *cis-trans* isomerism along the Dil-Dap peptide bond. That was further supported by a detailed molecular modeling and NMR analysis of the isomeric (6*R*)-isodolastatin 10 (*148*).

4.3 Synthesis of (6*R*)-Isodolastatin 10 and Other Chiral Modifications of Dolastatin 10

Because of the clinical development of dolastatin 10 and the need to further explore its mechanism of action (*150–153*) we required an X-ray crystal structure determination of the parent compound or a comparably active derivative. We were unable to obtain X-ray quality crystals from dolastatin 10 nor from eighteen of its configurational isomers (*118*, *139*, *151*). Most of these isomers showed IC_{50} values in the range 30–90 nM as compared to 1 nM for dolastatin 10. Fortunately, (6*R*)-isodolastatin 10 (**67**), with antineoplastic activity comparable to that of dolastatin 10, was successfully crystallized and its structure determined by X-ray analysis (*149*).

An all too frequent side-reaction in the synthesis of dolaphenine (*72*, *96*) was used to advantage for synthesis of (6*R*)-isodolastatin 10 (**67**). Manganese dioxide oxidation of the penultimate thiazolidine (**18**) often resulted in some isomerization. When the resulting optically impure dolaphenine [**19** and its (6*R*)-isomer] was coupled with dolaproine (Scheme 10), the product was a mixture of two diastereoisomers that were easily separated by column chromatography. The more polar amide (**68**) was found to possess the required stereochemistry for synthesis of (6*R*)-isodolastatin 10. The structure was confirmed by an X-ray analysis (*139*). Dipeptide **68** was deprotected with trifluoroacetic acid and coupled (DEPC) with tripeptide trifluoroacetate salt **64** to obtain (6*R*)-isodolastatin 10 (**67**, quantitative yield) (*149*). Crystallization from diethyl ether-dichloromethane-hexane yielded tiny crystals (mp 100–103°C) that proved suitable at low temperature for X-ray crystal structure determination.

4.3.1 Crystal Structure of (6R)-Isodolastatin 10

Because of the N...O distances usually found in N–H...O-type hydrogen bonds, none of the carbonyl groups or amide hydrogens in (6*R*)-isodolastatin 10 (**67**) appear to be involved in intramolecular hydrogen bonding. Instead, intermolecular hydrogen bonding occurs at N-7, O-16a,

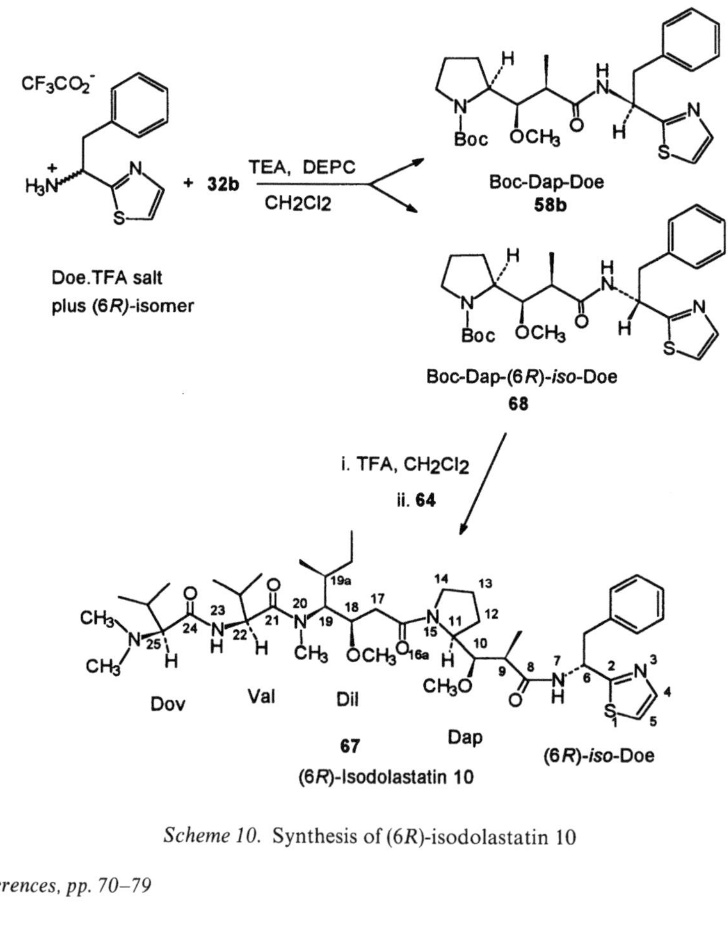

Scheme 10. Synthesis of (6*R*)-isodolastatin 10

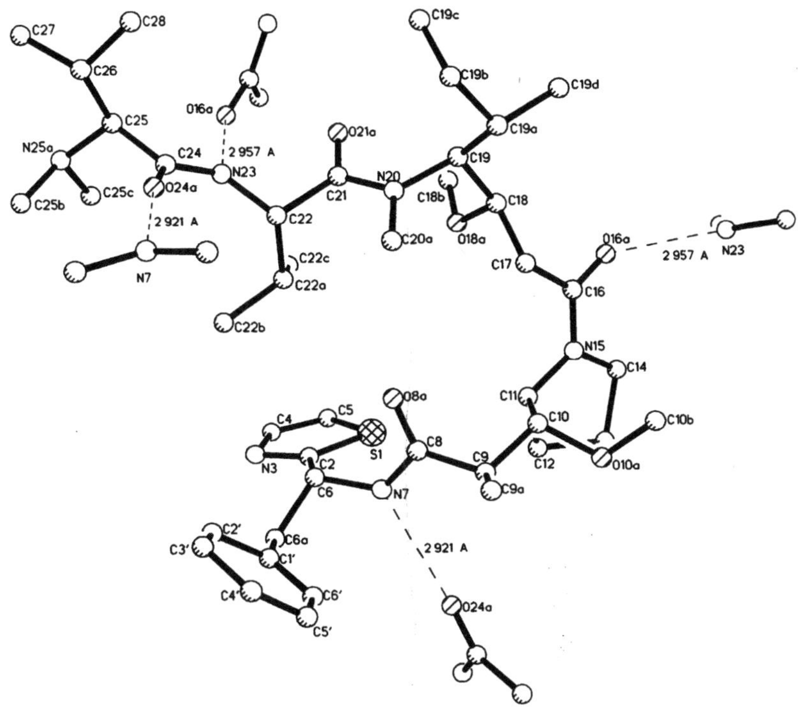

Fig. 9. Computer-generated drawing of (6*R*)-isodolastatin 10 (**67**)

N-23, and O-24a to adjacent molecules of isodolastatin 10 (see Fig. 9). Since (6*R*)-isodolastatin 10 leads to strong inhibition of tubulin assembly (*151–153*), these positions may be involved in molecular recognition of the active binding sites. Furthermore, it seems likely that H-bonding at O-16a has a role in positioning the molecule in the tubulin protein groove, producing a three-point attachment. The X-ray structure of (6*R*)-isodolastatin 10 proved to be very close conformationally to our working model that is based upon competitive tubulin binding experiments (*150–154*).

4.3.2 Molecular Modeling of (6R)-Isodolastatin 10

In the crystal structure of (6*R*)-isodolastatin 10 all amide bonds were found to be *trans* except for the Dil-Dap (N-5–N-16) peptide bond which proved to be *cis*. For the modeling studies (*149*), a distance-dependent dielectric constant of 4*R* was used throughout the calculations. The molecule was subjected to energy minimization using the ABNR algorithm, followed by simulated annealing (at 1000 K) and slow cooling to

50 K. During this process, the conformation of the peptide bond between Dil and Dap changed from *cis* to *trans*. Both conformers were first energy minimized (ABNR) and then subjected to dynamics at 300 K for 200 ps. Several low-energy structures were selected from the dynamics trajectory and minimized (ABNR). The potential energy of the various minimized

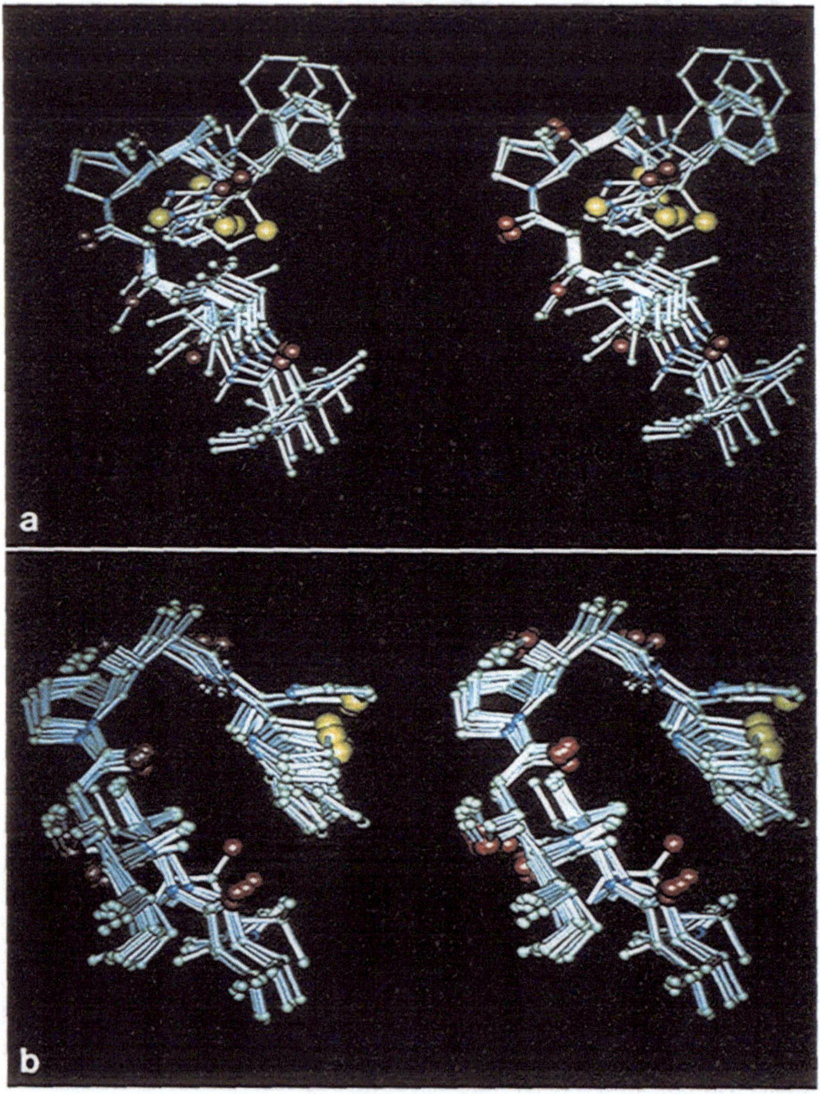

Fig. 10. Stereoview of the superimposed minimized dynamic datasets of (a) *cis*- and (b) *trans*-(6*R*)-isodolastatin 10 showing good convergence

conformers of both the *cis* and *trans* isodolastatins was found to be very close to that of the initial structures. A least-square superposition of the low-energy conformers of *cis* and *trans*-isodolastatins is given in Fig. 10. A plot of the potential energy *vs.* the corresponding dihedral angle showed the energy barrier between the *cis* and *trans* conformers to be about 21 kcal/mol. Both the *cis* and *trans* rotamers exhibited the characteristic backbone conformation over the entire dynamic trajectory. Although the *trans* conformer showed all three $\phi - \psi$ angles in the extended beta-sheet region (second quadrant), the *cis* conformer displayed the $\phi - \psi$ angle for Dil in the right-handed alpha-helix region (third quadrant). Both result in a bent structure for (6*R*)-isodolastatin 10 (**67**).

4.4 Structural Modifications of Dolastatin 10

Our initial approach (*146*) to exploring structural requirements for the extraordinary anticancer activity of dolastatin 10 involved replacement of the dolaphenine unit with easily accessible arylalkyl amides **69a–g** (Scheme 11). Here we further utilized our original synthetic approach to dolastatin 10. Thus, the required Boc-dolaproine amides **70a–g** were synthesized in high yields by the coupling of Boc-dolaproine (**32b**) and the required amines with use of diethyl phosphorocyanidate (DEPC) (*142*) as the amide-forming reagent. The Boc protecting group was removed with trifluoroacetic acid to afford the corresponding trifluoroacetates **71a–g**. Condensation of each amide with tripeptide **64** (*144*) was performed with DEPC to yield the new dolastatin 10 analogues **72a–g**.

Preparation of the new dipeptides **71h–k** and tripeptides **73a–d** was required for synthesis of pentapeptides **74a–h**. Each was synthesized following the general methods used for obtaining dolastatin 10. Dipeptides **71h–k** were prepared in good yields by the DEPC-assisted coupling of Boc-Dap (**32b**) with the appropriate amino acids. Synthesis of tripeptides **73a, b** involved peptide bond formation between Dil and Z-Ile or Z-Leu, respectively. In these reactions the coupling was accomplished by use of bromo(trisdimethylamino)phosphonium hexafluorophosphate (BrOP) (*141*). The resulting dipeptides were deprotected by way of hydrogen transfer (*144*) and coupled (DEPC) with dolavaline to provide the required tripeptides. For tripeptides **73c, d**, dipeptide Val-Dil-COOBu^t (**61**) (*144*) was coupled with the appropriate protected amino acids. The di- (**71h–k**) and tripeptides (**73a–d**) were then deprotected with trifluoroacetic acid and converted (DEPC) to peptides **74a–h**. The preceding dolastatin 10 structural changes provide an illustration of over one hundred such modifications we have made in dolastatin 10.

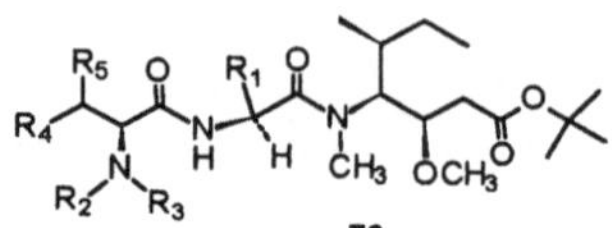

73

a, R_1 = *s*-Bu, $R_2 = R_3 = R_4 = R_5 = CH_3$ (Dov-Ile-Dil)

b, R_1 = *i*-Bu, $R_2 = R_3 = R_4 = R_5 = CH_3$ (Dov-Leu-Dil)

c, R_1 = *i*-Pr, $R_2 = R_4$ = H, R_3 = Z, R_5 = (

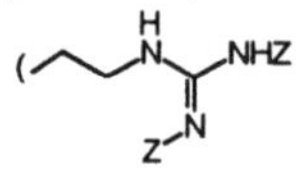

d, R_1 = *i*-Pr, $R_2 = R_4$ = H, R_3 = Z, $R_5 = (CH_2)_3NHZ$

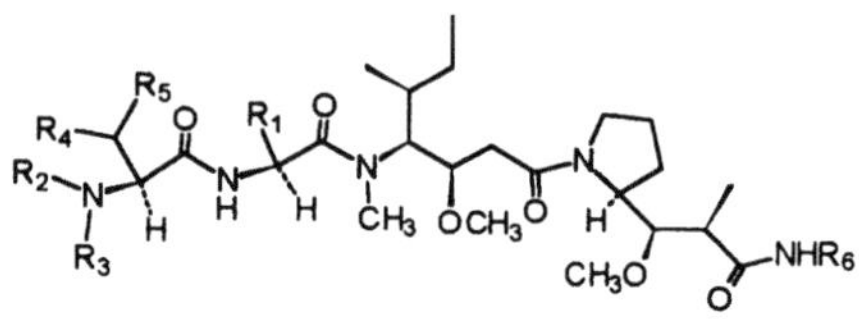

74a-h

a, R_1 = *i*-Pr, $R_2 = R_3 = R_4 = R_5 = CH_3$, R_6 = (

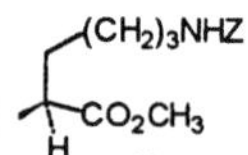

b, R_1 = *i*-Pr, $R_2 = R_3 = R_4 = R_5 = CH_3$, R_6 = (

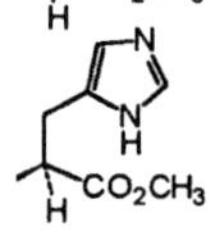

c, R_1 = *s*-Bu, $R_2 = R_3 = R_4 = R_5 = CH_3$, R_6 = (

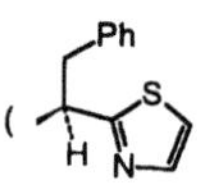

d, R_1 = *s*-Bu, $R_2 = R_3 = R_4 = R_5 = CH_3$, R_6 = (

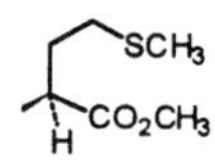

e, R_1 = *i*-Bu, $R_2 = R_3 = R_4 = R_5 = CH_3$, R_6 = (

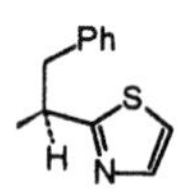

f, R_1 = *i*-Pr, $R_2 = R_4$ = H, R_3 = Z, $R_5 = (CH_2)_3NHZ$, R_6 = (

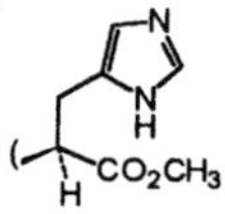

g, R_1 = *i*-Pr, $R_2 = R_4$ = H, R_3 = Z, R_5 =(

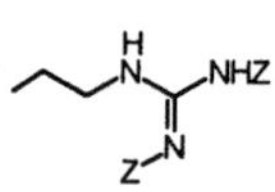

, R_6 = (

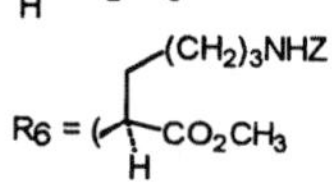

h, R_1 = *i*-Pr, $R_2 = R_4$ = H, R_3 = Z, R_5 =(

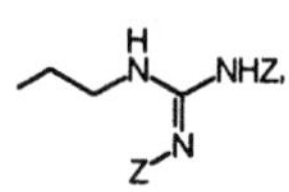

, R_6 = (

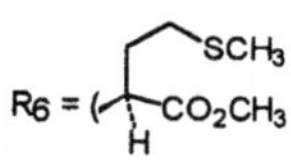

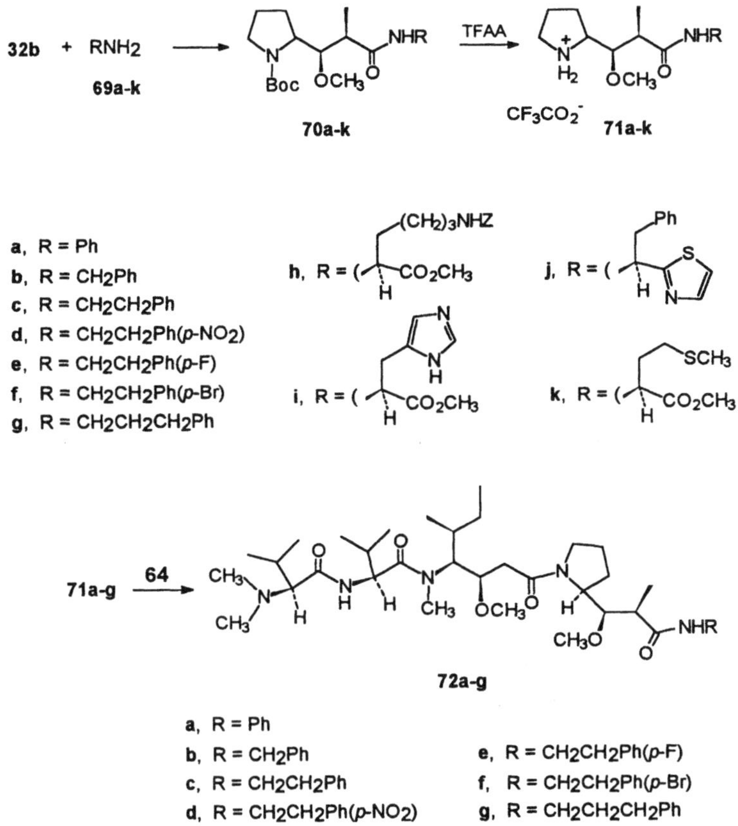

Scheme 11. Synthesis of analogues of dolastatin 10

4.5 Synthesis of Dolastatin 15

Structurally, the most interesting unit of the depsipeptide dolastatin 15 (**9**) is (*S*)-dolapyrrolidone (Dpy). Such modified amino acids are presumably derived biosynthetically from phenylalanine through a two-carbon condensation. Natural products containing a glycine-derived pyrrolidone C-terminus have previously been found in *Streptomyces* (*e.g.*, the antibiotic althiomycin) (*155*) and include the blue-green algae components malyngamide (*156*) and pukeleimide (*157*). Dysidin (*158*), a constituent of both sponges and blue-green algae, contains a valine-derived pyrrolidone C-terminus. A hexachloro metabolite, dysidamide (*159*), has been isolated from a *Dysidea* species of sponge. Interestingly, Hiva has been found to be incorporated in the Hip unit of the potent tunicate didemnins (*103*, *104*).

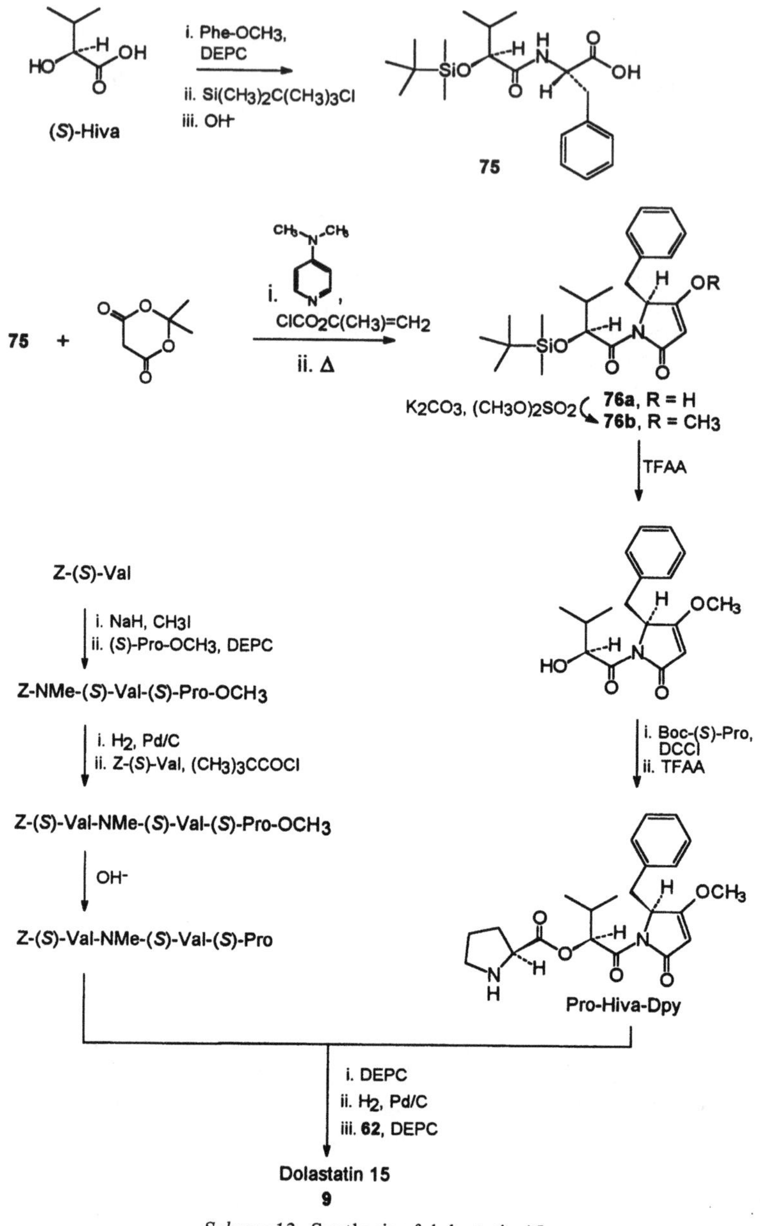

Scheme 12. Synthesis of dolastatin 15

Except for synthesis of the dolapyrrolidine moiety which was accomplished *via* acylation (*160*) of Meldrum's ester, total synthesis of dolastatin 15 proceeded smoothly as outlined in Scheme 12 (*161*). Isopropenyl chloroformate was found superior to several other reagents for generation of reactive mixed carbonic anhydrides derived from carboxylic acid **75** and was used with 4-dimethylaminopyridine. After removal of base with 10% aqueous $KHSO_4$, the Meldrum's ester adduct was heated in refluxing toluene to afford (in 68% yield) pyrrolidone **76a**. Methylation of the tautomeric mixture (**76a**) with dimethylsulfate and potassium carbonate in tetrahydrofuran gave methyl vinyl ether **76b**, without any detectable *C*-alkylated product. Some racemization was detected at the Phe center to give (*S*,*S*)- and (*S*,*R*)-*tert*-butyldimethylsilyl-Hiva-Dpy (**76b**) in an approximate ratio of 7.4:1. The yield of pyrrolidone **76a** fell dramatically to 21% (and lower) when the Meldrum's ester adduct was cyclized in methanol (*162*). In 1992 PONCET and co-workers reported a similar synthesis of dolastatin 15 (*163*).

4.6 Synthesis of Dolastatin C

Isolation and structure determination of the marginally cytotoxic dolastatin C (**10**) by YAMADA and colleagues (*54*) proved to be interesting.

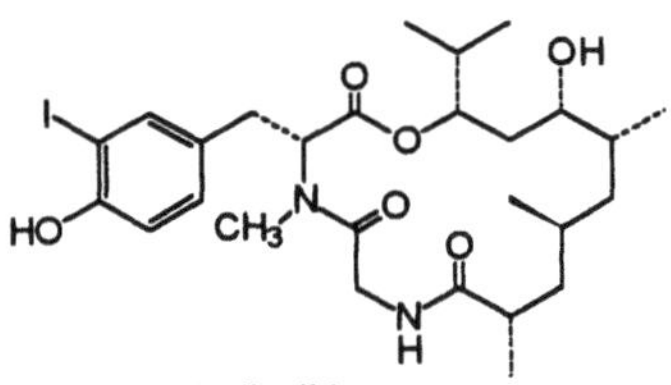

Doliculide

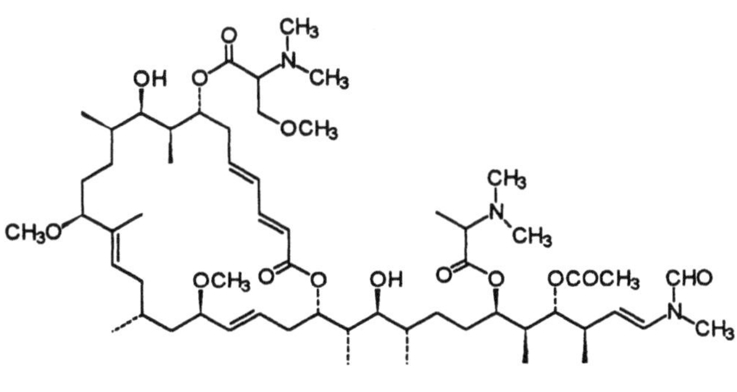

Aplyronine A

Because this depsipeptide appears to be part of an early step in the biosynthesis of dolastatin 10 that may have been retained for a subsequent different role in Japanese *D. auricularia*, it is useful to outline the first synthesis (Scheme 13) (*54*). YAMADA and co-workers (*55*) have also completed (1994) workable syntheses of the cyclic depsipeptide dolastatin D (**11**) and doliculide (*58*, *164*) (from the same sea hare). The Nagoya University group directed by Professor Yamada has also been pursuing the antineoplastic (*165*, *166*) and other (*167*) constituents of another Japanese sea hare, namely *Aplysia kurodai*. Here the isolation, structure determination, and first total synthesis (*166*) of the strongly active antineoplastic agent aplyronine A is especially noteworthy.

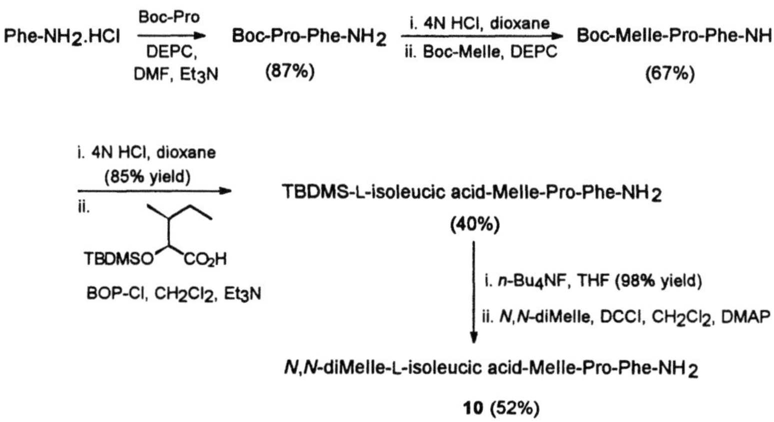

Scheme 13. Synthesis of dolastatin C

5. Spectral and Analytical Characterizations

5.1 Dolastatin 10

Extensive applications of high-resolution EI, FAB and tandem mass spectrometry, combined with results of rigorous two-dimensional ^{1}H- and ^{13}C-NMR, NOE, and HMBC NMR experiments, allowed structural assignments for dolastatins 10–15. As none of these complex peptides could be induced to crystallize, confirmation of the structure by crystallographic techniques was precluded. However, the dolastatin 10 sequence was unequivocally assigned (*49*) on the basis of SP-SIMS measurements

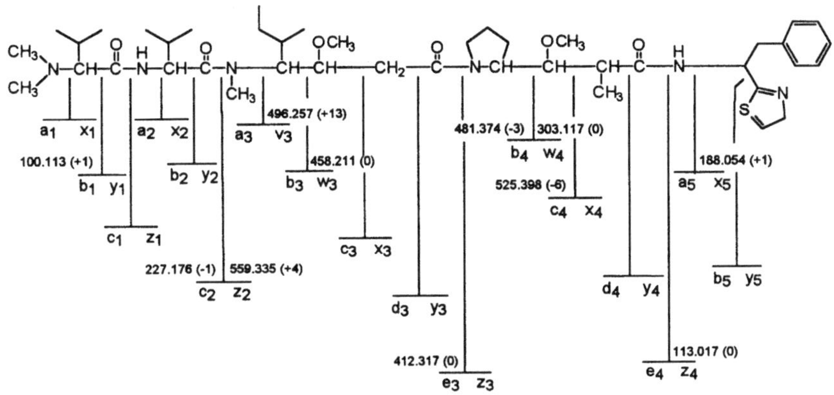

Fig. 11. Dolastatin 10 mass (M^+ *m/e* 784.489) spectral fragmentation

Table 2. *Collision-Induced Dissociations of Selected Higher Masses from the SP-SIMS Fragmentation of Dolastatin 10*

	Parent Ions		
785	559	458	412
M + H	$z_2 + 2H$	w_3	e_3
	Daughter Ions[a,b]		
753	188	426	100
$M + H - CH_3OH$	x_5	$w_3 - CH_3OH$	b_1
100	205	86	214
b_1 and/or $(z_2b_3) + H$	z_4	(w_3/e_3)	$(y_2/e_3) + H$
188[c]	374	188	380
x_5	$z_3 + 2H$	x_5	$e_3 - CH_3OH$
559	100	356	352
$z_2 + 2H$	$(z_2/b_3) + H$		$d_3 - CH_3OH$
154	154	205	311
$(z_2/e_3) + H - CH_3OH$	$(z_2/e_3) + H - CH_3OH$	$z_4 + 2H$	$(y_1/e_3) - H$
205	170	170	253
$z_4 + 2H$	$(z_3/e_4) + H$	$(z_3/e_4) + H$	
458	138	138	227
w_3	$(z_3/e_4) + H - CH_3OH$	$(z_3/e_4) + H - CH_3OH$	c_2
170	527	303	368
$(z_3/e_4) + H$	$z_2 + 2H - CH_3OH$	w_4	(x_1/e_3)

[a] See Fig. 11 for the letter designations.
[b] The eight strongest derivative ions are listed in order of decreasing abundances.
[c] The very abundant *m/z* 188 ion seems to mask a *m/z* 186 ($(z_2/e_3) + H$), but its derived ion appears at *m/z* 154.

Table 3. *Dolastatin 10 Correlated ^{13}C- and ^{1}H-NMR Assignments in Dichloromethane-d_2 Solution*[a]

Structure assignment	^{13}C (mult)	Chemical shift, ppm ^{1}H (mult, *J* (Hz), integration)	NOE
2	170.51 (w)		
4	142.77 (d)	7.717 (d, 3.3, 1 H)	
5	118.76 (d)	7.254 (d, 3.3, 1 H)	
6	53.02 (d)	5.516 (ddd, 5.7, 7.2, 9.3, 1 H)	
6a	41.48 (t)	3.399 (dd, 5.7, 14, 1 H)	
		3.256 (dd, 5.7, 14, 1 H)	
6b1	137.74 (s)		
6b2, 6b6	128.74 (d) × 2	7.243 (d, 7.9, 2 H)	
6b3, 6b5	129.80 (d) × 2	7.214 (dd, 7.9, 9.2, 2 H)	
6b4	127.02 (d)	7.194 (t, 9.2, 1 H)	
7		7.256 (t, 7.6, 1 H)	9, 9a, 10, 10ab, 11
8	175.67 (s)		
9	44.79 (d)	2.282 (quintet, 7.2, 1 H)	7
9a	14.49 (q)	1.085 (d, 7.1, 3 H)	
10	82.05 (d)	3.845 (dd, 2.0, 8.2, 1 H)	7
10ab	60.89 (q)	3.309 (s, 3 H)	6, 7
11	59.86 (d)	3.985 (m, 1 H)	
12	25.00 (t)	1.804 (ddd, 5.5, 7.0, 19, 1 H)	
		1.600 (ddd, 7, 9.2, 19, 1 H)	
13	25.45 (t)	1.446 (ddd, 4.7, 7, 19, 1 H)	
		1.75 (ddd, 4.7, 7.8, 12.7, 1 H)	
14	48.03 (t)	3.401 (dd, 7.8, 10, 1 H)	17
		3.390 (m, 1 H)[b]	
16	174.01 (s)		
17	38.11 (t)	2.394 (ABq, 9.0, 2 H)	14
18	78.86 (d)	4.122 (br t, 8.7, 1 H)	
18ab	58.16 (q)	3.313 (s, 3 H)	22
19	54.11 (d)	3.26–3.39 (H)[c]	
19a	33.62 (d)	1.680 (1 H)[b]	
19b	26.25 (t)	1.370 (br m, 1 H)	
		1.000 (br m, 1 H)[b]	
19c	10.92 (q)	0.832 (t, 7.4, 3 H)	
19d	19.82 (q)	1.003 (d, 6.8, 3 H)	
20a	30.90 (q)	3.012 (s, 3 H)	22
21	171.39 (s)		
22	54.20 (d)	4.761 (dd, 6.5, 8.8, 1 H)	18, 18ab, 20, 19
22a	31.42 (d)	1.983 (sextet, 6.7, 1 H)	
22b	18.18 (q)	0.941 (d, 6.8, 3 H)	
22c	16.09 (q)	0.977 (d, 6.8, 3 H)	
23		6.861 (d, 8.9, 1 H)	25, 25bc
24	172.44 (s)		
25	76.77 (d)	2.454 (d, 6.9, 1 H)	23
25bc	49.92 (q) × 2	2.262 (s, 6 H)	23
26	28.08 (d)	2.073 (sextet, 6.7, 1 H)	
27	20.24 (q)	0.964 (d, 6.8, 3 H)	
28	18.18 (q)	0.902 (d, 6.8, 3 H)	

[a] Residual $CHDCl_2$ as internal reference (5.32 ppm).
[b] Overlapping signal.
[c] Signal assigned from NOE data.

in conjunction with the collision-activated (-induced) decomposition (CAD or MS/MS or tandem mass spectrometry with a Kratos MS-50-3F) of the SP-SIMS ions (and fragments resulting from decarbonylation, see Fig. 11 and Table 2). Of the two final sequences, namely Dov-Val-Dap-Dil-Doe or Dov-Val-Dil-Dap-Doe, the latter was found to be correct by CAD in conjunction with HREIMS (Fig. 11 and Table 2) experiments using the intact peptide. Final confirmation of structure **4** for dolastatin 10 was obtained by H-[^{1}H]-NOE difference NMR experiments as summarized in Table 3. As noted above (Section 4.0), the structure of dolastatins 10 and 15, including stereochemical assignments, were substantiated by total syntheses.

5.2 Dolastatins 11 and 12

Examination of the proton and carbon NMR spectra indicated that dolastatin 11 was primarily a peptide (*50*). By means of a combination of ^{1}H,^{1}H-COSY, ^{1}H,^{13}C-COSY, and ^{1}H,^{1}H-relayed COSY, all of the spin systems were derived. The presence of two glycine and two alanine units and one unit each of *N*-methylleucine (MeLeu), *O*,*N*-dimethyltyrosine (*O*,*N*-diMeTyr), and *N*-methylvaline (MeVal) units was deduced. Glycine and alanine were also detected by amino acid analysis of the dolastatin 11 acid hydrolysis (6 N HCL; 110°C; 24 h) product. Three other structural units were found: isobutyric acid (Ibu), 2-hydroxy-3-methylpentanoic acid (Hmp), and the *β*-amino acid, 2-methyl-3-aminopentanoic acid (Map). The sequence of dolastatin 11 was first deduced by NOE experiments and further established by a sensitivity-enhanced heteronuclear multiple bond correlation experiment (HMBC). Analysis of the collision-activated spectra of selected FAB-produced ions from dolastatin 11 and from its saponification (0.1 N NaOH; 5.5 h, rt) product afforded the same conclusion. Furthermore, the all-L pentapeptide hydrochloride Gly-MeLeu-Gly-MeVal-*O*,*N*-diMeTyr-OMe was synthesized and its NMR characteristics were found quite comparable with that segment of dolastatin 11.

Spectral analyses of dolastatins 11 and 12 gave almost identical results. Inspection of the 2D NMR spectra (^{1}H and ^{13}C) suggested differences in only two units: the N-8 amide proton of the dolastatin 12 Ala unit was found to be replaced by a methyl group, and the Tyr methoxyl group by a hydrogen to give the phenylalanine analogue. These minor structural modifications were in accord with the observed mass of dolastatin 12 which was 16 amu lower than that of dolastatin 11 (**5**) and led to structure **6** for dolastatin 12 (*50*).

5.3 Dolastatin 13

From results of detailed high-field (400-MHz) ^{1}H- and ^{13}C-NMR, and high-resolution FABMS peak-matching experiments, molecular formula $C_{46}H_{63}N_7O_{12}$ was deduced for dolastatin 13 (*51*). A combination of ^{1}H,^{1}H COSY, ^{1}H,^{13}C COSY, and ^{1}H,^{1}H relayed COSY experiments indicated eight discrete spin-coupled systems of which four corresponded to the well-known amino acids threonine, *N*-methyl-phenylalanine, and valine (two units). Threonine and the two valine units were also detected by amino acid analyses of the products from acid-catalyzed (6 N HCl, 110°C, 24 h) hydrolysis. Assignment of the fifth and sixth units as an N,N-disubstituted phenylalanine derivative was realized by NMR spectrometry. From a series of double and triple relayed coherence transfer experiments (homonuclear relay) and from sensitivity-enhanced HMBC, the sixth unit was found to be the new cyclic hemiacetal Ahp (3-amino-6-hydroxy-2-piperidone), presumably derived from a Phe-Glu dipeptide precursor (Gly-γ-carboxyl-γ-aldehyde). Continuation of the NMR experiments led to assignment of the seventh unit as the rare dehydro amino acid α,β-dehydro-2-aminobutanoic acid (*cis*-Abu), presumably from dehydration of Thr, and the eighth, as 2-*O*-methylglyceric acid (MeGlc).

The Δ-Abu olefin was only tentatively assigned. The correct sequence of units was achieved by using HMBC and by combining the results from experiments in two different solvents (Table 4). With dichloromethane-d_2

Table 4. *^{1}H- and ^{13}C-NMR Assignments and Selected NOE's for Dolastatin 13 in CD_2Cl_2* Solution and HMBC Correlations from CD_2Cl_2 *(0.031 M) and C_5D_5N (0.015 M) Solutions*[a]

	C no.	^{13}C	^{1}H (mult, *J* (Hz), integration)	NOEs	HMBC[b]
Val_2	2	174.38			
	3	59.05	4.18 t, 7.5		2, 3a, 3b, 3c, 5
	3a	30.93	2.02 oct, 6.8	4	3, 3b, 3c
	3b	19.26	0.92 d, 6.7, 3 H		3, 3a, 3c
	3c	19.59	0.95 d, 6.9, 3 H		3, 3a, 3b
	4		7.42	3a	5, 6
MePhe	5	171.86			
	6	62.39	5.28		5, 6a, 6b
	6a	34.14	3.85 dd, 14.0, 2.8	6	
			2.81 dd, 14.0, 11.5	6, 6c	6, 6b, 6c
	6b	137.96			
	6c	129.72	7.35		6a, 6d, 6e
	6d	129.31	7.36		6b, 6c
	6e	127.55	7.26		6d
	7a	31.69	2.89 s, 3 H		6, 8

Table 4 (*continued*)

	C no.	^{13}C	^{1}H (mult, J (Hz), integration)	NOEs	HMBC[b]
Phe	8	172.49[c]			
	9	51.46	5.07 dd, 11.2, 5.0	9a, 9c, 15	8, 9a, 9b, 11, 15
	9a	35.27	2.93 dd, 14.6, 11.2	9c, 15	8, 9, 9b
			2.09 dd, 14.6, 4.8		9, 9b
	9b	136.28			
	9c	129.55	6.84 dd, 7.7, 2.1	9, 9a, 15	9a, 9d, 9e
	9d	128.69	7.24		9b, 9c
	9e	127.30	7.23		9d
Ahp	11	171.86			
	12	49.52	4.05		11
	13	22.21	2.21 dq, 3.0, 14.6	15, 16	11, 12, 14, 15
			1.68		
	14	29.80	1.85 ddd, 14.0, 5.0, 2.6		12
			1.59		
	15	75.85	5.22	9, 9a, 9c, 12	11, 13
	15a		3.98		
	16		7.12 d, 9.5	19	12, 17
ΔAbu	17	163.78			
	18	129.53			
	18a	134.15	6.76 br q, 7.0		17, 18b
	18b	13.24	1.66 d, 7.0, 3 H	19	17, 18, 18a, 20
	19		7.99 br s	16, 21, 23, 29	20
Thr	20	169.94			
	21	55.40	4.90 d, 10.4	22b	20, 22, 22a, 24
	22	73.96	5.24	3	
	22a	21.06	1.41 d, 6.8, 3 H	21	21, 22
	23		6.72 d, 10.4	19, 26	24
Val_1	24	172.56[c]			
	25	61.93	4.01		24, 25a, 25b
	25a	29.59	2.41 oct, 6.9	23, 26	24, 25, 25b, 25c
	25b	19.42	1.13 d, 6.9, 3 H		25, 25a, 25c
	25c	19.82	1.11 d, 6.9, 3 H		25, 25a, 25c
	26		7.38	25a, 25b	25, 25a, 27
MeGlc	27	172.82			
	28	83.43	3.74 t, 1.8	28b	27, 28a
	28a	58.23	3.38 s, 3 H	28	28
	29	63.09	4.01		
			3.63 ddd, 12.0, 4.7, 1.8	29b	27, 28
	29a		4.42 t, 5.1	19, 29	29

[a] Where no multiplicity is noted, it could not be determined owing to overlapping signals.
[b] Underlined numbers are carbonyl carbons.
[c] Chemical shift values are interchangeable.

Table 5. *Dolastatin 14 Correlated* ^{13}C- *and* ^{1}H-*NMR Assignments in Dichloromethane-*d_2 *Solution*[a]

Structure assignment	^{13}C	Chemical shift, ppm, ^{1}H mult (J, Hz)	Structure assignment	^{13}C	Chemical shift, ppm, ^{1}H mult (J, Hz)
2	171.0		18b	137.7	
3	55.8	4.82 q (7.0)	18c	129.9 × 2	7.22 m
3a	16.1	1.45 d (7.0)	18d	128.6 × 2	7.21 m
4a	29.9	2.75 s	18e	126.9	7.19 m
5	196.4		19a	31.1	3.11 s
6	49.7	5.77 dd (9.3, 5.3)	20	172.9	
6a	35.5	2.37 dd (15.3, 5.3) 2.99 dd (15.3, 9.3)	21	55.2	4.65 t (9.9)
6b	172.1		21a	30.8	1.95 d heptet (9.9, 6.6)
6c		5.25 brs, 5.75[b]	21b	18.6	0.90 d (6.7)
7a	30.5	2.93 s	21c		0.64 d (6.5)
8	171.6		22		6.01 d (9.9)
9	57.6	4.66 dd (8.8, 4.0)	23	169.3	
9a	28.7	1.67 m, 2.02 m	24	128.4	
9b	25.1	1.67 m, 1.87 m	24a	12.9	1.88 (2.2)
9c	47.7	3.52 m, 3.68 m	25	134.7	6.70 dq (11.2, 2.2)
11	168.6		26	127.8	6.35 dd (14.9, 11.2)
12	60.1	4.89 d (10.8)	27	138.0	5.99 ddd (14.9, 8.8, 6.1)
12a	27.2	2.08 d heptet (10.8, 6.7)	28	37.1	2.30 m, 2.48 m
12b	18.1	0.50 d (6.8)	29	80.0	3.27 m
12c	19.8	0.88 d (6.5)	29a	56.9	3.32 s
13	30.4	2.64 s	30	34.4	1.47 m
14	170.3		31	29.7	2.05 m
15	58.9	5.01 d (10.7)	32	131.2	5.44 dt (15.1, 6.1)
15a	27.9	2.15 d heptet (10.7, 6.7)	33	130.8	5.38 dt (15.1, 6.1)
15b	18.5	0.72 d (6.8)	34	33.1	2.02 m
15c	19.7	0.78 d (6.5)	35	26.6	1.30 m
16a	29.9	2.67 s	36	35.6	1.37 m, 163 m
17	169.6		37	73.1	4.79 m
18	55.0	5.74 dd (8.4, 7.0)	37a	19.8	1.20 d (6.1)
18a	35.6	2.83 dd (13.7, 7.0) 3.25 dd (13.7, 8.4)			

[a]Residual $CHDCl_2$ as internal reference (δ 5.32).
[b]Overlapping signal.

as solvent, several segments of dolastatin 13 were established, but owing to overlapping signals in the carbonyl region it was necessary to use data obtained in pyridine-d_5, where only two carbonyl chemical shifts overlapped. The structure of dolastatin 13 (**7**) was confirmed by a sequence determination using FAB combined with tandem (MS/MS) mass spectrometry. Upon standing in chloroform solution, dolastatin 13 selectively lost 1 mol equiv of water to yield dehydrodolastatin 13.

5.4 Dolastatin 14

Unfortunately, collision-activated decomposition (MS/MS) of HREI ions from the mass spectral fragmentation of dolastatin 14 gave results that were difficult to interpret. An unequivocal structure for dolastatin 14 (**8**) was eventually deduced (*52*) by a combination of various high-field (400-MHz) NMR techniques that included $^{1}H,^{1}H$-COSY, $^{1}H,^{13}C$-COSY (consult Table 5) and $^{1}H,^{1}H$-relayed COSY. By means of these methods the new hydroxyhexadecenoic acid (named dolatrienoic acid) and amino acid units were ascertained. Two of the olefin segments of dolatrienoic acid were clearly *E* as the respective vinyl protons were coupled with $J = 15$ Hz. The *E* configuration for the trisubstituted double bond was derived from the very large value of the ^{1}H-[^{1}H] NOE difference experiment involving the C-24a methyl protons and the proton at C-26. The sequence of dolastatin 14 components was established primarily by results of COSY, HETCOR, NOE, and HMBC NMR experiments in deuteromethylene chloride. Lack of an N-10 to C-11 connection in the HMBC spectra was circumvented by the strong NOEs observed in both directions between the protons at C-9c and C-12. In deuteromethylene chloride the carbonyl group chemical shifts of C-23 and C-5 fell within 0.1 ppm of one another and did not provide an unambiguous linkage assignment in the ring system. However, in deuteroacetone the difference increased to 0.3 ppm, and their connections were then safely established using HMBC methods. When the C-6a proton absorbing at δ 2.99 was irradiated, strong NOEs were observed for both amide protons at N-6c, thereby locating this otherwise elusive amino group. Based on our complete structure determinations for dolastatins 3 (**3**) (*47*), 10 (**4**) (*49*), and 15 (**9**) (*53*), the Phe, Pro, and Val amino acid units of dolastatin 14 may bear the usual *S* configuration.

5.5 Dolastatin 15

Initially the presence of valine, *N*-methylvaline (MeVal), and dolavaline (Dov) was ascertained by high-field (400-MHz) NMR using $^{1}H,^{1}H$-COSY, $^{1}H,^{13}C$-COSY and $^{1}H,^{1}H$-relayed COSY. By these 2D NMR techniques, followed by a homonuclear Hartman-Hahn 2D experiment

(HOHAHA), we were able to unambiguously establish the presence of two proline units and a 2-hydroxyisovaleric acid (Hiva) unit. By subtraction of the elemental composition of the six units hereby derived from the molecular formula of dolastatin 15, the seventh unit was found to correspond to $C_{12}H_{12}NO_2$. By NMR this segment contained a phenyl ring, an isolated olefinic methine, a methoxyl group, and an isolated —CH_2CH— as recognizable spin systems. Further structural deductions required NOE and HMBC NMR procedures. A series of ^1H–[^1H] NOE difference experiments indicated that the phenyl ring was attached to the methylene group, forming part of a phenylalanine system, and that the methoxyl group was located near the olefinic methine proton. Results of the HMBC experiment revealed a new phenylalanine biosynthetic product designated dolapyrrolidone (Dpy). Presumably, Dpy might originate biosyn-

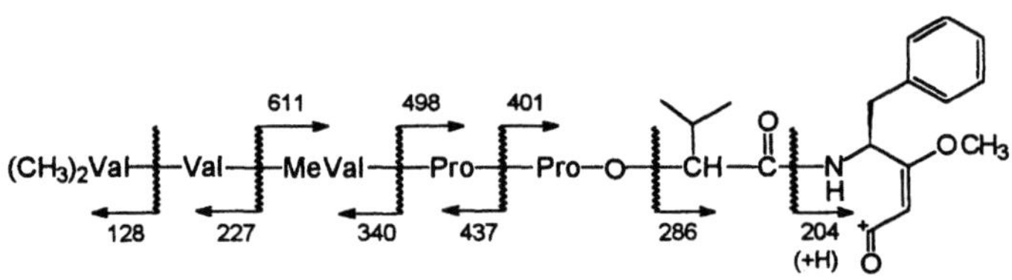

Scheme 14. Mass spectral fragmentation of dolastatin 15

Table 6. *Dolastatin 15* (**9**) *NOE and HMBC Correlations*

^{1}H-position	NOE	HMBC
2	3b	1, 4
3b	2	3
4		2, 3
4a	4c	3
7		6, 9
10		9
13	10, 12, 16	
16	13, 18, 19	15
18	16, 19	
19	22	
22	19	21, 24
22b	26	
23a	22a, 22b, 25	22, 24
25		24
26	28, 29ab	27
28	25, 26	27
29ab		28

thetically from *N*-acetylphenylalanine methyl ester by intramolecular condensation. Sequence determination of dolastatin 15 (**9**) was achieved using ^{1}H–[^{1}H] NOE and long-range proton carbonyl coupling information from the HMBC experiment (Table 6). The assigned structure **9** (*53*) was further supported by collision-activated decomposition (MS/MS) of the SP-SIMS ions combined with HREI mass spectral results (see Scheme 14). The absolute configuration was established by combination of an X-ray analysis and synthesis (see Section 4.5).

Interestingly, dolastatins 10–15 (*48*) appear only remotely related. Hence, this sea hare's facility for selecting and concentrating cytostatic and antineoplastic constituents and/or synthesizing such substances is a truly virtuoso performance.

6. Cytostatic and Antineoplastic Activities

6.1 Dolastatins 10–15

The powerful activity of dolastatin 10 against animal and human cancer cell lines is well illustrated by viewing the U. S. National Cancer Institute's (NCI) human cancer cell line panel results (Table 7) for this remarkable peptide. The following results from a comparison of dolastatin 10 with the next most powerful *D. auricularia* constituent, dolastatin 15, against the NCI human cancer cell line panel are also quite revealing (*48*). The averaged individual $-\log_{10} GI_{50}$ values obtained for dolastatin 10 and dolastatin 15, respectively, from each of the 60 cell lines are provided, with each cell identifier: CCRF-CEM (10.21, 9.60), HL-60TB (10.96, 10.00); K-562 (10.43, 9.70), MOLT-4 (10.00, 9.39), RPMI-8226 (10.25, 9.51), SR (9.38, 9.44); A549/ATCC (9.77, 8.31), EKVX (8.82, 8.38), HOP-18 (8.00, 8.00), HOP-62 (9.92, 8.70), HOP-92 (9.17, 8.59), HCI-H266 (9.92, 8.49), NCI-H23 (9.96, 8.85), NCI-H322M (9.08, 8.08), NCI-H460 (10.82, 9.28), NCI-H522 (10.49, 8.89), LXFL 529 (10.07, 9.27); DMS 114 (10.00, 9.37), DMS 273 (10.55, 9.64); COLO 205 (10.54, 9.20); DLD-1 (10.00, 9.27), HCC-2998 (10.17, 9.15), HCT-116 (10.21, 9.30), HCT-15 (9.68, 8.68), HT29 (11.02, 9.42), KM12 (10.36, 8.57), KM20L2 (10.85, 9.19), SW-620 (10.54, 9.48), SF-268 (9.47, 8.51), SF-295 (10.00, 8.77), SF-539 (10.19, 9.33), SNB-19 (9.00, 8.46), SNB-75 (10.12, 8.68), SNB-78 (9.39, 8.85), U251 (9.80, 8.72), XF-498 (9.92, 8.72); LOX IMVI (10.07, 9.38), MALME-3M (10.25, 8.96), M14 (10.39, 8.96), M19-MEL (10.11, 9.41), SK-MEL-2 (10.23, 9.28), SK-MEL-28 (8.89, 8.20), SK-MEL-5 (10.62, 9.06), UACC-257 (9.25, 8.40), UACC-62 (9.68, 9.00); IGROVI (9.92, 9.17), OVCAR-3 (10.80, 9.70), OVCAR-4 (9.19, 8.46), OVCAR-5 (9.55, 8.00), OVCAR-8

Table 7. *National Cancer Institute Development Therapeutic Program Mean Graph:*

Panel/Cell Line	Log_{10} GI50
Leukemia	
CCRP-CEM	−10.21
HL-60(TB)	−10.96
K-562	−10.43
MOLT-4	−10.00
RPMI-8226	−10.25
SR	−9.38
SR	−11.13
Non-Small Cell Lung Cancer	
A549/ATCC	−9.77
EKVX	−8.82
HOP-13	−8.00
HOP-62	−9.92
HOP-92	−9.17
NCI-H226	−9.92
NCI-H23	−9.96
NCI-H322M	−9.08
NCI-H460	−10.82
NCI-H522	−10.49
LXFL529	−10.07
Small Cell Lung Cancer	
DMS 114	−10.00
DMS 273	−10.55
Colon Cancer	
COLO 205	−10.54
DLD-1	−10.00
HCC-2998	−10.17
HCT-116	−10.21
HCT-15	−9.68
HT29	−11.02
KM12	−10.36
KM20L2	−10.85
SW-620	−10.54
CNS Cancer	
SP-268	−9.47
SP-295	−10.00
SP-539	−10.19
SNB-19	−9.00
SNB-75	−10.12
SNB-78	−9.39
U251·	−9.80
XP 498	−9.92
Melanoma	
LOX IMVI	−10.07
MALME-3M	−10.25
M14	−10.39
M19-MEL	−10.11
SK-MEL-2	−10.23
SK-MEL-28	−8.89
SL-MEL5	−10.62
UACC-257	−9.25
UACC-62	−9.68
Ovarian Cancer	
IGROVI	−9.92
OVCAR-3	−10.10
OVCAR-4	−9.19
OVCAR-5	−9.55
OVCAR-8	−9.24
SK-OV-3	−10.25
Renal Cancer	
786-0	−10.24
A498	−9.27
ACHN	−9.25
CAKI-1	−9.27
RXF-393	−10.02
SN12C	−10.24
TK-10	−8.39
UO-31	−9.20
MG_MID	−9.91
Delta	1.22
Range	3.13

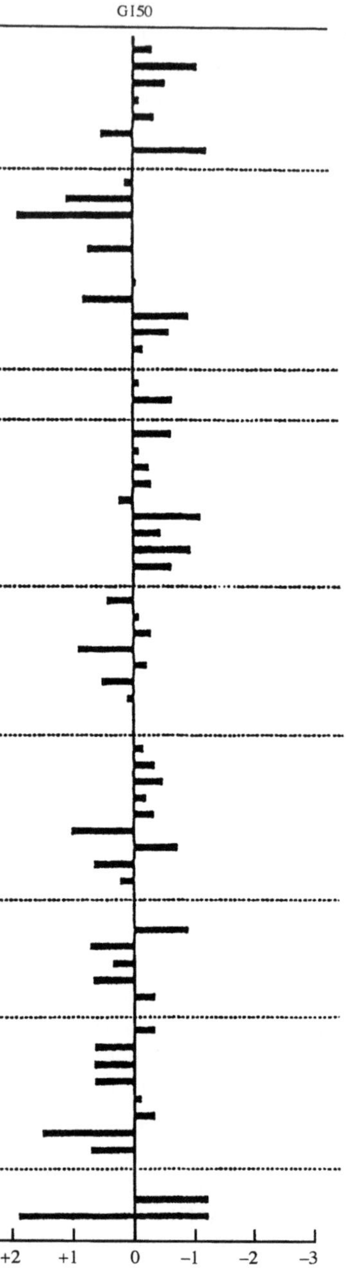

References, pp. 70–79

Dolastatin 10, NSC-376128; Units: Molar; Test date: July 16, 1992

Log_{10} TGI	Log_{10} LC50
-8.00	-8.00
-8.33	-8.03
-8.00	-8.00
-8.00	-8.00
-8.09	-8.00
-8.00	-8.00
-8.46	-8.00
-8.00	-8.00
-8.00	-8.00
-8.00	-8.00
-8.00	-8.00
-8.19	-8.05
-9.06	-8.00
-8.34	-8.34
-8.00	-8.00
-9.49	-8.49
-10.16	-8.44
-8.11	-8.00
-8.64	-8.00
-8.92	-8.11
-9.59	-8.92
-8.00	-8.00
-9.09	-8.32
-8.31	-8.00
-8.44	-8.11
-9.37	-8.38
-8.15	-8.00
-10.08	-8.68
-8.12	-8.00
-8.00	-8.00
-8.55	-8.00
-9.64	-8.37
-8.00	-8.00
-8.10	-8.00
-8.00	-8.00
-8.35	-8.00
-8.85	-8.00
-8.19	-8.08
-8.00	-8.00
-8.96	-8.00
-8.00	-8.00
-8.66	-8.00
-8.00	-8.00
-8.82	-8.08
-8.22	-8.00
-8.00	-8.00
-8.06	-8.00
-8.64	-8.00
-8.00	-8.00
-8.01	-8.00
-8.00	-8.00
-8.44	-8.00
-8.00	-8.00
-8.41	-8.00
-8.00	-8.00
-8.00	-8.00
-8.21	-8.01
-8.19	-8.00
-8.00	-8.00
-8.00	-8.00
-8.39	-8.07
1.77	0.85
2.16	0.92

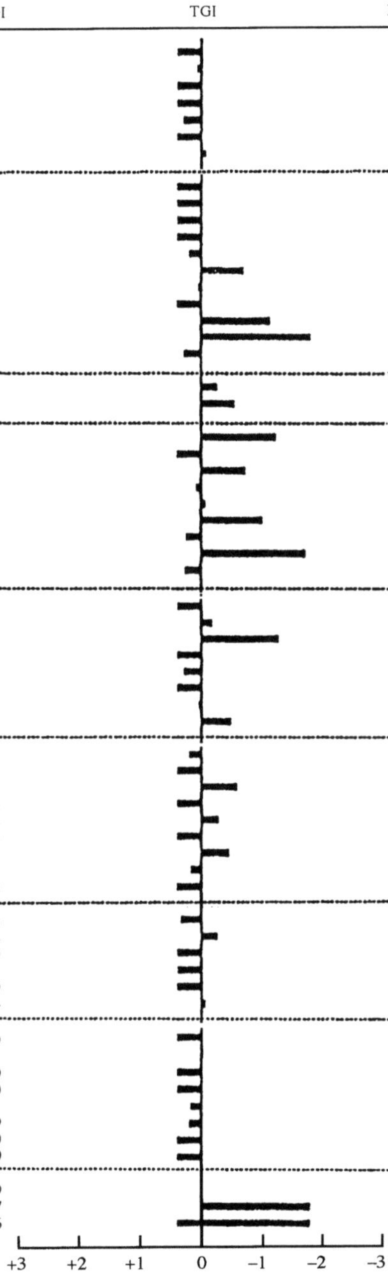

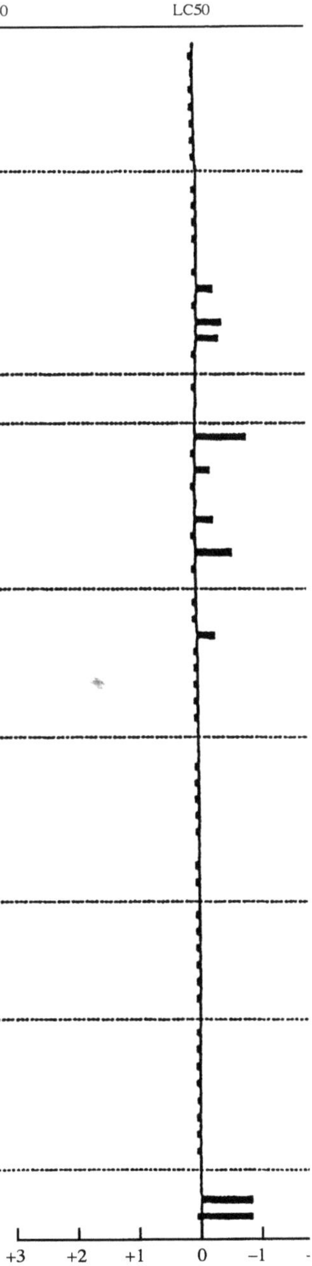

Table 8. *Comparative Activities (GI_{50}; μg/ml) of Dolastatins 10–15 Against a Human Tumor Cell Line Minipanel from the NCI Primary Screen*

Compound	Cell Line					
Dolastatin	OVCAR-3[a]	SF-295[b]	A498[c]	NCI-H460[d]	KM20L2[e]	SK-MEL-5[f]
10	9.5×10^{-7}	7.6×10^{-6}	2.6×10^{-5}	3.4×10^{-6}	4.7×10^{-6}	7.4×10^{-6}
11	0.04	0.031	0.023	1.9×10^{-4}	0.037	0.034
12	0.14	0.37	0.78	0.088	0.29	0.32
13	2.5	3.9	4.5	3.5	2.4	2.7
14	4.1×10^{-3}	2.0×10^{-3}	0.038	2.7×10^{-3}	0.020	6.4×10^{-5}
15	1.3×10^{-4}	4.3×10^{-4}	2.7×10^{-4}	2.8×10^{-4}	1.1×10^{-4}	1.7×10^{-4}

[a]Ovarian adenocarcinoma; [b]Brain (glioma); [c]Renal carcinoma; [d]Lung large cell carcinoma; [e]Colon carcinoma; [f]Melanoma.

(9.24, 8.80), SKOV-3 (10.25, 8.82); 786-0 (10.24, 9.37), A498 (9.27, 8.31), ACHN (9.25, 8.30), CAKI-1 (9.27, 8.72), RXF-393 (10.02, 8.74), SN12C (10.24, 8.70), TK-10 (8.39, 8.21), UO-31 (9.20, 8.32).

A direct comparison of dolastatins 10–15 against a mini-panel of the NCI human cancer cell lines clearly shows the extraordinary potency of dolastatin 10 followed by dolastatins 15 and 14 (Table 8).

In a series of collaborative *in vitro* studies of dolastatin 10 coordinated by the NCI and our institute with other university laboratories, concentrations of this peptide that cause total growth inhibition of the more sensitive human cancer lines were all in the 0.08–0.9 ng/mL range. With a human cancer colony-forming assay, 12 of 23 cell lines were sensitive at 1 ng/mL and, *e.g.*, afforded IC_{50} values of less than 10 ng/mL for the colon lines WiDr and LoVo and less than 1 ng/mL for DLD-1 and HCT-15. Against human leukemia cells the IC_{50} proved to be 0.02–0.1 nM and with human lymphoma cells 0.1–1 pg/mL. Results already published (*168–171*) involving human leukemia and lymphoma cell lines (Table 9) highlight some of the very impressive cytostatic activities of dolastatins 10 and 15 and lack of cytotoxicity to peripheral blood mononuclear cells (PBMNC) and tonsillar mononuclear cells (tonsillar MNC). The cancer cells were not killed even at maximum growth inhibition.

The IC_{50} of dolastatin 10 for both PHA- and SAC-stimulated tonsillor MNC was two to ten times higher than that found for leukemia cell lines. Enriched T- and B-cell populations of the same origin yielded similar results, but dolastatin 10 proved to be a very potent cytostatin for PHA-stimulated T-cells. Even at high concentrations of dolastatin 10, growth suppression of SAC-stimulated B-cells was not as complete as the inhibition of PHA-stimulated T-cells. Importantly, dolastatin 15, even at higher concentrations, did not inhibit growth of PHA- or SAC-stimulated

References, pp. 70–79

Table 9. *Dolastatin 10 and 15 versus Human Leukemia Cell Lines*

Cell Line	IC_{50}(M) D-10	IC_{50}(M) D-15
BV-173[a]	4.7×10^{-11}	5.9×10^{-10}
KM-3[a]	8.7×10^{-11}	8.9×10^{-10}
K-562[b]	12×10^{-11}	1.2×10^{-9}
U-937[c]	1.8×10^{-11}	9.5×10^{-11}
HL-60[d]	5.5×10^{-11}	2.5×10^{-11}
KARPAS-299[e]	9.8×10^{-11}	2.8×10^{-11}
B-Cells	1.0×10^{-9}	$>2.0 \times 10^{-9}$
T-Cells	3.7×10^{-10}	$>2.0 \times 10^{-9}$

[a] Pre B-cell leukemia.
[b] Erythroleukemia.
[c] Monocytic leukemia.
[d] Myeloid leukemia.
[e] T-cell lymphoma.

MNC or appropriately stimulated T- and B-cell populations but was highly cytostatic for the leukemia cell lines. With both dolastatins 10 and 15, growth inhibition was reversed when treatment of the growth medium was terminated, which gave further evidence of the cytostatic activity (*168*). In general, dolastatin 10 was found (Table 9) to be 4 to 12 times more cytostatic than dolastatin 15. Furthermore, both dolastatins 10 and 15 were found to inhibit potently (0.02–0.08 nM) freshly explanted acute myeloid leukemia from patients (*170*). Again the dolastatins proved to affect such dividing cells and not interfere with the viability of resting cells. Also, both dolastatins at 100 nM completely inhibited DNA synthesis, cell proliferation, and immunoglobulin production of the human chronic B-leukemia (most common leukemia in adults) cell lines JVM-2 and EHEB and led to a major number of treated cells residing in the S and G2/M phase (cytostatic rather than cytotoxic effects) with rapid (but temporary) expression of C-myc and bcl-2 RNA. The cytostatic activity was not accompanied by induction of differentiation. Neither high nor low concentrations of dolastatin 10 or dolastatin 15 induced maturation or differentiation of HL-60 cells. Very importantly, both dolastatins showed low toxicities to resting or proliferating nonleukemic cells, and incubation with these anticancer drugs did not cause an increase in the amount of dead cells in comparison of morphological or cytochemical criteria to untreated controls. In our Institute we have also found dolastatin 10 to exhibit similar strong cell growth inhibition of human pancreatic (BXPC-3, ED_{50} $<10.0 \times 10^{-7}$; PANC-1, ED_{50} 2.9×10^{-5}) and squamous cell (FADU, ED_{50} $<10.0 \times 10^{-7}$) carcinomas.

Directly following the first isolation of dolastatin 10 (*49*) in May, 1984, it was used against a series of NCI early stage murine tumor models employing *i.p.* daily or intermittent treatment regions, which led to excellent results (Table 10). Long-term survivors, considered cures, were observed in the 3–40 μg/kg does range with *i.p.* implants and *i.p.* treatment schedules. When tumor implant sites and drug injection were separated it was found necessary to use single high (300–450 μg/kg) *i.v.* bolus doses. For example, in experiments directed by Dr. J. Plowman (NCI) using dolastatin 10 to treat (*i.p.*) the HL-60 promyelocytic leukemia in doses of

Table 10. *Experimental Anticancer Activity of Dolastatin 10 in Murine* In Vivo *Systems T/C (μg/kg)*[a] *with i.p. Tumor and i.p. Dose Regimens*

P388 Lymphocytic Leukemia	*M5076 Ovary Sarcoma*
toxic (13.0)	toxic (26)
155 and 17% cures (6.5)	166 (13)
146 and 17% cures (3.25)	142 (6.5)
137 (1.63)	151 (3.25)
B16 Melanoma	
238 and 40% cures (11)	*OVCAR-3 Human Ovary Xenograft*
182 (6.67)	*(Nude Mouse)*
205 (4.0)	300 (40)
171 (3.4)	
142 (1.44)	
	MX-1 Human Mammary Xenograft
LOX Human Melanoma Xenograft	*(Nude Mouse)*
(Nude Mouse)	toxic (26)
toxic (52)	137 (13)
301 and 67% cures (26)	178 (6.25)
301 and 50% cures (13)	
206 and 33% cures (6.5)	*MX-1 Human Mammary Xenograft*
170 and 17% cures (3.25)	*(Tumor Regression)*
LOX in separate experiments	14 (52)
340 and 50% cures (43)	50 (26)
181 and 33% cures (26)	61 (13)
192 (15)	69 (6.25)
138 and 17% cures (9.0)	
L1210 Lymphocytic Leukemia	
152 (13)	
135 (6.5)	
139 (3.25)	
120 (1.63)	

[a] T/C = Test/Control, both bearing tumor, expressed in time of survival except for the MX-1 system which is recorded as tumor regression (% of the tumor remaining).

References, pp. 70–79

337–450 μg/mL on days 1 and 13, it produced 50–80% day-91 survivors. Comparable activity was observed with the advanced stage *s.c.* lung cancer NCI-H522, LOX human melanoma and the OVCAR-3 human ovary cancer. In short, dolastatin 10 was found to demonstrate good activity against human tumor xenografts implanted at distant sites. Dolastatin 15 also displays impressive *in vivo* activity. Surprisingly, in some cancer xenograft systems it can be used very effectively at doses exceeding 30 mg/kg! Both dolastatin 10 and the dolastatin 15 structural modification LU-103793 began phase 1 human clinical trials in 1995. Presently dolastatin 15 is in advanced preclinical developments. Hopefully the outstanding preclinical promise will be confirmed in the clinical trials.

In February, 1995 dolastatin 11 was selected by the NCI Division of Cancer Treatment for preclinical development based on its cancer cell growth inhibitory activity and especially because of its unusual biochemical effects on actin and tubulin. Dolastatin 14 continues to appear promising, but further development is still at a fairly early stage. Furthermore, structural modifications of dolastatins 10–15 continue to offer outstanding promise of allowing important improvements in antineoplastic activity and therapeutic index.

6.2 Dolastatin 10 Structural Modifications

6.2.1 Chiral Isomers

In order to probe the chiral and other structural parameters for such remarkable antineoplastic activity, we first explored the synthesis and biological effects of 18 (of 128 possible) chiral isomers of dolastatin 10 when one to five asymmetric carbons in the Dil-Dap sequence were reversed. We found (*118*) that isodolastatins **78–94** (Table 11) are less cytostatic than the parent pentapeptide (**4**, P388 ED_{50} 10^{-4} μg/mL) against cell growth of the P388 lymphocytic leukemia, with ED_{50} values $\geqslant 10^{-3}$ μg/mL, and that (19*aR*)-isodolastatin 10 (**77**) was comparable to the natural product, as was also (6*R*)-isodolastatin 10 (*149*) (**67**, see Section 4.3).

6.2.2 Structural Substitutions

From the onset of our synthetic approaches to dolastatin 10 and its first total synthesis, we explored structural modifications of this potentially important anticancer drug. A description of some early results from these studies directed at defining structure/activity relationships among

Table 11. *Dolastatin 10 Chiral Isomers*

No.	Isodolastatin 10	P388 ED_{50}, μg/mL
77	19a*R*	4.9×10^{-5}
78	9*S*	1.6×10^{-3}
79	10*S*	3.9×10^{-3}
80	18*S*, 19*R*	3.2×10^{-3}
81	9*S*, 19a*R*	5.6×10^{-3}
82	18*S*, 19*R*, 19a*R*	3.2×10^{-3}
83	9*S*, 10*S*	3.9×10^{-2}
84	18*S*	6.0×10^{-1}
85	9*S*, 18*S*	6.2×10^{-1}
86	9*S*, 18*S*, 19*R*	2.7×10^{-1}
87	9*S*, 18*S*, 19a*R*	4.1×10^{-1}
88	9*S*, 18*S*, 19*R*, 19a*R*	1.5×10^{-1}
89	10*S*, 18*S*, 19*R*, 19a*R*	2.0×10^{-1}
90	9*S*, 10*S*, 18*S*, 19*R*, 19a*R*	6.3×10^{-1}
91	9*S*, 19*S*	$> 1.0 \times 10^{0}$
92	9*S*, 19*R*, 19a*R*	$> 1.0 \times 10^{0}$
93	9*S*, 10*S*, 18*S*	$> 1.0 \times 10^{0}$
94	9*S*, 10*S*, 19*R*, 19a*R*	$> 1.0 \times 10^{0}$

such powerful antineoplastic peptides now follows. As noted above (Sections 4.2.5 and 4.4), we also developed a useful way to improve synthesis of the dolastatin 10 right terminal unit, dolaphenine (Doe), and later we designed analogues that completely lack this thiazole derivative. These new anticancer drug candidates that are most promising and closely related structurally to dolastatin 10 have been designated auristatins (derived from *Dolabella auricularia*, the source of natural dolastatin 10).

All of the new amides, peptides, and dolastatin 10 modifications summarized in Section 4.4 were evaluated against the murine P388 and L1210 leukemia cell lines and a series of human cancer cell lines (see Tables 12–14). The dolastatin 10 derivatives arising from the arylalkyl amides were found to display exceptionally high activity. In fact, compound **72c**, designated as auristatin PE, showed cancer cell, human cancer xenograft, and tubulin inhibition comparable (see Table 14) to that of natural dolastatin 10. Incorporation of more polar amino acids into the carboxyl or N-terminal ends of the molecule significantly decreased activity. Both sets of dolastatin 10 structural modifications provided very useful structure/activity information that has formed the basis for a series

Table 12. *Human Cancer Cell Line and Murine P388 Lymphocytic Leukemia Inhibitory Activities of Selected Dolastatin 10 Structural Modifications*

	Cell type	Cell line	D-10	**72a**	**72b**	**72c**	**72d**	**72e**	**72f**	**72g**
GI-50* (ng/ml)	Ovarian	OVCAR-3	0.001	<0.1	>0.1	0.004	0.24	0.016	0.15	0.18
	CNS	SF-295	0.0081	<0.1	>0.1	0.014	0.35	0.046	0.43	0.28
	Renal	A498	0.049	<0.1	>0.1	0.013	0.64	0.059	0.46	0.39
	Lung-NSC	NCI-H460	0.0033	0.21	>0.1	0.005	0.28	0.025	0.27	0.31
	Colon	KM20L2	0.0042	<0.1	>0.1	0.005	0.3	0.33	0.32	0.33
	Melanoma	SK-MEL-5	0.0085	<0.1	>0.1	0.012	0.12	0.044	0.38	0.31
TGI* (ng/ml)	Ovarian	OVCAR-3	0.0069	<0.1	>0.1	0.027	0.65	0.050	0.5	0.54
	CNS	SF-295	0.048	1.0	>0.1	0.042	>10	>10	>10	0.79
	Renal	A498	0.54	1.3	>0.1	>0.1	>10	2.8	>10	>10
	Lung-NSC	NCI-H460	0.04	1.3	>0.1	>0.1	1.2	0.14	1.2	>10
	Colon	KM20L2	0.028	0.91	>0.1	>0.1	1.3	2.9	4.1	>10
	Melanoma	SK-MEL-5	>10	2.9	>0.1	>0.1	1.3	>10	>10	>10
ED_{50}* (ng/ml)	Mouse Leukemia	P388	0.046	4.7	3.78	0.231	45.9	3.72	5.15	0.44

Table 12 (*continued*)

	Cell type	Cell line	**74a**	**74b**	**74c**	**74d**	**74e**	**74f**	**74g**	**74h**
	Ovarian	OVCAR-3	6.9	0.71	0.0000091	0.067	0.031	>1000	120	74
	CNS	SF-295	>10	>10	0.000025	0.25	0.27	>1000	670	130
GL-50* (ng/ml	Renal	A498	>10	>10	0.000058	0.27	0.26	>1000	>1000	600
	Lung-NSC	NCI-H460	>10	>10	0.0000058	0.12	0.22	>1000	270	250
	Colon	KM20L2	>10	0.99	0.0000072	0.034	0.034	>1000	290	280
	Melanoma	SK-MEL-5	3.0	0.8	0.0000048	0.044	0.058	>1000	80	380
	Ovarian	OVCAR-3	>10	>10	0.000060	0.67	0.51	>1000	670	740
	CNS	SF-295	>10	>10	0.00025	>1000	>10	>1000	>1000	>1000
TGI* ng/ml	Renal	A498	>10	>10	17	35	6.3	>1000	>1000	>1000
	Lung-NSC	NCI-H460	>10	>10	0.000065	0.13	0.7	>1000	>1000	620
	Colon	KM20L2	>10	>10	0.0001	1.3	1	>1000	>1000	>1000
	Melanoma	SK-MEL-5	>10	>10	0.00015	22	>10	>1000	>1000	>1000
ED_{50}* (ng/ml)	Mouse Leukemia	P388	>10	>10	0.0248	0.32	0.0312	513	>1000	329

*ED_{50} (P388) and GL-50 (HTCL) are the drug dosages (ng/ml) needed to inhibit the cancer cell growth by 50%. The TGI represents 100% cancer cell growth in ng/ml. There is no mathematical difference between ED_{50} and GI-50 which are both calculated using the same formula. The only difference is historical usages. The values are given here in ng/ml rather than μg/ml.

Table 13. *U.S. NCI Mean Panel GI_{50} Value, and TGI-Based COMPARE Correlation Coefficients Using Dolastatin 10 as the Benchmark, for Peptide Amides* **72a–d, g**

Peptide	GI_{50} ($\times 10^{-8}$)M	Coefficient
4	0.012	1.00
72a	1.00	0.828
72b	0.174	0.851
72c	0.049	0.901
72d	0.182	0.805
72g	0.166	0.883

Table 14. *Inhibition of Growth of the Murine L1210 Leukemia Cell Line and Tubulin Polymerization by Dolastatin 10 and Peptide Amides* **4, 72a–g,** *and* **74a–h**

Peptide	Inhibition of L1210 Leukemia Cell Growth (IC_{50}, μM)	Inhibition of Tubulin Polymerization (IC_{50}, μM)
4	0.5	1.3 ± 0.2
72a	60	1.8 ± 0.2
72b	80	1.6 ± 0.2
72c	0.6	1.3 ± 0.2
72d	20	1.3 ± 0.05
72e	2	1.6 ± 0.3
74f	4	1.8 ± 0.3
72g	20	1.9 ± 0.2
74a	200	2.7 ± 0.6
74b	20	1.2 ± 0.08
74c	0.1	1.8 ± 0.2
74d	0.3	1.6 ± 0.1
74e	0.2	2.7 ± 0.5
74f	> 1000	3.2 ± 0.1
74g	600	> 40
74h	300	> 40

of extensive structural studies still under way. Analogous structure/activity investigations of dolastatins 11–15 are being pursued.

Recently, some extensive structural modifications of dolastatin 10 have been recorded (*172*) and auristatin PE is now in phase I clinical trials.

7. Biochemical Mechanisms of Action

7.1 Inhibition of Tubulin Assembly and Mitosis

The microtubule components of cells perform necessary steps in a variety of processes such as mitosis. Dolastatin 10 was found by E. Hamel at the US NCI to be powerfully effective at binding to tubulin, inhibiting polymerization (*153*) (Table 15) and causing cells to accumulate in metaphase arrest (*151–153, 173*). At higher concentration levels, intracellular microtubules completely disappear (*154*). Although dolastatin 10 noncompetitively inhibits the binding of vincristine to tubulin, it does not inhibit the binding of colchicine and certain other plant-derived tubulin inhibitors (*153*). Dolastatin 10 strongly affects nucleotide-tubulin interactions and nearly eliminates exchange of GTP at its tubulin site without displacing GTP already bound to tubulin. When this nucleotide site is depleted, formation of a cross-link between two cysteine thiol groups and *N,N'*-ethylenebis(iodoacetamide) can be readily accomplished, but dolastatin 10 strongly inhibited this reaction (*154, 173*). Furthermore, dolastatin 10 has been shown to cause the tubulin assembly process to truncate into unique aggregates (*174*).

The chiral isomers of dolastatin 10 prepared to date, except for the two with reversal of configuration at the 6- (**67**) or 19a-positions (**77**), were all found to be less effective than the natural peptide in inhibiting microtubule assembly. Indeed, most were considerably less active (*151*). Considering the structural modifications of dolastatin 10 currently in hand, it is clear that the Doe unit can be replaced with certain amides such as those derived from phenethylamine (see Section 6). These analogues still inhibit microtubule assembly at the extraordinarily low concentrations (ca. 1.2. μM) typical of the parent peptide (dolastatin 10).

Table 15. *Inhibition of Tubulin Assembly by Anticancer Drugs Acting on Areas Near the* Vinca *Site (for* Vinca *Alkaloids and Maytansine)*

Compound	IC_{50} (μM)
Dolastatin 10	1.2 ± 0.08
Phomopsin A	1.4 ± 0.2
Vinblastine	1.5 ± 0.1
Maytansine	3.5 ± 0.1
Rhizoxin	6.8 ± 0.5

The IC_{50} values for dolastatin 15 against murine leukemia L1210 and human Burkitt lymphoma were found by HAMEL (*175*) to be 3 nM each, and when the toxic level was reached these cells were arrested in mitosis. With Chinese hamster ovary cells, both dolastatins 10 and 15 caused moderate loss of microtubules at the IC_{50} concentrations (0.5 nM and 5 nM respectively), and at ten times higher concentrations the microtubules disappeared. In contrast, dolastatin 15 *in vitro* only weakly reacts with tubulin with an IC_{50} of 23 μM (compared to 1.2 μM for dolastatin 10 and 1.5 μM for vinblastine) for inhibition of glutamate-induced tubulin polymerization. Of the presently known interactions of dolastatin 10 with tubulin, dolastatin 15 was found only to inhibit tubulin-dependent GTP hydrolysis.

7.2 Effects on Actin

Although dolastatin 11 causes unusual morphological changes during tubulin assembly, its effects on actin are even more dramatic. The actin proteins play a key role in the ability of cells to advance (creep), and the unique effects of dolastatin 11 on these processes have been observed (NCI unpublished) by K. DUNCAN and E. HAMEL. Currently dolastatin 11 is entering preclinical development (NCI), and the significance of its interactions with actin proteins in cancer patients will eventually be evaluated.

8. Pharmacology and Toxicology

The development of dolastatins 10 and 15 to clinical trial have necessitated a considerable number of formulation, pharmacological, and toxicological studies by the NCI and its European collaborators that will eventually be published. Some of the most important current conclusions using dolastatin 10 as an example now follow.

Dolastatin 10 was radiolabeled (^3H) by tritium exchange and found stable in murine, canine and human plasma (80–90% protein bound) for ⩾8 hr at 37°C. Development of a radioimmunoassay for dolastatin 10 provided the best means for detecting extremely small concentrations. A very convenient phosphate buffer formulation based on our earlier HPLC techniques (*148*) proved quite workable and was employed for delivery of dolastatin 10. The murine and canine animal safety studies were uneventful and suggested 65 μg/m^2 as an appropriate starting dose for the phase 1 clinical trials.

9. Conclusion

From the first observations that the Indian Ocean sea hare *Dolabella auricularia* contained potentially important anticancer constituents, 23 years of intense research have already passed and we are still only at the beginning of appreciating the many scientific and medical opportunities offered by the dolastatin discoveries. With dolastatins 11 and 15 in preclinical development, and dolastatin 10, auristatin PE, and LU-103793 already (1995) in the first human clinical trials, the prospects for this unprecedented series of structurally unique anticancer drug lead compounds (and/or future structural modifications) in contributing to future improvements in human cancer treatments seem promising. Because of their unusual influence on the chemistry of certain important cellular proteins and nucleic acids, the dolastatins will also find an increasing role as important biochemical probes.

Acknowledgements

Grateful appreciation is expressed to everyone in the author's research group and others over the past 23 years who contributed to advancing the dolastatin discoveries to the present clinical trials. In this regard special thanks are extended to Drs. MICHAEL R. BOYD, ERNEST HAMEL, CHERRY L. HERALD, FIONA HOGAN, YOSHIAKO KAMANO, JEAN M. SCHMIDT, SHEO BUX SINGH, JAYARAM SRIRANGAM, MATTHEW SUFFNESS, and MICHAEL D. WILLIAMS. I also wish to acknowledge the expert contributions of Dr. FIONA HOGAN and Mrs. MARIE BAUGHMAN to the final stages of manuscript preparation, Mrs. THERESA THORNBURGH for assistance with the Tables, and the financial support provided by Outstanding Investigator Grant OIG CA44344-01A1-07 awarded by the US National Cancer Institute, DHHS.

References

1. HALSTEAD, B.W.: Poisonous and Venomous Marine Animals of the World, Vols. 1, 2. Washington, D.C.: US Government Printing Office. 1965, 1967.

2. WOODWARD, R.B., and J.Z. GOUGOUTAS: The Structure of Tetrodotoxin. J. Amer. Chem. Soc., **86**, 5030 (1964).

3. GOTO, T., Y. KISHI, S. TAKAHASHI, and Y. HIRATA: Tetrodotoxin. Tetrahedron, **21**, 2059 (1965).

4. NIGRELLI, R.F., S. JAKOWSKA, and I. CALVENTI: Ectyonin, an Antimicrobial Agent from the Sponge *Microciona prolifera* Verrill. Zoologica, **44**, 173 (1959).

5. JAKOWSKA, S., and R.F. NIGRELLI: Antimicrobial Substances from Sponges. Ann. NY Acad. Sci., **90**, 913 (1960).

6. BURKHOLDER, P.R., and L.M. BURKHOLDER: Antimicrobial Activity of Horny Corals. Science, **127**, 1174 (1958).

7. CIERESZKO, L.S., D.H. SIFFORD, and A.J. WEINHEIMER: Chemistry of Coelenterates. I. Occurrence of Terpenoid Compounds in Gorgonians. Ann. NY Acad. Sci., **90**, 917 (1960).

8. CIERESZKO, L.S.: Chemistry of Coelenterates, III: Occurrence of Antimicrobial Terpenoid Compounds in the Zooxanthellae of Alcyonarians. Trans. NY Acad. Sci., **24**, 502 (1962).
9. MARDEROSIAN, A.H.D.: Marine Pharmaceuticals. J. Pharm. Sci., **58**, 1 (1969).
10. NIGRELLI, R.F., M.F. STEMPIEN, C.D. RUGGIERI, V.R. LIGOURI, and J.T. CECIL: Substances of Potential Biomedical Importance from Marine Organisms. Fed. Proc., **26**, 1197 (1967).
11. SIGEL, M.M., L.L. WELLHAM, W. LICGTER, L.E. DUDECK, J.L. GARGUS, and A.H. LUCAS: Anticellular and Antitumor Activity of Extracts from Typical Marine Invertebrates. In: Food-Drugs from the Sea Proceedings 1969 (H.W. YOUNGKEN, Jr., ed.), p. 281. Washington, D.C.: Marine Technology Society. 1969.
12. WHITSON, D., and H.W. TITUS: The Use of Starfish Meal in Chick Diets. Bull. Bingham Oceanog. Coll., **9**, 24 (1945).
13. LEE, C.F.: US Fish Wildlife Service. Fish Leaflet, 391 (1951).
14. HEILBRUNN, L.V., A.B. CHAET, A. DUNN, and W.L. WILSON: Antimitotic Substances from Ovaries. Biol. Bull., **106**, 158 (1954).
15. BALTZER, F.: Über die Giftwirkung weiblicher Bonellia-Gewebe auf das Bonellia-Männchen und andere Organismen und ihre Beziehung zur Bestimmung des Geschlechts der Bonellienlarve. Mitt. Naturf. Ges. Bern., **8**, 98 (1925).
16. BALTZER, F.: Über die Giftwirkung der weiblichen Bonellia und ihre Beziehung zur Geschlechtsbestimmung der Larve. Rev. Suisse Zool., **32**, 87 (1925).
17. FRIESS, S.L., F.G. STANDAERT, E.R. WHITCOMB, R.F. NIGRELLI, J.D. CHANLEY, and H. SOBOTKA: Some Pharmacologic Properties of Holothurin A, a Glycosidic Mixture from the Sea Cucumber. Ann. NY Acad. Sci., **90**, 893 (1960).
18. SCHMEER, M.R., and C.V. HUALA: Mercenene: *In Vivo* Effects of Mollusk Extracts on the Sarcoma 180. Ann. NY Acad. Sci., **118**, 605 (1965).
19. LI, C.P., B. PRESCOTT, and W.B. JAHNES: Antiviral Activity of a Fraction of Abalone Juice. Proc. Soc. Exp. Biol. Med., **109**, 534 (1962).
20. PETTIT, G.R., J.F. DAY, J.L. HARTWELL, and H.B. WOOD: Antineoplastic Components of Marine Animals. Nature, **227**, 962 (1970).
21. PETTIT, G.R., J.L. HARTWELL, and H.B. WOOD: Arthropod Antineoplastic Agents. Cancer Res., **28**, 2168 (1968).
22. PETTIT, G.R.: The Bryostatins. In: Progress in the Chemistry of Organic Natural Products, No. 57. Founded by Zechmeister, L. (HERZ, W., G.R. KIRBY, W. STEGLICH, and CH. TAMM, eds.), p. 153–195. Wien-New York: Springer-Verlag. 1991.
23. PETTIT, G.R.: Marine Animal and Terrestrial Plant Anticancer Constituents. Pure and Appl. Chem., **66**, 2271 (1994).
24. HYMAN, L.H.: The Invertebrates, Mollusca. American Museum of Natural History. New York: McGraw-Hill. 1967.
25. EALES, N.B.: L.M.B.C. Memoirs, Vol. 24. on "Typical British Marine Plants and Animals, Aplysia" (HARDMAN, W.A., and J. JOHNSTONE, eds.), Liverpool University Press. 1921.
26. PLINY: Historia Naturalis, Lib. 9, Lib. 32, *ca.* 60 A.D.
27. DONATI, G., and B. PORFIRIO: Marine Pharmacology and Toxicology. The Dolastatins. La Conchiglia 16, May, 1984.
28. ENGEL, H.: The genus *Dolabella.* Zool. Mededeel. Leyden, **24**, 197 (1942).
29. See Ref. 1, Vol. 1, p. 709. 1965.
30. PETTIT, G.R., R.H. ODE, C.L. HERALD, R.B. VON DREELE, and C. MICHEL: The Isolation and Structure of Dolatriol. J. Amer. Chem. Soc., **98**, 4677 (1976).
31. BEBBINGTON, A.: Aplysiid Species from East Africa with Notes on the Indian Ocean Aplysiomorpha Gastropoda Opisthobranchia. Zool. J. Linn. Soc., **54**, 63 (1974).

32. KUWASAWA, K., and K. MATSUI: Postjunctional Potentials and Cardiac Acceleration in a Mollusc *Dolabella auricularia*. Experientia, **26**, 1100 (1970).
33. KUWASAWA, S.: Heartbeat Inhibiting Effect of Inhibitory Connective Potential in *Dolabella auricularia* Heart. J. Physiol. Soc. Jap., **34**, 453 (1972).
34. KUWASAWA, K.: Localization of Acetylcholine Response in the Heart of *Dolabella auricularia*. J. Physiol. Soc. Jpn., **35**, 403 (1973).
35. HILL, R.B.: Effects of 5-Hydroxytryptamine on Action Potentials and on Contractile Force in the Ventricle of *Dolabella auricularia*. J. Exp. Biol., **61**, 529 (1974).
36. HILL, R.B.: Effects of Acetylcholine on Resting and Action Potentials and on Contractile Force in the Ventricle of *Dolabella auricularia*. J. Exp. Biol., **61**, 629 (1974).
37. SUZUKI, S.: Localization of Intracellular Calcium and its Translocation During Mechanical Activity in the Smooth Muscle of a Mollusk *Dolabella auricularia*. J. Electron Microsc., **26**, 253 (1977).
38. SUGI, H., and S. SUZUKI: Ultrastructural and Physiological Studies on the Longitudinal Body Wall Muscle of *Dolabella auricularia*. Part 1. Mechanical Response and Ultrastructure. J. Cell. Biol., **79**, 454 (1978).
39. MAKINO, N.: Hemocyanin from *Dolabella auricularia*. Part 4. Dissociation by DEAE Cellulose. J. Biochem. (Tokyo), **72**, 29 (1972).
40. WATSON, M.: Ph.D. Thesis, University of Hawaii. Some Aspects of the Pharmacology, Chemistry and Biology of the Midgut Gland Toxins of Some Hawaiian Sea Hares, especially *Dolabella auricularia* and *Aplysia plumonica*. Ann Arbor, Michigan: University Microfilms, Inc. 1969.
41. SCHEUER, P.J.: Recent Developments in the Chemistry of Marine Toxins (DEVRIES, A., and E. KOCHVA, eds.). Toxins of Animal and Plant Origin. Vol. 2. Proceedings of the 2nd International Symposium on Animal and Plant Toxins. Tel Aviv, Israel, Feb. 22–28, 1970. New York, London: Gordon and Breach Science Publishers.
42. WATSON, M.: Midgut Gland Toxins of Hawaiian Sea Hares. Part 1. Isolation and Preliminary Toxicological Observations. Toxicon, **11**, 259 (1973).
43. PETTIT, G.R., C.L. HERALD, R.H. ODE, P. BROWN, D.J. GUST, and C. MICHEL: The Isolation of Loliolide from an Indian Ocean Opisthobranch Mollusc. J. Nat. Prod., **43**, 752 (1980).
44. PETTIT, G.R., Y. KAMANO, Y. FUJI, C.L. HERALD, M. INOUE, P. BROWN, D. GUST, K. KITAHARA, J.M. SCHIDT, D.L. DOUBEK, and C. MICHEL: Marine Animal Biosynthetic Constituents for Cancer Chemotherapy. J. Nat. Proc., **44**, 482 (1981).
45. PETTIT, G.R., Y. KAMANO, R. AOYAGI, C.L. HERALD, D.L. DOUBEK, J.M. SCHMIDT, and J.J. RUDLOE: Antineoplastic Agents 100. The Marine Bryozoan *Amathia convoluta*. Tetrahedron, **41**, 985 (1985).
46. PETTIT, G.R., Y. KAMANO, P. BROWN, D. GUST, M. INOUE, and C.L. HERALD: Structure of the Cyclic Peptide Dolastatin 3 from *Dolabella auricularia*. J. Amer. Chem. Soc., **104**, 905 (1982).
47. PETTIT, G.R., Y. KAMANO, C.W. HOLZAPFEL, W.J. VAN ZYL, A.A. TUINMAN, C.L. HERALD, L. BACZYNSKYJ, and J.M. SCHMIDT: The Structure and Synthesis of Dolastatin 3. J. Amer. Chem. Soc., **109**, 7581 (1987).
48. PETTIT, G.R., Y. KAMANO, C.L. HERALD, Y. FUJI, H. KIZU, M.R. BOYD, F.E. BOETTNER, D.L. DOUBEK, J.M. SCHMIDT, J.-C. CHAPUIS, and C. MICHEL: Isolation of Dolastatins 10–15 from the Marine Mollusc *Dolabella auricularia*. Tetrahedron, **49**, 9151 (1993).
49. PETTIT, G.R., Y. KAMANO, C.L. HERALD, A.A. TUINMAN, F.E. BOETTNER, H. KIZU, J.M. SCHMIDT, L. BACZYNSKYJ, K.B. TOMER, and R.J. BONTEMS: The Isolation and Structure of a Remarkable Marine Animal Antineoplastic Constituent: Dolastatin 10. J. Amer. Chem. Soc., **109**, 6883 (1987).

50. PETTIT, G.R., Y. KAMANO, H. KIZU, C. DUFRESNE, C.L. HERALD, R. BONTEMS, J.M. SCHMIDT, F.E. BOETTNER, and R.A. NIEMAN: Isolation and Structure of the Cell Growth Inhibitory Depsipeptides Dolastatins 11 and 12. Heterocycles, **28**, 553 (1989).
51. PETTIT, G.R., Y. KAMANO, C.L. HERALD, C. DUFRESNE, R.L. CERNY, D.L. HERALD, J.M. SCHMIDT, and H. KIZU: Isolation and Structure of the Cytostatic Depsipeptide Dolastatin 13 from the Sea Hare *Dolabella auricularia*. J. Amer. Chem. Soc., **111**, 5015 (1989).
52. PETTIT, G.R., Y. KAMANO, C.L. HERALD, C. DUFRESNE, R.E. BATES, and J.M. SCHMIDT: Antineoplastic Agents 190. Isolation and Structure of the Cyclodepsipeptide Dolastatin 14. J. Org. Chem., **55**, 2989 (1990).
53. PETTIT, G.R., Y. KAMANO, C. DUFRESNE, R.C. CERNY, C.L. HERALD, and J.M. SCHMIDT: Isolation and Structure of the Cytostatic Linear Depsipeptide Dolastatin 15. J. Org. Chem., **54**, 6005 (1989).
54. SONE, H., T. NEMOTO, M. OJIKA, and K. YAMADA: Isolation, Structure, and Synthesis of Dolastatin C, a New Depsipeptide from the Sea Hare *Dolabella auricularia*. Tetrahedron Lett., **34**, 8445 (1993).
55. SONE, H., T. NEMOTO, H. ISHIWATA, M. OJIKA, and K. YAMADA: Isolation, Structure, and Synthesis of Dolastatin D, a Cytotoxic Cyclic Depsipeptide from the Sea Hare *Dolabella auricularia*. Tetrahedron Lett., **34**, 8449 (1993).
56. YAMADA, K.: Private communication.
57. SONE, H., T. NEMOTO, and K. YAMADA: Doliculols A and B, the Non-halogenated C_{15} Acetogenins with Cyclic Ether from the Sea Hare *Dolabella auricularia*. Tetrahedron Lett., **34**, 3461 (1993).
58. ISHIWATA, H., H. SONE, H. KIGOSHI, and K. YAMADA: Enantioselective Total Synthesis of Doliculide, a Potent Cytotoxic Cyclodepsipeptide of Marine Origin, and Structure-Cytotoxicity Relationships of Synthetic Doliculide Congeners. Tetrahedron, **50**, 12853 (1994).
59. YAMAZAKI, M: Antitumor and Antimicrobial Glycoproteins from Sea Hares. Comp. Biochem. Physiol., **105C**, 141 (1993).
60. PETTIT, G.R., D. KANTOCI, D.L. DOUBEK, B.E. TUCKER, W.E. PETTIT, and R.M. SCHROLL: Isolation of the Nickel-Chlorin Chelate Tunichlorin from the South Pacific Sea Hare *Dolabella auricularia*. J. Nat. Prod., **56**, 1981 (1993).
61. PETTIT, G.R., P.S. NELSON, and C.W. HOLZAPFEL: Synthesis of the *cyclo*-[(gly)Thz- (*R*)- and *cyclo*-[(gly)Thz-(*S*)-(gln)Thz-L-Val-L-Leu-L-Pro] Isomers of Dolastatin 3. J. Org. Chem., **50**, 2654 (1985).
62. PETTIT, G.R., and C.W. HOLZAPFEL: Synthesis of the Modified Dolastatin 3 Sequence *cyclo*-[L-Val-L-Leu-L-Pro-(*R*,*S*)-(gln)Thz-(gly)Thz]. J. Org. Chem., **51**, 4586 (1986).
63. PETTIT, G.R., and C.W. HOLZAPFEL: Synthesis of the Dolastatin 3 Isomer *cyclo*-[L-Pro-L-Leu-L-Val-(*R*,*S*)-(gln)Thz-(gly)Thz]. J. Org. Chem., **51**, 4580 (1986).
64. HOLZAPFEL, C.W., and G.R. PETTIT: Synthesis of the Dolastatin Thiazole Amino Acid Component (gln)Thz. J. Org. Chem., **50**, 2323 (1985).
65. ECKART, K., U. SCHMIDT, and H. SHWARZ: Amino Acids and Peptides; 60. Synthesis of Biologically Active Cyclopeptides; 10. Synthesis of 16 Structural Isomers of Dolastatin 3; II: Synthesis of the Linear Educts and the Cyclopeptides. Liebigs Ann. Chem., 1940 (1986).
66. SCHMIDT, U., and R. UTZ: Synthetic Studies on the Elucidation of the Structure and Configuration of Dolastatin 3. Angew Chem., Int. Ed. Engl., **23**, 725 (1984).
67. HAMADA, Y., K. KOHDA, and T. SHIOIRI: Proposed Structure of the Cyclic Peptide Dolastatin 3, A Powerful Cell Growth Inhibitor, Should Be Revised! Tetrahedron Lett., **25**, 5303 (1984).

68. Schmidt, U., and D. Weller: Total Synthesis of Ulithiacyclamide. Tetrahedron Lett., **27**, 3495 (1986).
69. Schmidt, U., R. Utz, A. Lieberknecht, H. Griesser, B. Potzolli, J. Bahr, K. Wagner, and P. Fischer: Sequenzierung von Thiazolaminosäuren enthaltenden Cyclopentapeptiden ("Dolastatin 3") durch FAB/MSMS. Synthesis, 236 (1987).
70. Bernier, J.-L., R. Houssin, and J.-P. Henichart: Analog of Dolastatin 3. Synthesis, ^{1}H-NMR Studies and Spatial Conformation. Tetrahedron, **42**, 2695 (1986).
71. Holzapfel, C.W., G.R. Pettit, and G.M. Cragg: Biosynthetic Product Molecular Weight Determimations by Solution Phase Secondary Ion Mass Spectrometry Employing Group 1A Metal Salts. J. Nat. Prod., **48**, 513 (1985).
72. Pettit, G.R., S.B. Singh, F. Hogan, P. Lloyd-Williams, D.L. Herald, D.D. Burkett, and P.J. Clewlow: The Absolute Configuration and Synthesis of Natural (−)-Dolastatin 10. J. Amer. Chem. Soc., **111**, 5463 (1989).
73. Hayashi, K., Y. Hamada, and T. Shioiri: Synthetic Study on Dolastatin 10, an Antineoplastic Pentapeptide of Marine Origin. In: Peptide Chemistry (Yanaihara, N., ed.), p. 291. Osaka: Protein Research Foundation. 1990.
74. Kano, S., Y. Yuasa, and S. Shibuya: Highly Diastereoselective Synthesis of *N*-Boc Dolaisoleuine, Unusual Amino Acid in Dolastatin 10. Heterocycles, **31**, 1597 (1990).
75. Maugras, I., J. Poncet, and P. Jouin: Stereocontrolled Synthesis of *N,O*-Dimethyl-γ-Amino-β-Hydroxy Acids: Analogues of the (*R*)-MeIle-Ψ(CHOMe)-Gly Residue of the Cytotoxic Marine Pseudopeptide Dolastatin 10. Tetrahedron, **46**, 2807 (1990).
76. Bredenkamp, M.W., C.W. Holzapfel, R.M. Snyman, and W.J. van Zyl: Observations on the Hantzsch Reaction: Synthesis of *N*-^{t}Boc-*S*-Dolaphenine. Synth. Commun., **22**, 3029 (1992).
77. Hamada, Y., K. Hayashi, and T. Shioiri: Efficient Stereoselective Synthesis of Dolastatin 10, an Antineoplastic Peptide from a Sea Hare. Tetrahedron Lett., **32**, 931 (1991).
78. Tomioka, K., M. Kanai, and K. Koga: An Expeditious Synthesis of Dolastatin 10. Tetrahedron Lett., **32**, 2395 (1991).
79. Shioiri, T., K. Hayashi, and Y. Hamada: Stereoselective Synthesis of Dolastatin 10 and Its Congeners. Tetrahedron, **49**, 1913 (1993).
80. Roux, F., I. Maugras, J. Poncet, G. Niel, and P. Jouin: Synthèse de la Dolastatine 10 et de la [*R*-Doe]-Dolastatine 10. Tetrahedron, **50**, 5345 (1994).
81. Ireland, C.M., and P.J. Scheuer: Ulicyclamide and Ulithiacyclamide, Two New Small Peptides from a Marine Tunicate. J. Amer. Chem. Soc., **102**, 5688 (1980).
82. Ireland, C.M., A.R. Durso, Jr., R.A. Newman, and M.P. Hacker: Antineoplastic Cyclic Peptides from the Marine Tunicate *Lissoclinum Patella*. J. Org. Chem., **47**, 1807 (1982).
83. Biskupiak, J.E., and C.M. Ireland: Absolute Configuration of Thiazole Amino Acids in Peptides. J. Org. Chem., **48**, 2302 (1983).
84. Hamamoto, Y., M. Endo, M. Nakagawa, T. Nakanishi, and K. Mizukawa: A New Cyclic Peptide, Ascidiacyclamide, Isolated from Ascidian. J. Chem. Soc., Chem. Commun., 323 (1983).
85. Wasylyk, J.M., J.E. Biskupiak, C.E. Costello, and C.M. Ireland: Cyclic Peptide Structures from the Tunicate *Lissoclinum Patella* by FAB Mass Spectrometry. J. Org. Chem., **48**, 4445 (1983).
86. Hamada, Y., S. Kato, and T. Shioiri: New Methods and Reagents in Organic Synthesis. 51. A Synthesis of Ascidiacyclamide, a Cytotoxic Cyclic Peptide from Ascidian – Determination of Its Absolute Configuration. Tetrahedron Lett., **26**, 3223 (1985).

87. HAMADA, Y., M. SHIBATA, and T. SHIOIRI: New Methods and Reagents in Organic Synthesis. 58. A Synthesis of Patellamide A, a Cytotoxic Cyclic Peptide from a Tunicate. Revision of Its Proposed Structure. Tetrahedron Lett., **26**, 6501 (1985).
88. HAMADA, Y., M. SHIBATA, and T. SHIOIRI: New Methods and Reagents in Organic Synthesis. 55. Total Syntheses of Patellamides B and C, Cytotoxic Cyclic Peptides from a Tunicate. 1. Their Proposed Structures Should Be Corrected. Tetrahedron Lett., **26**, 5155 (1985).
89. HAMADA, Y., M. SHIBATA, and T. SHIOIRI: New Methods and Reagents in Organic Synthesis. 56. Total Syntheses of Patellamides B and C, Cytotoxic Cyclic peptides from a Tunicate. 1. Their Real Structures have been Determined by Their Syntheses. Tetrahedron Lett., **26**, 5159 (1985).
90. SCHMIDT, U., R. UTZ, and P. GLEICH: What is the Structure of the Patellamides? Tetrahedron Lett., **26**, 4367 (1985).
91. SCHMIDT, U., and H. GRIESSER: Total Synthesis and Structure Determination of Patellamide B. Tetrahedron Lett., **27**, 163 (1986).
92. KATO, S., Y. HAMADA, and T. SHIOIRI: Total Synthesis of Ulithiacyclamide, a Strong Cytotoxic Cyclic Peptide. Tetrahedron Lett., **27**, 2653 (1986).
93. SCHMIDT, U., and P. GLEICH: Total Synthesis of Ulicyclamide. Angew. Chem., Int. Ed. Engl., **24**, 569 (1985).
94. BREDENKAMP, M., C. HOLZAPFEL, and W. VAN ZYL: The Chiral Synthesis of Thiazole Amino Acid Enantiomers. Synth. Commun., **20**, 2235 (1990).
95. KELLY, R.C., I. GEBHARD, and N. WICNIENSKI: Synthesis of (*R*)- and (*S*)-(glu)Thz and the Corresponding Bisthiazole Dipeptide of Dolastatin 3. J. Org. Chem., **51**, 4590 (1986).
96. PETTIT, G.R., F. HOGAN, D.D. BURKETT, S.B. SINGH, D. KANTOCI, J. SRIRANGAM, and M.D. WILLIAMS: The Dolastatins 16. Synthesis of Dolaphenine. Heterocycles, **39**, 81 (1994).
97. PARIKH, J.R., and W. VON E. DOERING: Sulfur Trioxide in the Oxidation of Alcohols by Dimethyl Sulfoxide. J. Amer. Chem. Soc., **89**, 5505 (1967).
98. FATIADI, A.J.: Active Manganese Dioxide Oxidation in Organic Chemistry. Synthesis, **65**, 133 (1976).
99. HAYASHI, K., Y. HAMADA, and T. SHIOIRI: Stereoselective Total Synthesis of Dolastatin 10 Utilizing the Evans-Aldol Reaction. In: Third Symposium on the Chemistry of Natural Products. Osaka. 1991.
100. IRAKO, N., Y. HAMADA, and T. SHIOIRI: A New Efficient Synthesis of (*S*)-Dolaphenine ((*S*)-2-Phenyl-1-(2-thiazolyl)ethylamine), the C-Terminal Unit of Dolastatin 10. Tetrahedron, **48**, 7251 (1992).
101. SCHMIDT, U., P. GLEICH, H. GRIESSER, and R. UTZ: Synthesis of Optically Active 2-(1-Hydroxyalkyl)-thiazole-4-carboxylic Acids and 2-(1-Aminoalkyl)-thiazole-4- carboxylic Acids. Synthesis, 992 (1986).
102. RINEHART, K.L., R. SAKAI, V. KISHORE, D.W. SULLINS, and K.-M. LI: Synthesis and Properties of the Eight Isostatine Stereoisomers. J. Org. Chem., **57**, 3007 (1992).
103. RINEHART, K.L., B. KISHORE, S. NAGARAJAN, R.J. LAKE, J.B. GLOER, F.A. BOZICH, K.-M. LI, R.E. MALECZKA, JR., W.L. TODSEN, M.H.G. MUNRO, D.W. SULLINS, and R. SAKAI: Total Synthesis of Didemnins A, B, and C. J. Amer. Chem. Soc., **109**, 6846 (1987).
104. SCHMIDT, U., M. KRONER, and H. GRIESSER: Total Synthesis of the Didemnins – 2. Synthesis of Didemnin A, B, C and Prolyldidemnin A. Tetrahedron Lett., **29**, 4407 (1988).
105. HAMADA, Y., Y. KONDO, M. SHIBATA, and T. SHIOIRI: Efficient Total Synthesis of Didemnins A and B. J. Amer. Chem. Soc., **111**, 669 (1989).

106. MAIBAUM, J., and D.H. RICH: Inhibition of Porcine Pepsin by Two Substrate Analogues Containing Statine. The Effect of Histidine at the P_2 Subsite on the Inhibition of Aspartic Proteinases. J. Med. Chem., **31**, 625 (1988).

107. WAGNER, I., and H. MUSSO: New Naturally Occurring Amino Acids. Angew. Chem., Int. Ed. Engl., **22**, 816 (1983).

108. EWING, W.R., B.D. HARRIS, K.L. BHAT, and M.M. JOUILLIÉ: Synthetic Studies of the Detoxin Complex. 1. Total Synthesis of (−)-Detoxinine. Tetrahedron, **42**, 2421 (1986).

109. HAGIWARA, H., K. KIMURA, and H. UDA: High Diastereoselection in the Aldol Reaction of the Bistrimethylsilyl Enol Ether of Methyl Acetoacetate with 2-Benzyloxyhexanal. Synthesis of (−)-Pestalotin. J. Chem. Soc., Perkin Trans. 1, 693 (1992).

110. FRANCK-NEUMANN, M., P.-J. COLSON, P. GEOFFROY, K.M. TABA: The Aldol Condensation Reaction of Diene (Tricarbonyl)iron Complexes. From Metal Coordinated Trimethylsilyl Enol Ethers to Polyols of Known Configuration. Tetrahedron Lett., **33**, 1903 (1992).

111. KOBAYASHI, S., and I. HACHIYA: The Aldol Reaction of Silyl Enol Ethers with Aldehydes in Aqueous Media. Tetrahedron Lett., **33**, 1625 (1992).

112. DEVANT, R.M., and H.-E. RADUNZ: A Short Novel and Efficient Asymmetric Synthesis of Statine Analogues. Tetrahedron Lett., **29**, 2307 (1988).

113. WUTS, P.G.M., and S.R. PUTT: Synthesis of *N*-Boc-Statine and *epi*-Statine. Synthesis, 951 (1989).

114. PETTIT, G.R., S.B. SINGH, D.L. HERALD, P. LLOYD-WILLIAMS, D. KANTOCI, D.D. BURKETT, J. BARKÓCZY, F. HOGAN, and T.R. WARDLAW: The Dolastatins. 17. Synthesis of Dolaproine and Related Diastereoisomers. J. Org. Chem., **59**, 6287 (1994).

115. SMITH, A.B., III, T.L. LEENAY, H.-J. LIU, L.A.K. NELSON, and R.G. BALL: A Caveat on the Swern Oxidation. Tetrahedron Lett., **29**, 49 (1988).

116. BRAUN, M., and R. DEVANT: Effective Synthetic Equivalents of a Chiral Acetate Enolate. Tetrahedron Lett., **25**, 5031 (1984).

117. KANTOCI, D., G.R. PETTIT, and T.R. WARDLAW: Efficient Separation of Dolaproine Stereoisomers by Optimization of a Three Component Chromatographic Solvent System. J. Liq. Chromatogr., **13**, 3915 (1990).

118. PETTIT, G.R., S.B. SINGH, F. HOGAN, and D.D. BURKETT: Chiral Modifications of Dolastatin 10: The Potent Cytostatic Peptide (19a*R*)-Isodolastatin 10. J. Med. Chem., **33**, 2177 (1990).

119. DIEM, M.J., D.F. BUROW, and J.L. FRY: Oxonium Salt Alkylation of Structurally and Optically Labile Alcohols. J. Org. Chem., **42**, 180 (1990).

120. WALKER, M.A., and C.H. HEATHCOCK: Extending the Scope of the Evans Asymmetric Aldol Reaction: Preparation of Anti and Non-Evans Syn Aldols. J. Org. Chem., **56**, 5747 (1991).

121. GAGE, J.R., and D.A. EVANS: Diastereoselective Aldol Condensation Using a Chiral Oxazolidinone Auxiliary: (2*S**,3*S**)-3-Hydroxy-3-Phenyl-2-Methylpropanoic Acid. Org. Synth., **68**, 83 (1989).

122. VAN DRAANEN, N.A., S. ARSENIYADIS, M.T. CRIMMINS, and C.H. HEATHCOCK: Protocols for the Preparation of Each of the Four Possible Stereoisomeric α-Alkyl-β-hydroxy Carboxylic Acids from a Single Chiral Aldol Reagent. J. Org. Chem., **56**, 2499 (1991).

123. HEATHCOCK, C.H.: Understanding and Controlling Diastereofacial Selectivity in Carbon–Carbon Bond-Forming Reactions. Aldrichimica Acta, **23**, 99 (1990).

124. DANDA, H., M.M. HANSEN, and C.H. HEATHCOCK: Reversal of Stereochemistry in the Aldol Reactions of a Chiral Boron Enolate. J. Org. Chem., **55**, 173 (1990).

125. PETTIT, G.R., S.B. SINGH, J.K. SRIRANGAM, F. HOGAN-PIERSON, and M.D. WILLIAMS: The Dolastatins. 19. Synthesis of Dolaisoleuine. J. Org. Chem., **59**, 1796 (1994).

126. PETTIT, G.R., M.D. WILLIAMS, J.K. SRIRANGAM, F. HOGAN, N.L. BENOITON, and D. KANTOCI: The dolastatins 25. Conformational isomerism of *N*-benzyloxycarbonyl-*N*-methylisoleucinol and related substances. J. Chem. Soc., Perkin Trans. 2, 919 (1995).
127. MCDERMOTT, J.R., and N.L. BENOITON: *N*-Methylamino Acids in Peptides Synthesis. III. Racemization During Deprotection by Saponification and Acidolysis. Can. J. Chem., **51**, 2555 (1973).
128. MCDERMOTT, J.R., and N.L. BENOITON: *N*-Methylamino Acids in Peptides Synthesis. IV. Racemization and Yields in Peptide-Bond Formation. Can. J. Chem., **51**, 2562 (1973).
129. MCDERMOTT, J.R., and N.L. BENOITON: *N*-Methylamino Acids in Peptides Synthesis. II. A New Synthesis of *N*-Benzyloxycarbonyl *N*-Methylamino Acids. Can. J. Chem., **51**, 1915 (1973).
130. EVANS, D.A., J. BARTROLI, and T.L. SHIH: Enantioselective Aldol Condensations. 2. Erythro-Selective Chiral Aldol Condensations *via* Boron Enolates. J. Amer. Chem. Soc., **103**, 2127 (1981).
131. BOGER, D.L., and T.T. CURRAN: Synthesis of the Lower Subunit of Rhizoxin. J. Org. Chem., **50**, 1830 (1985).
132. GUTIERREZ, C.G., and L.R. SUMMERHAYS: Organotin-Mediated Selective Desulfurization: Tri-*N*-butyltin Hydride Reduction of Unsymmetric Sulfides. J. Org. Chem., **49**, 5206 (1984).
133. ANDERSON, G.W., and F.M. CALLAHAN: *t*-Butyl Esters of Amino Acids and Peptides and Their Use in Peptide Synthesis. J. Amer. Chem. Soc., **82**, 3359 (1960).
134. COOK, B., R.R. HILL, and G.E. JEFFS: Efficient One-Step Synthesis of Diastereoisomeric Cyclic Dipeptides from Amino Acids: Three Diastereoisomers of Cyclo-L-isoleucyl-L-isoleucine. J. Chem. Soc., Perkin Trans. 1, 1199 (1992).
135. ABRAHAM, R.J., J.T. JACKSON, and W.A. THOMAS: The Fixed Conformation of the Leucyl Side-Chain in a Tripeptide. Org. Mag. Res., **14**, 543 (1980).
136. TORII, K., and Y. IITAKA: The Crystal Structure of L-Isoleucine. Acta Crystallogr., **B27**, 2237 (1971).
137. COGGINS, J.J.R., and N.L. BENOITON: Synthesis of *N*-Methylamino Acid Derivative from Amino Acid Derivatives Using Sodium Hydride-Methyl Iodide. Can. J. Chem., **49**, 1968 (1971).
138. STEINAUER, R., F.M.F. CHEN, and N.L. BENOITON: *N*-Methylamino Acids in Peptide Synthesis. Int. J. Peptide Protein Res., **26**, 109 (1985).
139. PETTIT, G.R., J. BARKOCZY, J.K. SRIRANGAM, S.B. SINGH, D.L. HERALD, M.D. WILLIAMS, D. KANTOCI, F. HOGAN, and T.L. GROY: The Dolastatins. 22. Synthesis of Boc-dolaproinyl-dolaphenine and Four Related Chiral Isomers. J. Org. Chem., **59**, 2935 (1994).
140. WENGER, R.M.: Synthesis of Cyclosporine and Analogues: Structural Requirements for Immunosuppressive Activity. Angew. Chem., Int. Ed. Engl., **24**, 77 (1985).
141. COSTE, J., M.-M. DUFOUR, A. PANTALONI, and B. CASTRO: BROP: A New Reagent for Coupling N-Methylated Amino Acids. Tetrahedron Lett., **31**, 669 (1990).
142. YAMADA, S., I. IKOTA, T. SHIOIRI, and S. TACHIBANA: Diphenyl Phosphorazidate (DPPA) and Diethyl Phosphorocyanidate (DEPC), Two New Reagents for Solid-Phase Peptide Synthesis and Their Application to the Synthesis of Porcine Motulin. J. Amer. Chem. Soc., **97**, 7174 (1975).
143. BOWMAN, R.E., and H.H. STROUD: N-Substituted Amino-acids. Part 1. A New Method of Preparation of Dimethylamino-acids. J. Chem. Soc., 1342 (1950).
144. PETTIT, G.R., J.K. SRIRANGAM, S.B. SINGH, M.D. WILLIAMS, D.L. HERALD, J. BARKOCZY, D. KANTOCI, and F. HOGAN: The Dolastatins. 24. Synthesis of (−)-dolastatin 10. J. Chem. Soc., Perkin Trans. 1, 859 (1996).

145. SHIOIRI, T.: Total Synthesis of Dolastatin 10, an Antitumor Marine Peptide Containing Unusual Amino Acids. Chem. Soc. Jpn., **52**, 392 (1994).

146. PETTIT, G.R., J.K. SRIRANGAM, J. BARKOCZY, M.D. WILLIAMS, K.P.M. DURKIN, M.R. BOYD, R. BAI, E. HAMEL, J.M. SCHMIDT, and J.-C. CHAPUIS: Antineoplastic Agents 337. Synthesis of dolastatin 10 structural modifications. Anti-Cancer Drug Design, **10**, 529 (1995).

147. AGUILAR, E., and A.I. MEYERS: Reinvestigation of a Modified Hantzsch Thiazole Synthesis. Tetrahedron Lett., **35**, 2473 (1994).

148. PETTIT, G.R., D. KANTOCI, D.L. HERALD, J. BARKOCZY, and J.A. SLACK: Procedures for the Analyses of Dolastatins 10 and 15 by High Performance Liquid Chromatography. J. Liq. Chromatogr., **17**, 191 (1994).

149. PETTIT, G.R., J.K. SRIRANGAM, D.L. HERALD, and E. HAMEL: The Dolastatins. 21. Synthesis, X-ray Crystal Structure, and Molecular Modeling of (6*R*)-Isodolastatin 10. J. Org. Chem., **59**, 6127 (1994).

150. HAMEL, E.: Interactions of Tubulin with Small Ligands. In: Microtubule Proteins (Avila, J., ed.), p. 89. Boca Raton: CRC Press. 1990.

151. BAI, R., G.R. PETTIT, and E. HAMEL: Structure-Activity Studies with Chiral Isomers and with Segments of the Antimitotic Marine Peptide Dolastatin 10. Biochem. Pharmacol., **40**, 1859 (1990).

152. BAI, R., G.R. PETTIT, and E. HAMEL: Binding of Dolastatin 10 to Tubulin at a Distinct Site for Peptide Antimitotic Agents Near the Exchangeable Nucleotide and Vinca Alkaloid Sites. J. Biol. Chem., **265**, 17141 (1990).

153. BAI, R., G.R. PETTIT, and E. HAMEL: Dolastatin 10, A Powerful Cytostatic Peptide Derived from a Marine Animal: Inhibition of Tubulin Polymerization Mediated through the Vinca Alkaloid Binding Domain. Biochem. Pharmacol., **39**, 1941 (1990).

154. BAI, R., M.C. ROACH, S.K. JAYARAM, J. BARKOCZY, G.R. PETTIT, R.F. LUDUENA, and E. HAMEL: Differential Effects of Active Isomers, Segments, and Analogs of Dolastatin 10 on Ligand Interactions with Tubulin. Biochem. Pharmacol., **45**, 1503 (1993).

155. KIRST, H.A., E.F. SZYMANSKI, D.E. DORMANN, J.L. OCCOLOWITZ, N.D. JONES, M.O. CHANEY, R.L. HAMILL, and M.M. HOEHN: Structure of Althiomycin. J. Antibiotics, **28**, 286 (1975).

156. CARDELLINA, J.H. II, F.-J. MARNER, and R.E. MOORE: Malyngamide A, a Novel Chlorinated Metabolite of the Marine Cyanophyte *Lyngbya majuscula*. J. Amer. Chem. Soc., **101**, 240 (1979).

157. CARDELLINA, J.H. II, and R.E. MOORE: The Structure of Pukeleimides A, B, C, D, E, F, and G. Tetrahedron Lett., 2007 (1979).

158. VON HOFHEINZ, W., and W.E. OBERHANSLI: Dysidin, ein neuartiger, chlorhaltiger Naturstoff aus dem Schwamm *Dysidea herbacea*. Helv. Chim. Acta, **60**, 660 (1977).

159. GEBREYESUS, T., T. YOSIEF, S. CARMELY, and Y. KASHMAN: Dysidamide, a Novel Hexachloro-metabolite from a Red Sea Sponge *Dysidea* sp. Tetrahedron Lett., **29**, 3863 (1988).

160. OIKAWA, Y., K. SUGANO, and O. YONEMITSU: Meldrum's Acid in Organic Synthesis. 2. A General and Versatile Synthesis of β-Keto Esters. J. Org. Chem., **43**, 2087 (1978).

161. PETTIT, G.R., T.J. THORNTON, J.T. MULLANEY, M.R. BOYD, D.L. HERALD, S.B. SINGH, and E.J. FLAHIVE: The Dolastatins 20. A Convenient Route to Dolastatin 15. Tetrahedron, **50**, 12097 (1994).

162. JOUIN, P., B. CASTRO, and D. NISATO: Stereospecific Synthesis of *N*-Protected Statine and Its Analogues *via* Chiral Tetramic Acid. J. Chem. Soc., Perkin Trans. 1, 1177 (1987).

163. PATINO, N., FREROT, N. GALEOTTI, J. PONCET, J. COSTE, M.-N. DUFOUR, and P. JOUIN: Total Synthesis of the Proposed Structure of Dolastatin 15. Tetrahedron, **48**, 4115 (1992).

164. Ishiwata, H., H. Sone, H. Kigoshi, and K. Yamada: Total Synthesis of Doliculide, a Potent Cytotoxic Cyclodepsipeptide from the Japanese Sea hare *Dolabella auricularia.* J. Org. Chem., **59**, 4712 (1994).
165. Ojika, M., H. Kigoshi, T. Ishigaki, and K. Yamada: Further Studies on Aplyronine A, an Antitumor Substance Isolated from the Sea Hare *Aplysia kurodai.* Tetrahedron Lett., **34**, 8501 (1993).
166. Kigoshi, H., M. Ojika, T. Ishigaki, K. Suenaga, T. Mutou, A. Sakakura, T. Ogawa, and K. Yamada: Total Synthesis of Aplyronine A, a Potent Antitumor Substance of Marine Origin. J. Amer. Chem. Soc., **116**, 7443 (1994).
167. Ojika, M., T. Yoshida, and K. Yamada: Aplysepine, a Novel 1,4-Benzodiazepine Alkaloid from the Sea Hare *Aplysia kurodai.* Tetrahedron Lett., **33**, 5307 (1993).
168. Quentmeier, H., S. Brauer, G.R. Pettit, and H.D. Drexler: Cytostatic Effects of Dolastatin 10 and Dolastatin 15 on Human Leukemia Cell Lines. Leukemia and Lymphoma, **6**, 245 (1992).
169. Steube, K.G., H. Quentmeier, G.R. Pettit, T. Pietsch, S. Brauer, D. Grunicke, S.M. Gignac, and H.G. Drexler: Inhibition of Cellular Proliferation by the Natural Peptides Dolastatin 10 and Dolastatin 15. Mol. Biol. Haematapoiesis, **2**, 567 (1992).
170. Steube, K.G., D. Grunicke, T. Pietsch, S.M. Gignac, G.R. Pettit, and H.G. Drexler: Dolastatin 10 and Dolastatin 15: Effects of Two Natural Peptides on Growth and Differentiation of Leukemia Cells. Leukemia, **6**, 1048 (1992).
171. Hu, Z.-B., S.M. Gignac, H. Quentmeier, G.R. Pettit, and H.G. Drexler: Effects of Dolastatins on Human B-Lymphocytic Leukemia Cell Lines. Leukemia Res., **17**, 333 (1993).
172. Kim, I.K., B.S. Lee, Y.T. Hong, H.S. Choi, and N.Y. Kwon: Synthesis of Dolastatin 10 Analogues. J. Korean Chem. Soc., **38**, 763 (1994).
173. Luduena, R.F., M.C. Roach, V. Prasad, and G.R. Pettit: Interaction of Dolastatin 10 with Bovine Brain Tubulin. Biochem. Pharmacol., **43**, 539 (1992).
174. Bai, R., G.F. Taylor, J.M. Schmidt, M.D. Williams, J.A. Kepler, G.R. Pettit, and E. Hamel: Interaction of Dolastatin 10 with Tubulin: Induction of Aggregation and Binding and Dissociation Reactions. Mol. Pharmacol. **47**, 965 (1995).
175. Bai, R., S.J. Friedman, G.R. Pettit, and E. Hamel: Dolastatin 15, a Potent Antimitotic Depsipeptide Derived from *Dolabella auricularia.* Interaction with Tubulin and Effects on Cellular Microtubules. Biochem. Pharmacol., **43**, 2637 (1992).

(Received September 25, 1995)

Acetogenins from Annonaceae

A. CAVÉ[1,*], B. FIGADÈRE[1], A. LAURENS[1], and D. CORTES[2],
[1] Laboratoire de Pharmacognosie, Faculté de Pharmacie, Université Paris-Sud, Châtenay-Malabry, France;
[2] Departamento de Farmacologia, Farmacognosia y Farmacodinamica, Faculdad de Farmacia, Universidad de Valencia, Burjasot, Spain

Contents

I. Introduction 81
II. Classification 82
Synonyms of acetogenins 84
Listing (structures and physical data) 89
Distribution 180
III. Extraction and isolation 188
IV. Structure elucidation 189
V. Biogenetic hypotheses 226
VI. Syntheses 231
VII. Biological activities 264
Acknowledgements 273
References 273

I. Introduction

In 1982, JOLAD *et al.* isolated uvaricin, a new antitumor agent, from the roots of *Uvaria acuminata* (Annonaceae), a bis-tetrahydrofuranoid fatty acid lactone (*1*) related to polyketides or acetogenins. However, it contained a number of original structural characteristics, particularly a linear acetogenin, a bis-tetrahydrofuran pattern flanked by hydroxyls and a terminal unsaturated lactone. Two years later, Dabrah and Sneden (*2*, *3*) and Cortes *et al.* (*4*) described four new products presenting the same structural characteristics. Because these products formed a new class of natural compounds, and are only found in species belonging to the family of Annonaceae, they are commonly called acetogenins from Annonaceae.

Annonaceous acetogenins constitute a series of C-35/C-37 natural products of polyketide origin derived from fatty acids. Their structure is characterized by a long alkyl chain bearing a terminal unsaturated γ-methyl-γ-lactone (sometimes rearranged to a γ-lactone containing an acetonyl α to the lactone carbonyl), one, two or three tetrahydrofuran rings and some oxygenated substituents along the chain, particularly α to a tetrahydrofuran, and in some cases double bonds and/or epoxides.

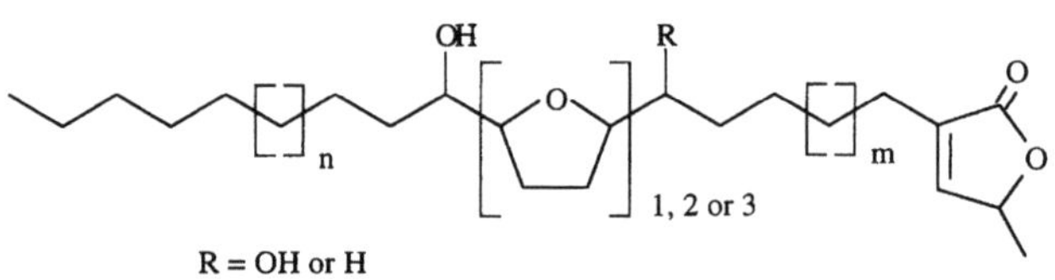

Because the acetogenins from Annonaceae exhibit a broad range of potential biological properties such as cytotoxic, antitumoral, antiparasitic, pesticidal, antimicrobial and immunosuppresive activities, research in the field expanded greatly with the number of isolated and reported acetogenins increasing rapidly. 31 acetogenins were described in 1990 (*5*) and 61 in 1993 (*6*) as part of two reviews, and they now number more than 160. At the same time, numerous synthetic studies were initiated by an ever-growing number of research teams, leading to a rapid increase in the number of papers dealing with isolation, structural elucidation, synthesis and biology. Between 1982 and 1992, about 80 papers appeared, and since 1993 they have reached about 130.

II. Classification

Acetogenins from Annonaceae can be classified into four main types according to the number and arrangement of the tetrahydrofuran rings along the alkyl chain.

Type A with one tetrahydrofuran nucleus. In most of the acetogenins of type A the tetrahydrofuran is flanked by two hydroxyls at the α and α′ positions.

Type B with two adjacent tetrahydrofurans. To date, four acetogenins of type B do not contain hydroxyl α of the tetrahydrofuran pattern.

Type C with two non-adjacent tetrahydrofurans separated by four methylenes. The tetrahydrofuran on the side of the lactone is flanked by only one hydroxyl.

Type D with three adjacent tetrahydrofurans. Currently, only one acetogenin belongs to this type.

TYPE

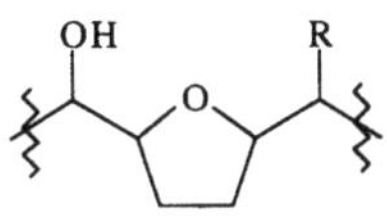

TYPE A

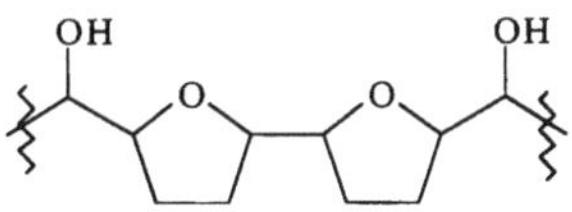

TYPE B

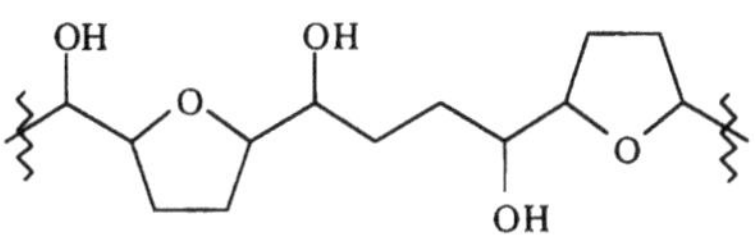

TYPE C

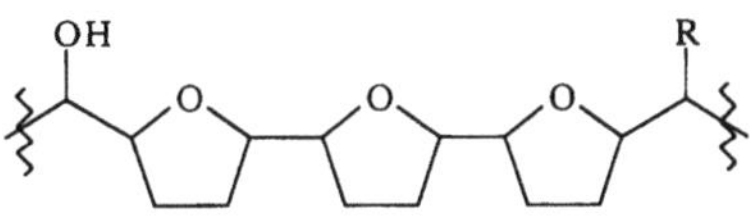

TYPE D

SUB-TYPE

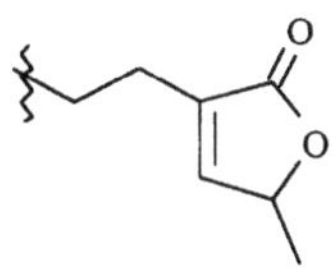

SUB-TYPE 1a

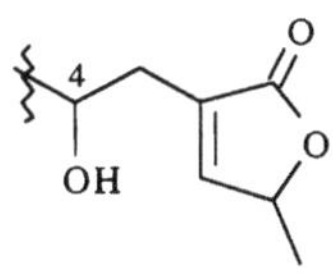

SUB-TYPE 1b

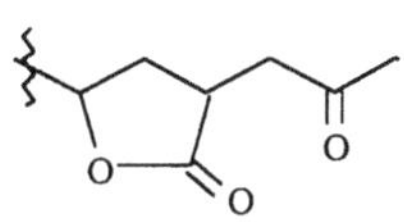

SUB-TYPE 2

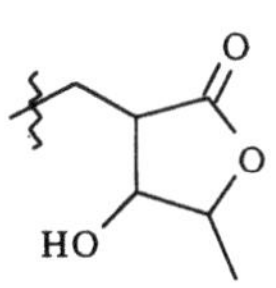

SUB-TYPE 3

These different types can be subdivided into three sub-types depending on the lactone:

Sub-type 1 with an unsaturated γ-methyl-γ-lactone. This is the most common sub-type. It can be divided into sub-type 1a and sub-type 1b depending on the absence or presence of a hydroxyl on the chain β to the lactone ring (position C-4). This subdivision is justified as the presence of the hydroxyl leads easily to sub-type 2 by translactonization.

Sub-type 2 (isoacetogenins (7)) with a saturated γ-lactone substituted by a propanone group. Members of this sub-type, in fact, are artefacts formed during extraction or purification processes (*8, 9*).

Sub-type 3 with a saturated β-hydroxy-γ-methyl-γ-lactone. Such a lactone can be considered as the precursor of the lactone of sub-type 1. It is rare and only four acetogenins belong to this sub-type.

To the four main types a fifth type can be added, type E. This type is characterized by absence of the tetrahydrofuran ring or rings. In their place, there is a double bond (or bonds) and/or epoxy groups. They are biogenetic precursors of the normal acetogenins. The acetogenins belonging to type E can lead to acetogenins belonging to type A, B or C by biomimetic hemisyntheses.

II.a. Names and Synonyms of Acetogenins (names which are set in italics have to be used)

Name of compound	Type
Almunequin or *Squamostatin-A*	C1a
Annogalene or *Xylomatenin*	A1b
Annoglaucin	B1b
Annohexocin	A1b
Annomonicin	A1b
Annomontacin	A1b
Annomontacin-10-one (hemisynthesis)	A1b
18,21-cis-Annomontacin-10-one (hemisynthesis)	A1b
Annomuricin-A	A1b
Annomuricin-B	A1b
Annomuricin-C	A1b
Annomutacin	A1b
Annonacin = Annonacin-1 = Howiicin-A	A1b
Annonacin-1 = *Annonacin*	A1b
Annonacin-2 = *Annonacin-A*	A1b
Annonacin-A = Annonacin-2	A1b
Annonacin-A-one = *Isoannonacin-A*	A2
Annonacinone = Annonacin-10-one	A1b
Annonareticin: probably Rolliniastatin or Asimicin	B1b
Annonastatin = *Asimicin*	A1b
Annonin I = *Squamocin* = Squamocin-A	B1a
Annonin III = *Motrilin*	B1a
Annonin IV = *Bullatanocin*	C1b
Annonin VI = *Rolliniastatin-2* = Bullatacin	B1b

Name of compound	Type
Annonin VIII = *Bullatalicin* = Cherimolin-1	C1b
Annonin XIV (wrong str.)	
Annonin XVI = *Squamostatin-A*	C1a
Annonsilin A	B1a
Annoreticuin	A1b
Annoreticuin-9-one	A1b
Annosenegalin	A1b
Annotemoyin-1	A1a
Annotemoyin-2	A1a
Asimicin = Squamocin-H = Annonastatin	B1b
Asimilobin	B1b
Asimin	B1a
Asiminacin or Squamocin-D	B1a
Asiminecin	B1a
Asiminenin A	A1b
Asiminenin B	A1b
Asitribin	B1a
Atemoyacin-A = *Parviflorin* = Squamocin-E	B1b
Atemoyacin-B = *Neoannonin* = Squamocin-J	B1a
Atemoyin or Squamocin-K	B1a
Bullacin	B1a
Bulladecinone	B2
Bullanin	B1a
Bullatacin or *Rolliniastatin-2*	B1b
Bullatacinone = Isorolliniastatin-2	B2
Bullatalicin or Cherimolin-1 = Squamostatin-B = Annonin VIII	B1b
C-12, 15-cis-Bullatalicin	C1b
Bullatalicinone or Isocherimolin-1	C2
Bullatanocin or Cherimolin-2 = Squamostatin-C = Crassiflorin = Purpureacin-1 = Annonin IV	C1b
C-12, 15-cis-Bullatanocin	C1b
Bullatanocinone	C2
C-12, 15-cis-Bullatanocinone	C2
Bullatencin = Bullatenin	A1a
Bullatin	B1a
Cherimolin-1 or *Bullatalicin* = Squamostatin-B = Annonin VIII	B1b
Cherimolin-2 = *Bullatanocin* = Squamostatin-C = Crassiflorin = Purpureacin-1 = Annonin IV	B1b
Corepoxylone	E1a
Coriacin	A1b
Coriadienin	E1b
Corossolin = Howiicin-C	A1a
Corossolone	A1a
Crassiflorin = Cherimolin-2 = *Bullatanocin* = Squamostatin-C = Purpureacin-1	C1b
Cyclogoniodenin C (hemisynthesis)	D1b
Cyclogoniodenin T (hemisynthesis)	D1b

Name of compound	Type
Cyclogonionenin C (hemisynthesis)	B1b
Cyclogonionenin T(hemisynthesis)	B1b
Densicomacin-1 = 13, 14-*erythro*-Densicomacin	A1b
Densicomacin-2 = 13, 14-*threo*-Densicomacin = Gigantetrocin (A or B)	A1b
4-Deoxyasimicin = *Isodesacetyluvaricin* = Squamocin-M	B1a
4-Deoxycoriacin	A1a
4-Deoxygigantecin	C1a
Desacetyluvaricin = Squamocin-L = Neodesacetyluvaricin	B1a
4-Desoxycherimolin-2 = *Squamostatin-E*	C1a
4-Desoxyhowiicin = *Solamin*	A1a
Diepomuricanin-A = Epoxyrollin-B	E1a
Diepomuricanin-B	E1a
Dieporeticanin-1 = Epoxyrollin-A	E1a
Dieporeticanin-2	E1a
Dieporeticenin	E1a
Diepoxymontin	E1a
Diepoxyrollin	E1a
Dihydrocherimolin = *Otivarin*	C3
Epomuricenin-A = Epoxymurin-A	E1a
Epomuricenin B	E1a
Epoxymurin-A = Epomuricenin-A	E1a
Epoxymurin-B	E1a
Epoxyrollin A (wrong str.) = *Dieporeticanin-1*	E1a
Epoxyrollin B (wrong str.) = *Diepomuricanin-A*	E1a
Giganenin	A1a
Giganin	E1b
Gigantecin	C1b
Gigantetrocin-A = Densicomacin-2 = Howiicin-F	A1b
Gigantetrocin-B = Howiicin-G?	A1b
Gigantetrocinone = Isodensicomacin	A2
Gigantetronenin	A1b
Gigantriocin = Howiicin-D	A1a
Gigantrionenin	A1a
Glaucanisin	B1b
Goniocin	D1b
Goniodenin	B1b
Gonionenin	A1b
Goniothalamicin	A1b
Goniothalamicinone	A2
Howiicin-A = *Annonacin*	A1b
Howiicin-B = *Murisolin*	A1b
Howiicin-C = *Corossolin*	A1a
Howiicin-D = *Gigantriocin*	A1a
Howiicin-E = *Muricatetrocin-A*	A1b
Howiicin-F = *Gigantetrocin-A*	A1b
Howiicin-G = *Gigantetrocin-B*	A1b

Name of compound	Type
8-Hydroxyannonacin	A1b
30-Hydroxybullatacin	B1b
31-Hydroxybullatacin	B1b
32-Hydroxybullatacin	B1b
10-Hydroxybullatacinone	B2
12-Hydroxybullatacinone	B2
28-Hydroxybullatacinone	B2
29-Hydroxybullatacinone	B2
30-Hydroxybullatacinone	B2
31-Hydroxybullatacinone	B2
32-Hydroxybullatacinone	B2
4-Hydroxy-25-desoxyneorollinicin (wrong str.) = *Rolliniastatin-1*	B1b
4-Hydroxy-25-desoxyrollinicin (wrong str.) = *Rolliniastatin-2*	B1b
Isoannonacin = Annonacin-A-one	A2
Isoannonacin-A = Annonacin-A-one	A2
Isoannonacinone	A2
Isoannoreticuin	A2
Isocherimolin-1 = *Bullatalicinone*	C2
Isodensicomacin = *Gigantetrocinone*	C2
Isodesacetyluvaricin = 4-Deoxyasimicin = Squamocin-M	B1a
Isomolvizarin-1	B2
Isomolvizarin-2	B2
Isomurisolin	A2
Isoneoannonacinone (probably Isoannonacinone)	A2
Isorolliniastatin-1	B2
Isorolliniastatin-2 = *Bullatacinone*	B2
Isorollinicin?	B1a
Isosylvaticin (hemisynthesis)	C2
Itrabin	B3
Jetein	A3
Laherradurin	B3
Longicin	A1b
Membranacin	B1a
Molvizarin	B1b
Montanacin	A1b
Motrilin or Annonin-III = Squamocin-C	B1a
Mucocin	A1b
Muricatacin	E
Muricatalin	A1b
Muricatetrocin-A = Howiicin-E	A1b
Muricatetrocin-B	A1b
Muricatin-A	A1b
Muricatin-B	A1b
Muricatin-C	A1b
Muricatocin-A	A1b
Muricatocin-B	A1b
Muricatocin-C	A1b

Name of compound	Type
Murihexocin A	A1b
Murihexocin B	A1b
Murisolin = Howiicin-B	A1b
16, 19-cis-Murisolin	A1b
Murisolin-A	A1b
Narumicin I	B1a
Narumicin II	B1a
Neoannonacin-10-one = *Annonacinone*	A1b
Neoannonacin-B = *Neoannonin* = Squamocin-J = Atemoyacin-B	B1a
Neoannonin = Squamocin-J = Neoannonacin-B = Atemoyacin-B	B1a
Neodesacetyluvaricin = *Desacetyluvaricin*	B1a
Neoisoannonacinone (probably *Isoannonacinone*)	A2
Neoisoannonacinone = *Isoannonacinone*	A2
Neoreticulatacin = *Reticulatain-1*	A1a
Otivarin = Dihydrocherimolin (wrong str.)	C3
Panalicin	B1a
Parvifloracin	C1b
Parviflorin or Squamocin-E	B1b
Plagionicin-A	A1a
Purpureacin-1 = *Bullatanocin* = Cherimolin-2	C1b
Purpureacin-2	B1b
Reticulacinone	A1b
Reticulatacin = Uvariamicin-II	A1a
Reticulatain-1 = Neoreticulatacin	A1a
Reticulatain-2	A1a
Reticulatamol	E1a
Reticulatamone	E1a
Rioclarine	B1b
Rolliniastatin-1 = 4-Hydroxy-25-desoxyneorollinicin	B1b
Rolliniastatin-2 or Bullatacin = Squamocin-G	B1b
Rollinicin = *Squamocin* = Annonin-1	B1a
Rollinone = *Isorolliniastatin-1*	B2
Senegalene	A1b
Solamin = 4-Desoxyhowiicin	A1a
Squamocin = Squamocin-A = Annonin-I = Rollinicin (wrong str.)	B1a
Squamocin-B	B1a
Squamocin-D or *Asiminacin*	B1a
Squamocin-E = *Parviflorin* = atemoyacin-A	B1b
Squamocin-F	B1a
Squamocin-G = *Rolliniastatin-2* = Bullatacin	B1b
Squamocin-H = *Asimicin*	B1b
Squamocin-I	B1a
Squamocin-J = *Neoannonin* = Neoannonacin-B = Atemoyacin-B	B1a
Squamocin-K or *Atemoyin*	B1a
Squamocin-L = *Desacetyluvaricin*	B1a
Squamocin-M = *Isodesacetyluvaricin* = 4-Deoxyasimicin	B1a
Squamocin-N	B1a

Name of compound	Type
Squamocinone	B1a
Squamone	A2
Squamosinin-A (doubtful)	D1b
Squamostanal-A	E
Squamostatin-A or Almunequin	C1a
Squamostatin-B = Cherimolin-1 = *Bullatalicin*	C1b
Squamostatin-B′ = *Squamostatin-A*?	C1a
Squamostatin-C = Cherimolin-2 = *Bullatanocin*	C1b
Squamostatin-D	C1a
Squamostatin-E = 4-Desoxycherimolin-2	C1a
Squamosten-A	A1b
Sylvaticin = Uleicin-C	C1b
12, 15-cis-Sylvaticin	C1b
Trieporeticanin = *tripoxyrollin*	E1a
Trilobacin	B1b
Trilobin	B1a
Tripoxyrollin = Trieporeticanin	E1a
Uleicin A structure to be revised	C1b
Uleicin B structure to be revised	C1b
Uleicin C = *Sylvaticin*	C1b
Uleicin D structure to be revised	C1b
Uleicin E structure to be revised	C1b
Uleirollin	B1a
Uvariamicin I	A1a
Uvariamicin II = *Reticulatacin*	A1a
Uvariamicin III	A1a
Uvariamicin IV	A1a
Uvaricin	B1a
Venezenin	E1b
Xylomatenin or Annogalene	A1b
Xylomaticin	A1b
Xylopiacin	A1b
Xylopianin	A1b
Xylopien	A1b

II.b. Listing of Acetogenins (Structures and Physical Data)

Almunequin or **Squamostatin-A** = Annonin XVI (inexact structure) (*10*), revised structure (*6, 11, 12*), = Squamostatin-B "bis" *,** (*13, 14*)

* There is no good argument for the relative configuration 19,20-*erythro* proposed for squamostatin-B "bis" which was based on a difference in mp

** The name squamostatin-B has also been given to an acetogenin (*20*) which is identical with bullatalicin

Annona squamosa (seeds) (*13, 14, 15*); *Annona cherimolia* (seeds) (*16*), (roots) (*17*)

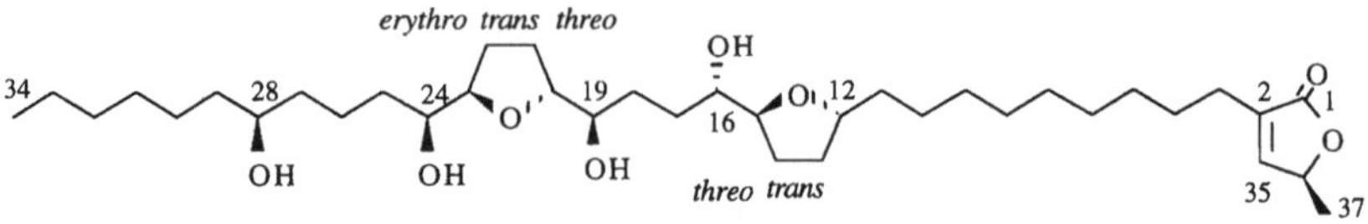

Squamostatin-A was first described without assignment of relative configuration (*15*) and with ^{13}C nmr data which apparently corresponded to the configuration *threo-threo-cis-erythro* (*12, 16*) while almunequin was assigned configuration *threo-threo-trans-erythro* (*16*). In a subsequent paper dealing with squamostatin-A the Japanese authors presented ^{13}C nmr data for squamostatin A identical with those of almunequin and proposed the configuration *trans-threo-threo-trans-erythro* (*20*).

$C_{37}H_{66}O_8$; M = 638
mp = 87–89 °C (*15*), mp = 105–106 °C (*13, 14*)
$[\alpha]_D = +11°$ (c = 0.4, MeOH) (*15*), $[\alpha]_D = +25°$ (c = 0.38, MeOH) (*16*), $[\alpha]_D = +13.5°$ (c = 0.14, $CHCl_3$) (*13, 14*)
IR (*13, 15, 16*); UV (*13, 15*)
^1H nmr (*10*), (500 MHz) (*13, 15*), (400 MHz) (*16*); ^{13}C nmr (*10*), (125 MHz) (*13, 14, 15*), (100 MHz) (*16*)
FABms (*13, 15, 16*), FABms/ms (*15*), CIms (*10, 15, 16*), EIms (*10, 11*)
HRms (*13*), HREIms (*13, 14*)
tetraacetate: ^1H nmr (*10*), (500 MHz) (*13, 14, 15*)
tetra-TMSi derivative: EIms (*11*)
mesitoate: ^1H nmr (C_6D_6, 500 MHz) (*13, 14*)
tetra-(+)-MTPA ester: ^1H nmr (*10*), (500 MHz) (*15*)
dihydroalmunequin: ^1H nmr (200 MHz) (*16*); ^{13}C nmr (50 MHz) (*16*); CIms (*16*)
16,19–24,28-bis-formaldehyde acetal derivative: HRFABms, TMSi derivative (EIms), ^1H nmr (500 MHz), ^{13}C nmr (125 MHz) → 16,19-*cis*, 24,28-*cis* (*18*)
16,19-mono-formaldehyde acetal derivative: HRFABms, TMSi derivative (EIms), ^1H nmr (500 MHz), ^{13}C nmr (125 MHz) → 16,19-*cis* (*18*)
(*R*)- and (*S*)-MTPA esters of 16,19-mono-formaldehyde acetal derivative: ^1H nmr (500 MHz) → 24*S* (*18*)
(*R*)- and (*S*)-MTPA esters ^1H nmr allows to fix the absolute stereochemistry at C-28 thank to the data of the terminal methyl. → absolute stereochemistry 28*S* by "advanced Mosher method" (*19*)
absolute stereochemistry: 12*R*, 15*S*, 16*S*, 19*R*, 20*R*, 23*R*, 24*S*, 28*S*, 36*S*

Annogalene (*21*, *22*) or **Xylomatenin** (*23*)

Annoglaucin

Annona glauca (roots) (*24*)

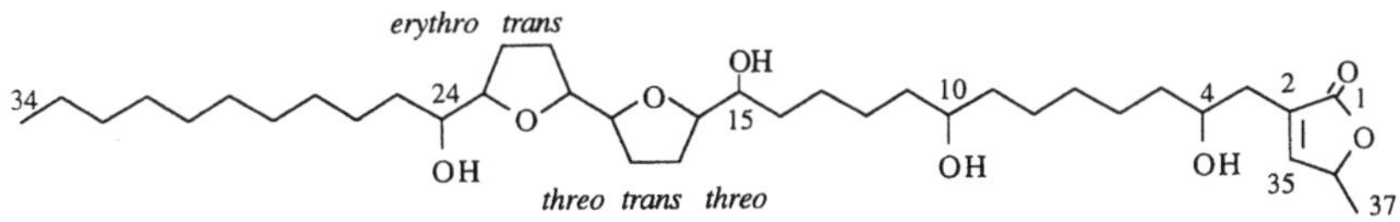

$C_{37}H_{66}O_8$; M = 638
$[\alpha]_D = +15.4°$ (c = 0.6, $CHCl_3$) (*24*)
IR (*24*); UV (*24*)
^1H nmr (200 MHz), ^{13}C nmr (50 MHz) (*24*)
CIms (*24*), FAB-Li ms (*24*)
tetracetate: ^1H nmr (200 MHz) (*24*), EIms (*24*)
tetra-TMSi derivative: EIms (*24*)

Annohexocin

Annona muricata (leaves) (*25*)

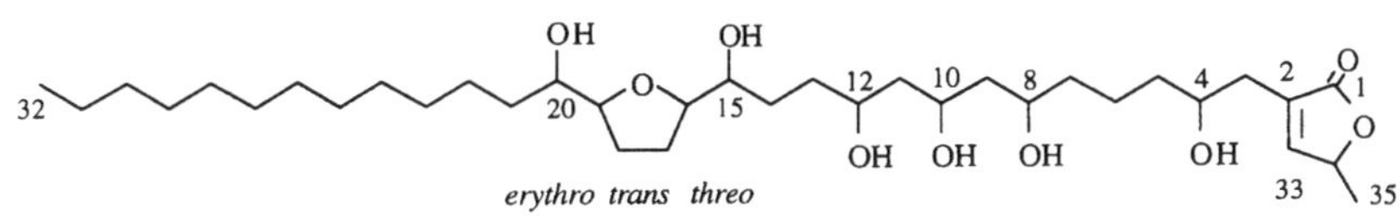

$C_{35}H_{64}O_9$; M = 628
$[\alpha]_D = +18.5°$ (c = 0.33, $CHCl_3$) (*25*)
UV (*25*)
^1H nmr (500 MHz) (*25*); ^{13}C nmr (125 MHz) (*25*)
FABms (*25*), HRFABms (*25*)
hexaacetate: ^1H nmr (500 MHz) (*25*), ^{13}C nmr (125 MHz) (*25*)
TMSi derivative: EIms (*25*), HREIms (*25*)

Annomonicin

Annona montana (seeds) (*26*); *Annona reticulata* (leaves) (*27*)

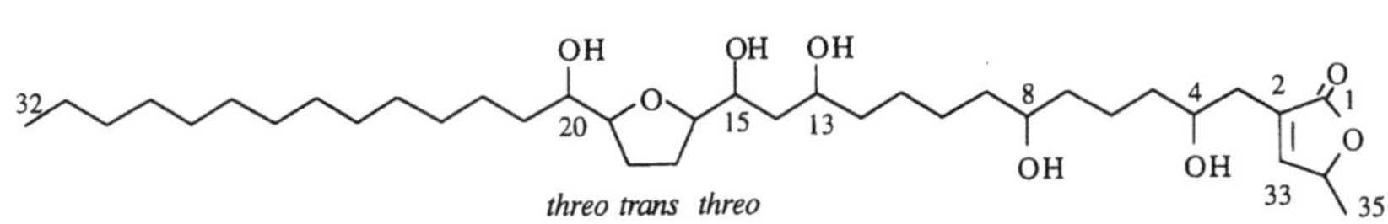

$C_{35}H_{64}O_8$; M = 612
mp = 45–48 °C (*26*), mp = 49–51° (*27*)
$[\alpha]_D = +4°$ (c = 1.0, MeOH) (*26*), $[\alpha]_D = +5°$ (c = 0.17, $CHCl_3$) (*27*)
IR (*26*); UV (*26*)
^{1}H nmr (300 MHz) (*26*); ^{13}C nmr (75 MHz) (*26*)
CIms (*26*)
pentaacetate: CIms (*26*); ^{1}H nmr (300 MHz) (*26*)

Most of the physical, chemical and spectroscopic data reported in (*26*) and (*27*) were essentially identical except the mp and the $[\alpha]_D$

Annomontacin

Annona montana (seeds) (*28*); *Goniothalamus giganteus* (bark) (*29*); *Xylopia aromatica* (*167*)

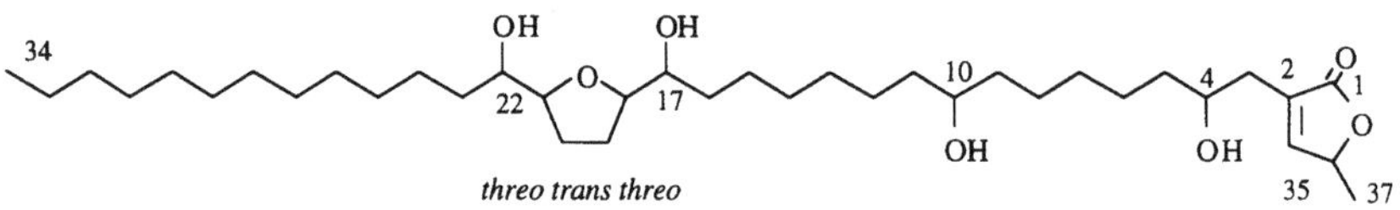

$C_{37}H_{68}O_7$; M = 624
mp = 34–36 °C (*28*), 70–72 °C (*29*)
$[\alpha]_D = +81°$ (c = 1, MeOH) (*28*), $[\alpha]_D = +18°$ (c = 0.45, $CHCl_3$ (*29*)
^{1}H nmr (300 MHz) (*28*); ^{13}C nmr (75 MHz) (*28*)
HRFABms (*28*), CIms (*28*)
tetraacetate: ^{1}H nmr (300 MHz) (*28*)

Annomontacin-10-one

hemisynthesis (*30*)

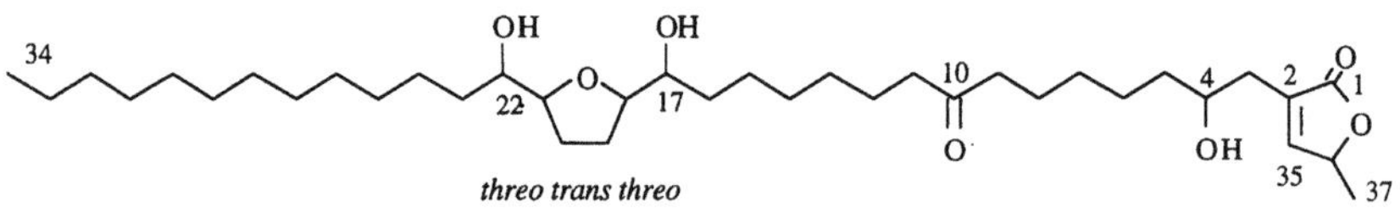

$C_{37}H_{66}O_7$; M = 622
mp = 44–46 °C (*30*)
^{1}H nmr (500 MHz) (*30*)
HRFABms (*30*), FABms (*30*)
tri-TMSi derivative: EIms (*30*)
hemisynthesis from venezenin (*m*-CPBA; $HClO_4$) (*30*)

18,21-*cis*-Annomontacin-10-one

hemisynthesis (*30*)

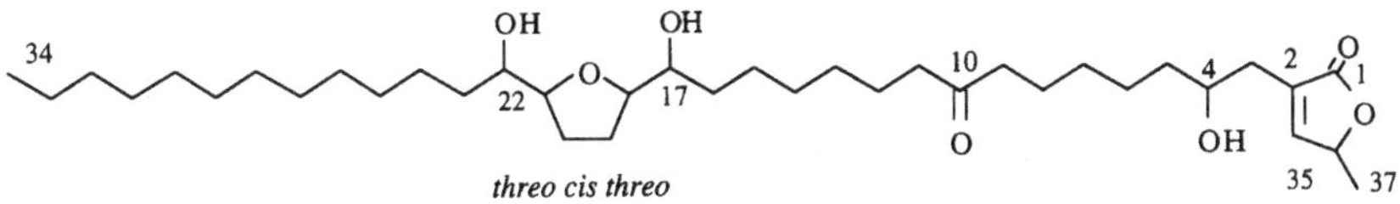

$C_{37}H_{66}O_7$; M = 622
mp = 42–44 °C (*30*)
^{1}H nmr (500 MHz) (*30*)
HRFABms (*30*), FABms (*30*)
tri-TMSi derivative: EIms (*30*)
hemisynthesis from venezenin (m-CPBA; $HClO_4$) (*30*)

Annomuricin-A

Annona muricata (leaves) (*31*)

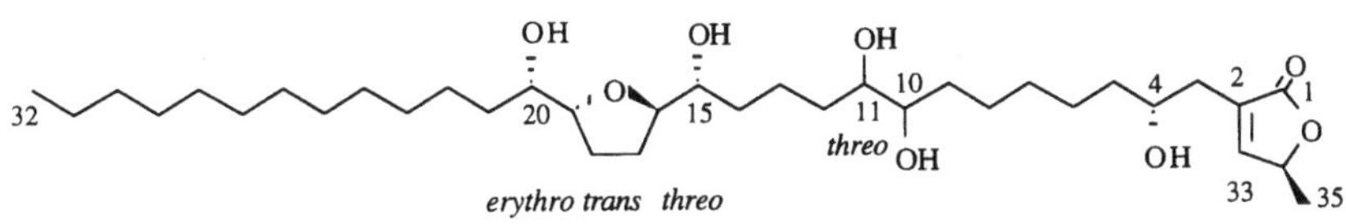

$C_{35}H_{64}O_8$; M = 612
$[\alpha]_D = -6.4°$ (c = 0.0025, solvent ?) (*31*)
IR (*31*); UV (*31*)
^{1}H nmr (500 MHz) (*31*); ^{13}C nmr (125 MHz) (*31*)
HRFABms (*31*), CIms (*31*)
pentaacetate: ^{1}H nmr (500 MHz) (*31*)
penta-TMSi derivative: EIms (*31*), HRCIms (*31*)
acetonide: ^{1}H nmr (500 MHz) (*31*)
(*R*)- and (*S*)-penta-MTPA esters: ^{1}H nmr (500 MHz) → 4*R*, 15*R*, 20*S* (*31*)
absolute stereochemistry: 4*R*, 15*R*, 16*R*, 19*R*, 20*S*, 34*S*

Annomuricin-B

Annona muricata (leaves) (*31*)

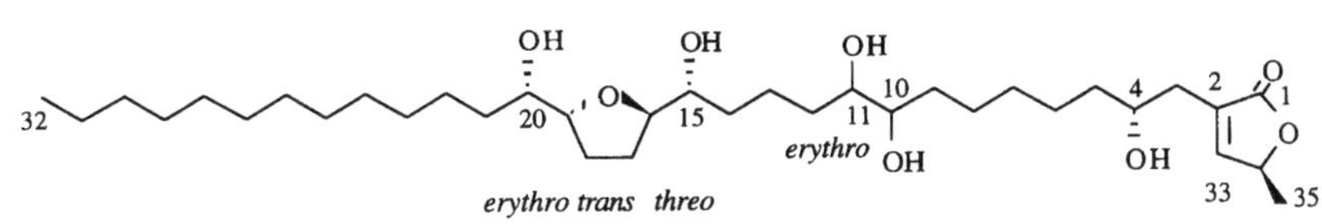

$C_{35}H_{64}O_8$; M = 612
$[\alpha]_D = -11.7°$ (c = 0.0064, solvent ?) (*31*)
IR (*31*); UV (*31*)
^{1}H nmr (500 MHz) (*31*); ^{13}C nmr (125 MHz) (*31*)
HRFABms (*31*), CIms (*31*)
pentaacetate: ^{1}H nmr (500 MHz) (*31*)
penta-TMSi derivative: EIms (*31*), HRCIms (*31*)
acetonide: ^{1}H nmr (500 MHz) (*31*)
(*R*)- and (*S*)-penta-MTPA esters: ^{1}H nmr (500 MHz) → 4*R*, 15*R*, 20*S* (*31*)
absolute stereochemistry: 4*R*, 15*R*, 16*R*, 19*R*, 20*S*, 34*S*

Annomuricin-C

Annona muricata (leaves) (*32*)

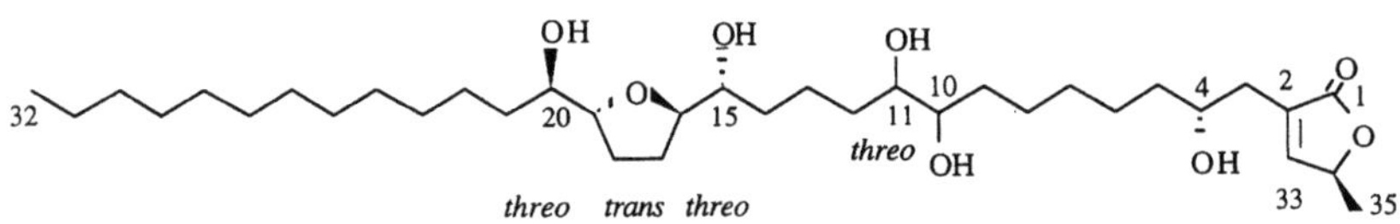

$C_{35}H_{64}O_8$; M = 612
$[\alpha]_D = +57.7°$ (c = 0.0005, EtOH) (*32*)
IR (*32*); UV (*32*)
^{1}H nmr (500 MHz) (*32*); ^{13}C nmr (125 MHz) (*32*)
HRFABms (*32*), CIms (*32*)
pentaacetate: ^{1}H nmr (500 MHz) (*32*)
penta-TMSi derivative: EIms (*32*),
acetonide: ^{1}H nmr (500 MHz) (*32*)
(*R*)- and (*S*)-penta-MTPA esters: ^{1}H nmr (500 MHz) (*32*) → 4*R*, 15*R*, 20*R*
absolute stereochemistry: 4*R*, 15*R*, 16*R*, 19*R*, 20*R*, 34*S*

Annomutacin

Annona muricata (leaves) (*33*)

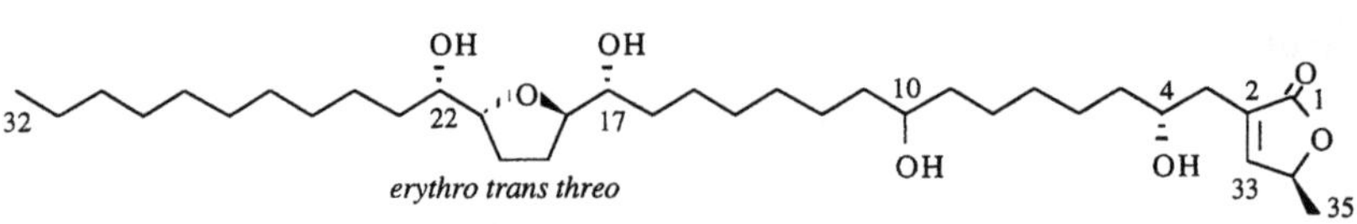

$C_{37}H_{68}O_7$; M = 624
$[\alpha]_D = +60.0°$ (c = 0.001, EtOH) (*33*)

IR (*33*); UV (*33*)
^{1}H nmr (500 MHz) (*33*); ^{13}C nmr (125 MHz) (*33*)
HRFABms (*33*), CIms (*33*)
tetraacetate: ^{1}H nmr (500 MHz) (*33*)
tetra-TMSi derivative: EIms (*33*)
(*R*)- and (*S*)-MTPA esters: ^{1}H nmr (500 MHz) (*33*), → 4*R*, 17*R*, 22*S* (*33*)
absolute stereochemistry: 4*R*, 17*R*, 18*R*, 21*R*, 22*S*, 36*S*

Annonacin = Annonacin-I (*10*), = Howiicin-A (*34*)

Annona densicoma (stem bark) (*35*); *Annona glabra* (bark) (*36*); *Annona montana* (seeds) (*26, 28*); *Annona muricata* (seeds) (*37, 38, 39, 40, 41*), (leaves) (*31*); *Annona senegalensis* (seeds) (*22*); *Annona squamosa* (seeds) (*10*); *Asimina longifolia* (leaves and twigs) (*42*); *Asimina parviflora* (twigs) (*43*); *Goniothalamus giganteus* (stem bark) (*44*); *Goniothalamus howii* (seeds) (*34*); *Xylopia aromatica* (*167*)

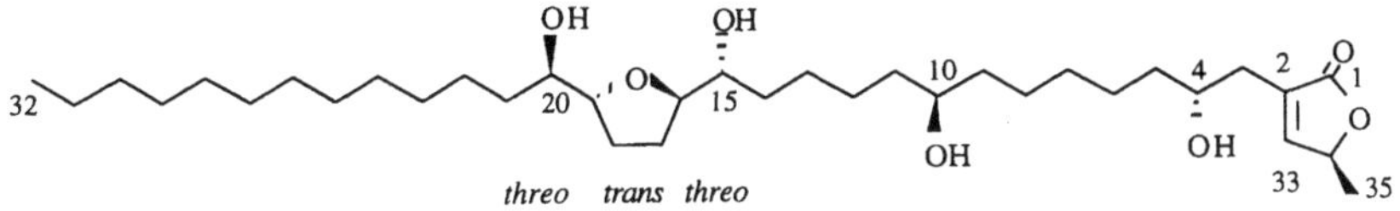

$C_{35}H_{64}O_7$; M = 596
mp = 57 °C (*44*), mp = 52–53 °C (*36*), mp = 65–67 °C (*34*), mp = 82–83.5 °C (*41*)
$[\alpha]_D = +1.4°$ (*44*), $[\alpha]_D = +11.4°$ (c = 0.04, MeOH) (*36*) $[\alpha]_D = +20.78$ (c = 5.05, $CHCl_3$) (*34*), $[\alpha]_D = +22.93°$ (c = 4.62, $CHCl_3$) (*41*)
IR (*22, 34, 35, 36, 41*); UV (*35, 36*)
^{1}H nmr (470 MHz) (*35*), (200 MHz) (*37*), (400 MHz) (*34, 41*), (600 MHz) (*36*); ^{13}C nmr (50 MHz) (*35*), (125 MHz) (*36*), (100 MHz) (*34, 41*)
CIms (*35*), FABms (*35, 36*), EIms (*34, 36, 41*)
tetraacetate: ^{1}H nmr (470 MHz) (*35*), (200 MHz) (*34, 37*), CIms (*35*), EIms (*34*)
tetra-TMSI and tetra-$TMSI_{d9}$ derivatives: HRCIms (*35*)
2,33-dihydroannonacin: EIms (*35*)
relative stereochemistry C(4)-C(36) *RS* (*45*)
oxidative degradation of annonacin → muricatacin: by comparison with a known synthetic analogue, the *R* stereochemistry at C-5 for hemisynthetic muricatacin was suggested (*39*)
(*R*)- and (*S*)-bis-MTPA esters: ^{1}H nmr (500 MHz) (*46*); ^{19}F nmr (*46*): → 4*R*, 15*R*, 20*R* (*46*) → 10*R* (*18*)
absolute stereochemistry: 4*R*, 10*R*, 15*R*, 16*R*, 19*R*, 20*R*, 34*S*

Muricatacin

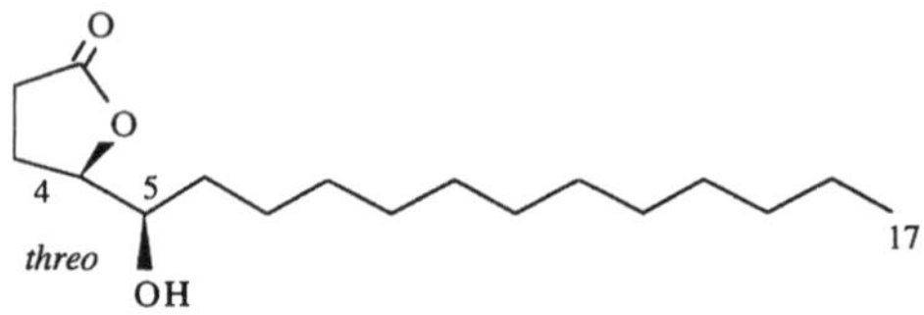

isolated as a mixture of 4*R*, 5*R* and 4*S*, 5*S*-muricatacin (*39*)

$C_{17}H_{32}O_3$; M = 284
$[\alpha]_D = -5.8°$ (c-?, solvent ?) (*39*)
F = 50 °C (*39*)
^1H nmr (500 MHz) (*39*); ^{13}C nmr (125 MHz) (*39*)
CIms (*39*), HRCIms (*39*)
hemisynthesis: annonacin → muricatacin $[\alpha]_D = -16.1°$ (c-?, solvent ?) (*39*)
Syntheses of muricatacin and isomers:
(−)-(4*R*, 5*R*)-muricatacin (*47, 48, 49, 50, 51, 52, 53, 54*)
(+)-(4*S*, 5*S*)-muricatacin (*48, 49, 50, 55, 56, 57, 58*)
(4*S*, 5*R*)-muricatacin (*49, 55, 56*)

Annonacin-A = Annonacin-2

Annona cherimolia (seeds) (*22*); *Annona muricata* (leaves) (*59*); *Annona squamosa* (seeds) (*10, 37*); *Annona senegalensis* (seeds) (*22*)

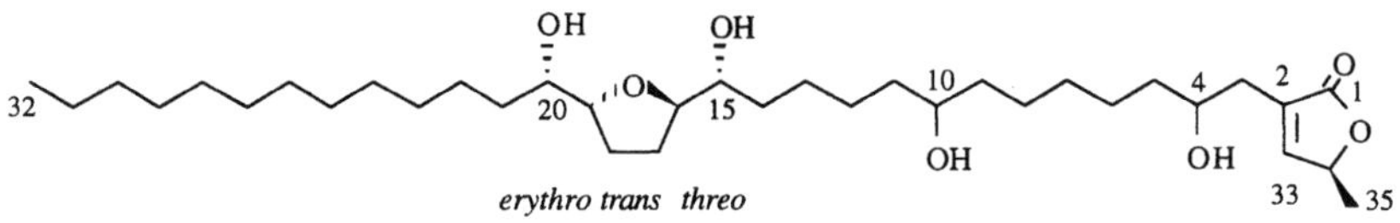

$C_{35}H_{64}O_7$; M = 596
$[\alpha]_D = +23.8°$ (c = 0.4, CH_2Cl_2) (*37*)
IR (*37*), (*22*)
^1H nmr (360 MHz) (*37*); ^{13}C nmr (90 MHz) (*37*)
ms (*37*), FAB-Li ms (*22*)
tetraacetate: ^1H nmr (360 MHz) (*37*)

Annonacin-A-one = Isoannonacin-A

Asimina triloba (stem bark) (*60*); *Annona muricata* (leaves) (*33*)

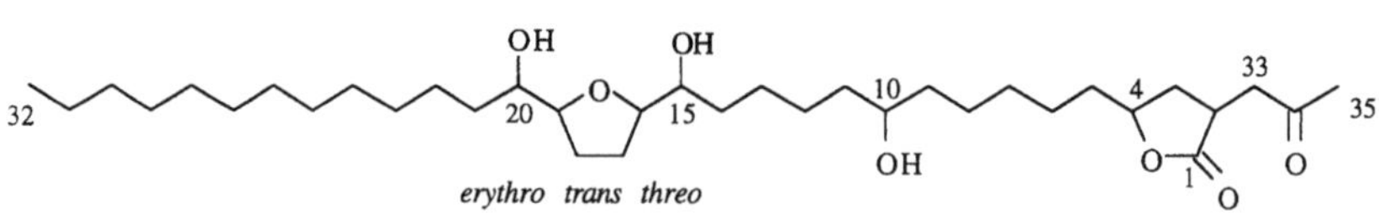

mixture of 2,4-cis- and 2,4-*trans*-isomers

$C_{35}H_{64}O_7$; M = 596
mp = 91–92 °C (*60*)
$[\alpha]_D = +20°$ (c = 0.2, $CHCl_3$) (*60*)
IR (*60*); UV (*60*)
^1H nmr (500 MHz) (*60*); ^{13}C nmr (125 MHz) (*60*)
FABms (glycerol) (*60*), HRFABms (glycerol) (*60*), CIms (*60*), EIms (*60*)
triacetate: ^1H nmr (500 MHz) (*60*)
tri-TMSi derivative: EIms (*60*)

The name annonacin-A-one is confusing; isoannonacin-A would be more appropriate; according to (*33*), Annonacin-A-one could be the 10-*S*-Annonacin-A-one!

10-*R*-Annonacin-A-one

Annona muricata (leaves) (*33*)

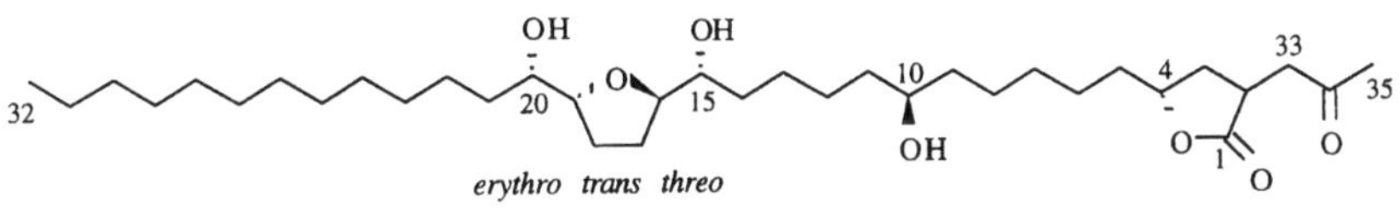

mixture of 2,4-*cis*- and 2,4-*trans*-isomers

$C_{35}H_{64}O_7$; M = 596
$[\alpha]_D = +15°$ (c = 0.002, ?) (*33*)
IR, (*33*); UV (*33*)
^1H nmr (500 MHz) (*33*); ^{13}C nmr (125 MHz) (*33*)
FABms (glycerol) (*33*), HRFABms (glycerol) (*33*), CIms (*33*)
triacetate: ^1H nmr (500 MHz) (*33*)
tri-TMSi derivative: EIms (*33*)
(*R*)- and (*S*)-MTPA esters: ^1H nmr (500 MHz) (*33*) → 4*R* (there is no other data)

The difference in C-10 stereochemistry between annonacin-A-one from *Asimina triloba* (*60*) and the compound isolated from *Annona muricata* is based on differences in some of the physical data and the multiplicity of the H-10 signals (3.40 and 3.60 ppm). The absolute configurations indicated in the formula are given without proof.

Annonacinone = Annonacin-10-one

Annona densicoma (stem bark) (*7*); *Annona montana* (seeds) (*28*); *Annona muricata* (seeds) (*38, 40*), (leaves) (*31*)

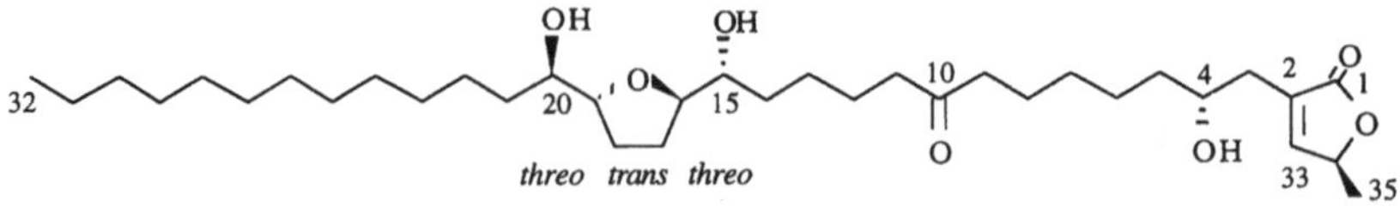

$C_{35}H_{62}O_7$; M = 594
mp = 73–75 °C (7)
$[\alpha]_D$ = + 31,1° (c = 0.06, MeOH) (7)
IR (7); UV (7)
^{1}H nmr (470 MHz) (7); ^{13}C nmr (67.5 MHz) (7), (75 MHz) (*28*)
CIms (7), HRCIms (7)
triacetate: ^{1}H nmr (470 MHz) (7), HRCIms (7)
reduction ($NaBH_4$) → annonacin (7)
(*R*)- and (*S*)-bis-MTPA esters: ^{1}H nmr (500 MHz) (*46*); ^{19}F nmr (*46*): → 4*R*, 15*R*, 20*R* (*46*), → C(4)–C(34) *R*, *S* (*45*)
absolute stereochemistry: 4*R*, 10*R*, 15*R*, 16*R*, 19*R*, 20*R*, 34*S*

Annonin-III or **Motrilin**

Annonsilin-A

Annona squamosa (seeds) (*61*)

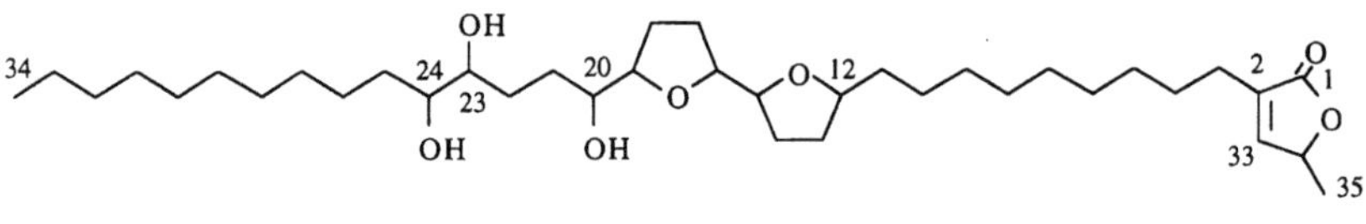

$C_{37}H_{66}O_7$; M = 622
mp = 107–109 °C (*61*)
$[\alpha]_D$ = + 18.65° (c = 1.67, $CHCl_3$) (*61*)
IR (*61*)
^{1}H nmr (300 MHz) (*61*); ^{13}C nmr (75 MHz) (*61*)
EIms (*61*)
triacetate: ^{1}H nmr (*61*); EIms (*61*)

Annoreticuin

Annona reticulata (leaves) (*62*)

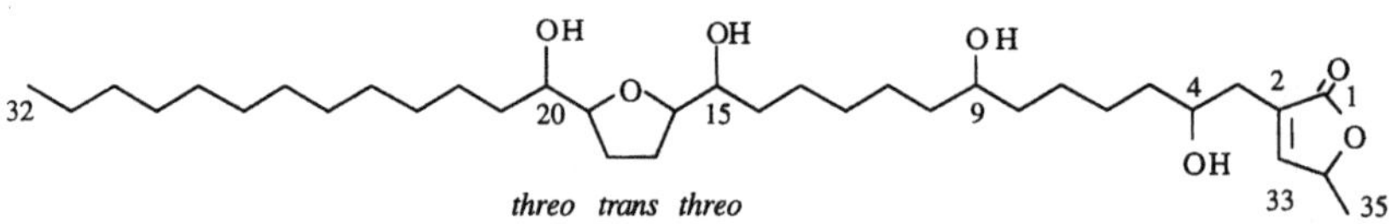

References, pp. 273–288

$C_{35}H_{64}O_7$; M = 596
$[\alpha]_D = +10.5°$ (c = 0.02, $CHCl_3$) (*62*)
IR (*62*); UV (*62*)
^1H nmr (200 MHz) (*62*); ^{13}C nmr (50 MHz) (*62*)
HRCIms (*62*), EIms (*62*)
tetraacetate: ^1H nmr (200 MHz) (*62*)
tetra-TMSi derivative: EIms (*62*)

Annoreticuin-9-one

Annona reticulata (leaves) (*27*)

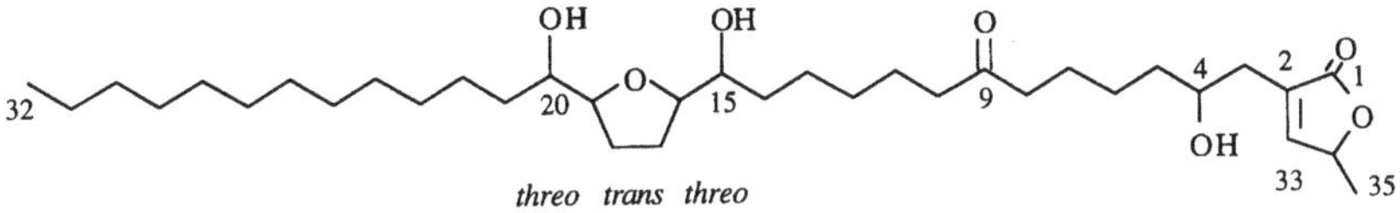

$C_{35}H_{62}O_7$; M = 594
$[\alpha]_D = +11.7°$ (c = 0.02, $CHCl_3$) (*27*)
IR (*27*); UV (*27*)
^1H nmr (200 MHz) (*27*); ^{13}C nmr (50 MHz) (*27*)
EIms (*27*), CIms (*27*), HRFAB-ms (*27*)
triacetate: UV (*27*); ^1H nmr (200 MHz) (*27*); EIms (*27*), CIms (*27*)
tri-TMSi derivative: EIms (*27*), FABms (*27*)
hemisynthesis of squamone (KOH in *t*-BuOH) (*27*) F = 87–89 °C, $[\alpha]_D = +7°$ (c = 0.12, $CHCl_3$)] (*27*)

Annosenegalin

Annona senegalensis (*21, 22*); *Annona cherimolia* (*22*)

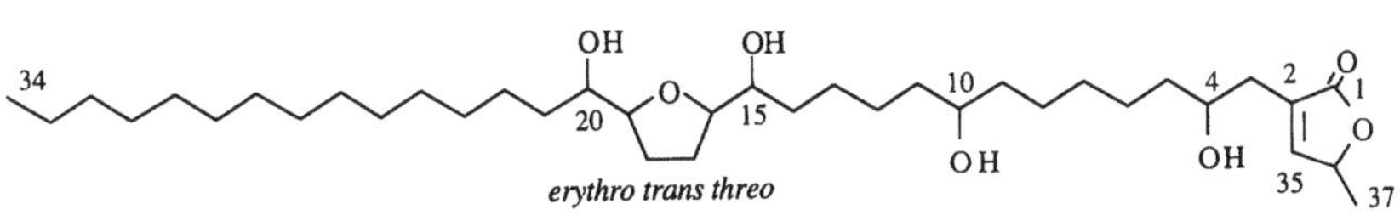

$C_{37}H_{68}O_7$; M = 624
$[\alpha]_D = +15°$ (c = 0.27, $CHCl_3$) (*22*)
IR (*22*)
^1H nmr (200 MHz) (*22*); ^{13}C nmr (50 MHz) (*22*)
EIms (*22*), FAB-Li ms (*22*)
tetraacetate: ^1H nmr (200 MHz) (*22*)

Annotemoyin-1

Annona atemoya (seeds) (*63*)

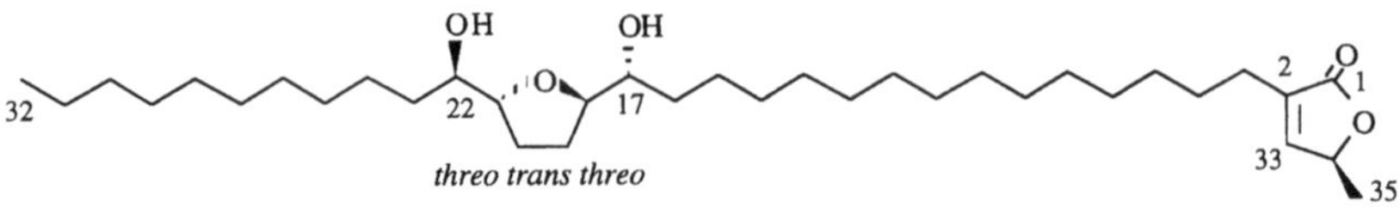

$C_{35}H_{64}O_5$; M = 564
$[\alpha]_D = 21°$ (c = 0.13, MeOH) (*63*)
IR (*63*); UV (*63*)
^{1}H nmr (200 MHz) (*63*); ^{13}C nmr (50 MHz) (*63*)
CIms (*63*), EIms (*63*)
By analogy to solamin, the absolute stereochemistry is suggested as 17*R*, 18*R*, 21*R*, 22*R*, 36*S* (*63*)

Annotemoyin-2

Annona atemoya (seeds) (*63*)

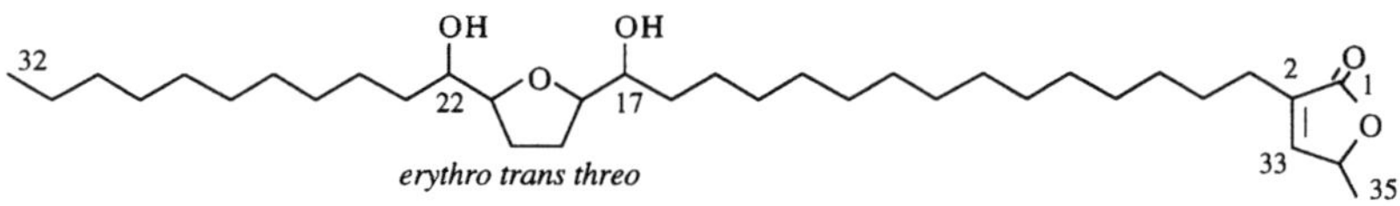

$C_{35}H_{64}O_5$; M = 564
$[\alpha]_D = +20°$ (c = 0.13, MeOH) (*63*)
IR (*63*); UV (*63*)
^{1}H nmr (200 MHz) (*63*); ^{13}C nmr (50 MHz) (*63*)
CIms (*63*), EIms (*63*)

Asimicin = Squamocin-H (*64*); = Annonastatin (*10, 37*), rev. struct. (*65*)

Annona cherimolia (seeds) (*66*); *Annona glabra* (seeds) (*67*); *Annona senegalensis* (seeds) (*68, 69*); *Annona squamosa* (seeds) (*10, 37, 64*); *Asimina parviflora* (twigs) (*70*); *Asimina triloba* (leaves and twigs) (*71*), (bark) (*72*), (seeds) (*73*); *Xylopia aromatica* (bark) (*30*); synthesis (*74*)

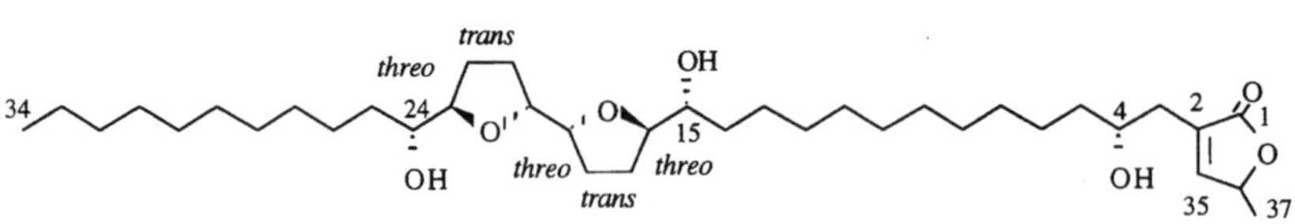

$C_{37}H_{66}O_7$; M = 622
mp = 68–69 °C (*71*); mp = 67–68 °C (*70*); mp = 70–72 °C (*66*), mp = 45–48 °C (*64*), mp = 68–68.5 °C (*74*)
$[\alpha]_D = +11°$ (c = 0.22, $CHCl_3$) (*68*), $[\alpha]_D = +21.8°$ (c = 0.61, MeOH) (*64*), $[\alpha]_D = +14.7$ (c = 0.31, $CHCl_3$) (*74*), $[\alpha]_D = +15°$ (c = 1.1, CH_2Cl_2) (*37*)
CD (*64*)
IR (*37, 70, 71*); UV (*71*)
^1H nmr ($CDCl_3$ and C_6D_6), (470 MHz) (*71*), (500 MHz) (*64, 70, 72*), (300 MHz) (*37*); ^{13}C nmr (50 MHz) (*37, 71*), (125 MHz) (*64, 72*)
EIms (*64*), HRFABms (nitrobenzylalcohol) (*64*), FABms (*71*), CIms (*70*)
tri-acetate: ^1H nmr (300 MHz) (*37, 71*)
relative stereochemistry C(4)–C(36) *RS* (*45*)
(*R*)-tri-MTPA ester: ^1H nmr (500 MHz) (*46*); ^{19}F nmr (*46*) → 15*R*, 24*R* (*46*); ^1H nmr (500 MHz) (*64*) → 15*R*, 16*R*, 19*R*, 20*R*, 23*R*, 24*R* (*64*)
absolute stereochemistry: 15*R*, 16*R*, 19*R*, 20*R*, 23*R*, 24*R*

Asimilobin*

Goniothalamus giganteus (bark) (75)

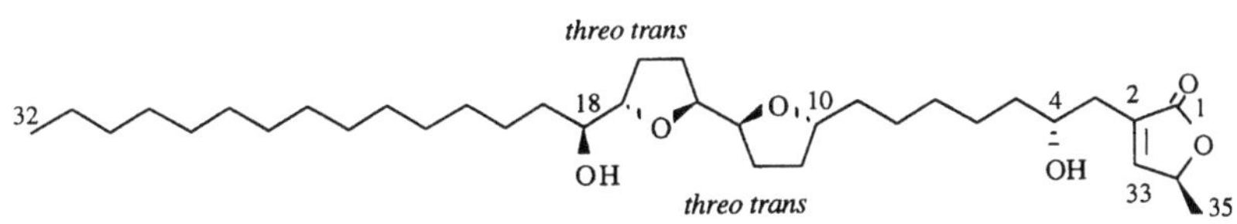

$C_{35}H_{62}O_6$; M = 578
$[\alpha]_D = +11.3°$ (c = 1.0, CH_2Cl_2) (75)
IR (75); UV (75)
^1H nmr [**]; ^{13}C nmr [**]
HRFABms (75), EIms [**]
diacetate: [**]
TMSi derivative: [**]
(*S*)- and (*R*)-MPTA esters: [**] → 4*R*, 18*S*, 36*S* (75)
absolute stereochemistry: 4*R*, 10*R*, 13*S*, 14*S*, 17*S*, 18*S*, 36*S*

---

* This name could be confused with asimilobine, an alkaloid isolated from the same plant and very common in the Annonaceae

** Spectroscopic data leading to the proposed structure were not reported, but were announced as being published elsewhere

Asimin

Asimina triloba (stem bark) (*76*)

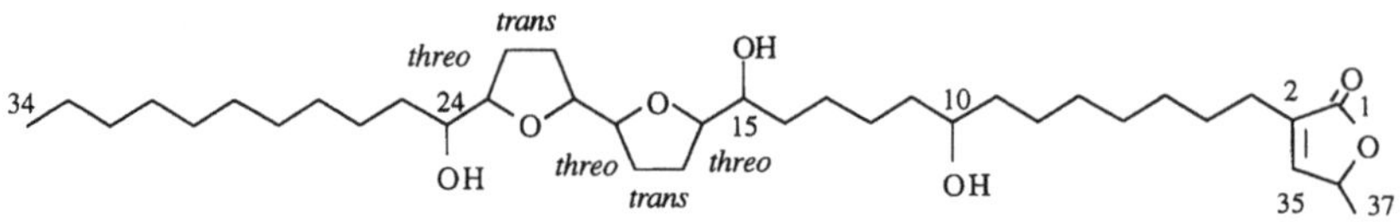

$C_{37}H_{66}O_7$; M = 622
$[\alpha]_D = +26°$ (c = 0.1, $CHCl_3$) (*76*)
IR (*76*); UV (*76*)
^{1}H nmr (500 MHz) (*76*); ^{13}C nmr (125 MHz) (*76*)
CIms (*76*), HRFABms (glycerol) (*76*)
triacetate: ^{1}H nmr (500 MHz) (*76*)
tri-TMSi and tri-TMSi$_{d9}$ derivatives: EIms (*76*)

Asiminacin or Squamocin-D (*64*)

Asimina triloba (stem bark) (*76*); *Annona squamosa* (seeds) (*64*)

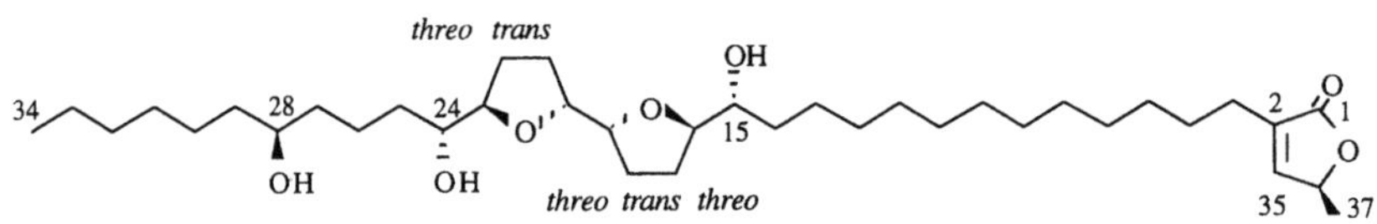

$C_{37}H_{66}O_7$; M = 622
$[\alpha]_D = +21.1°$ (c = 0.38, $CHCl_3$) (*76*), $[\alpha]_D = +30.1°$ (c = 0.58, MeOH) (*64*)
IR (*64*, *76*); UV (*76*)
^{1}H nmr (500 MHz) (*64*, *76*), δ value of terminal methyl allows to fix the position of hydroxyl between C-25 and C-34 (*76*); ^{13}C nmr (125 MHz) (*64*, *76*)
EIms (*64*), CIms (*76*), HRFABms (*64*, *76*)
tri-TMSi and tri-TMSi$_{d9}$ derivatives: EIms (*76*)
triacetate: ^{1}H nmr (500 MHz) (*64*, *76*)
tri-(*R*)-MTPA ester: ^{1}H nmr (500 MHz) (*64*): → 15*R*, 16*R* (*64*), → 28*S* (*64*)
absolute stereochemistry: 15*R*, 16*R*, 19*R*, 20*R*, 23*R*, 24*R*, 28*S*, 36*S*

Asiminecin

Asimina triloba (stem bark) (*76*)

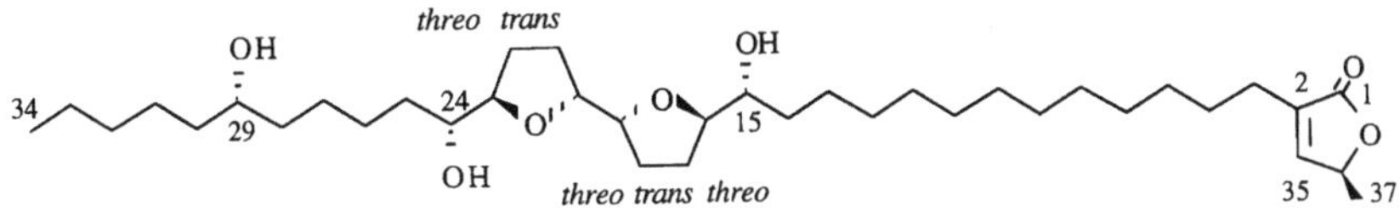

$C_{37}H_{66}O_7$; M = 622
$[\alpha]_D = +22°$ (c = 0.1, $CHCl_3$) (*76*)
IR (*76*); UV (*76*)
^{1}H nmr (500 MHz) (*76*); the δ value of the terminal methyl led to assignment of the hydroxyl between C-25 and C-34 (*76*); ^{13}C nmr (125 MHz) (*76*)
HRCIms, HRFABms (glycerol) (*76*)
tri-TMSi and tri-TMSi$_{d9}$ derivatives: EIms (*76*)
triacetate: ^{1}H nmr (500 MHz) (*76*)

Asiminenin-A

Asimina triloba (seeds) (*73*)

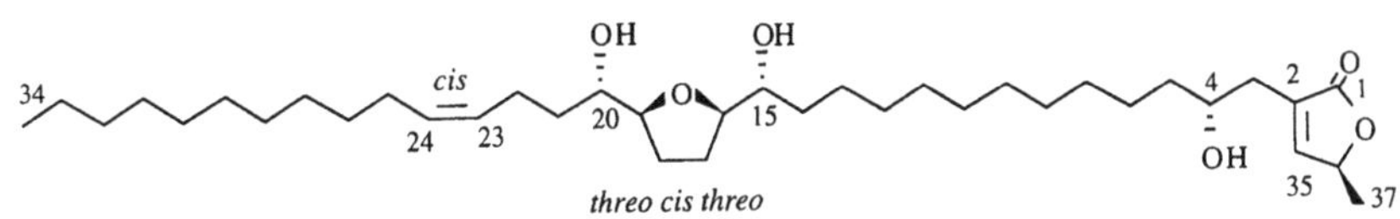

$C_{37}H_{66}O_6$; M = 606
mp = 58–59 °C (*73*)
$[\alpha]_D = +10°$ (c = 0.1, CH_2Cl_2) (*73*)
IR (*73*); UV (*73*)
^{1}H nmr (500 MHz) (*73*); ^{13}C nmr (125 MHz) (*73*)
HRCIms (*73*), CIms (*73*)
triacetate: EIms (*73*); ^{1}H nmr (500 MHz) (*73*)
tri-TMSi derivative: EIms (*73*)
tri-(*R*)- and tri-(*S*)-MTPA esters: ^{1}H nmr (500 MHz): → 4*R*, 15*R*, 20*S*, 36*S* (*73*)
absolute configuration: 4*R*, 15*R*, 16*R*, 19*S*, 20*S*, 36*S*

Asiminenin-B

Asimina triloba (seeds) (*73*)

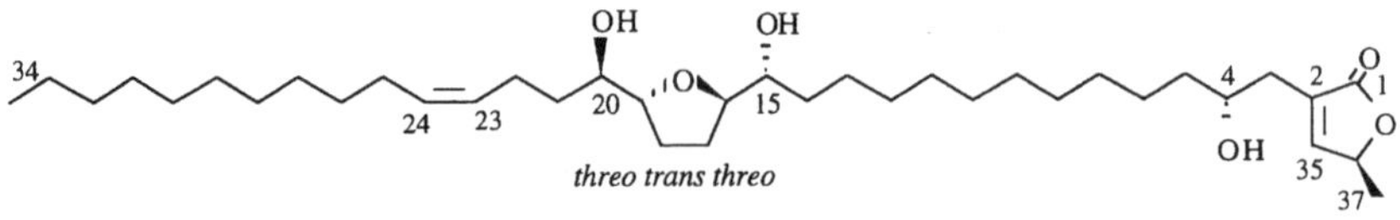

$C_{37}H_{66}O_6$; M = 606
mp = 54–55 °C (*73*)
$[\alpha]_D$ = + 17° (c = 0.1, CH_2Cl_2) (*73*)
IR (*73*); UV (*73*)
^{1}H nmr (500 MHz) (*73*); ^{13}C nmr (125 MHz) (*73*)
HRFABms (*73*), CIms (*73*)
triacetate: EIms (*73*); ^{1}H nmr (500 MHz) (*73*)
tri-TMSi derivative: EIms (*73*)
tri-(*R*)- and tri-(*S*)-MTPA esters: ^{1}H nmr (500 MHz): → 4*R*, 15*R*, 20*R*, 36*S* (*73*)
absolute configuration: 4*R*, 15*R*, 16*R*, 19*R*, 20*R*, 36*S*

Asitribin

Asimina triloba (seeds) (*73*)

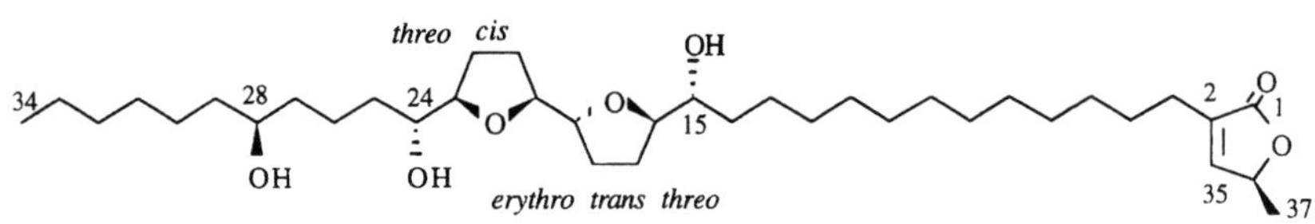

$C_{37}H_{66}O_7$; M = 622
mp = 71–72 °C (*73*)
$[\alpha]_D$ = + 15° (c = 0.1, CH_2Cl_2) (*73*)
CD (*73*) → 36*S*
IR (*73*); UV (*73*)
^{1}H nmr (500 MHz) (*73*); ^{13}C nmr (125 MHz) (*73*)
HRFABms (*73*), CIms (*73*)
triacetate: EIms (*73*); ^{1}H nmr (500 MHz) (*73*)
tri-TMSi derivative: EIms (*73*), CIms (*73*)
tri-(*R*)- and tri-(*S*)-MTPA esters: ^{1}H nmr (500 MHz): → 15*R*, 24*R*, 28*S* (*73*)
absolute stereochemistry: 15*R*, 16*R*, 19*R*, 20*S*, 23*R*, 24*R*, 28*S*

Atemoyin or **Squamocin-K** (*64*)

Annona atemoya (seeds) (*78*); *Annona squamosa* (seeds) (*64*)

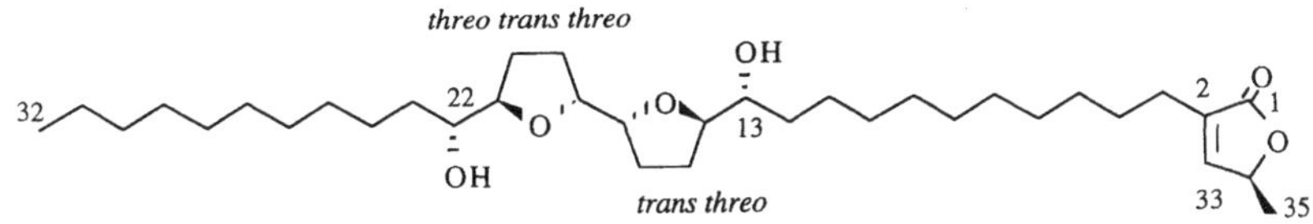

$C_{35}H_{62}O_6$; M = 578
$[\alpha]_D = +18°$ (c = 0.25, MeOH) (*78*), $[\alpha]_D = +20.5°$ (c = 0.53, MeOH) (*64*)
IR (*64*, *78*); UV (*64*, *78*)
^{1}H nmr (200 MHz) (*78*), (500 MHz) (*64*); ^{13}C nmr (50 MHz) (*78*), (125 MHz) (*64*)
FAB-Li ms (*78*), EIms (*64*, *78*), CIms (*78*), HRFABms (nitrobenzylalcohol) (*64*)
diacetate (*64*): ^{1}H nmr (500 MHz) (*64*)
oxidation → undecanal (*78*)
di-(*R*)-MPTA derivative: ^{1}H nmr (500 MHz) (*64*) → 13*R*, 22*R*, (*64*)
absolute stereochemistry: 13*R*, 14*R*, 17*R*, 18*R*, 21*R*, 22*R*, 34*S*

Bullacin

Annona bullata (bark) (*79*)

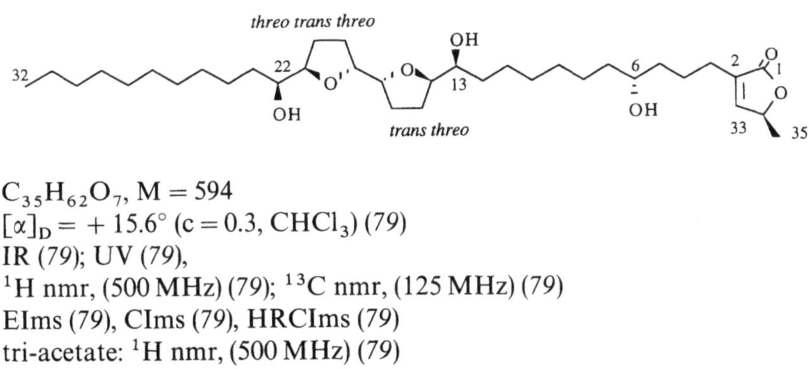

$C_{35}H_{62}O_7$, M = 594
$[\alpha]_D = +15.6°$ (c = 0.3, $CHCl_3$) (*79*)
IR (*79*); UV (*79*),
^{1}H nmr, (500 MHz) (*79*); ^{13}C nmr, (125 MHz) (*79*)
EIms (*79*), CIms (*79*), HRCIms (*79*)
tri-acetate: ^{1}H nmr, (500 MHz) (*79*)
tri-TMSi and tri-$TMSi_{d9}$ derivatives: EIms (*79*)
(*R*)- and (*S*)-MTPA esters: ^{1}H nmr (500 MHz) → 13*R*, 22*R* (*79*); positive data on the lactone side and negative data for H-7 suggested that the stereochemistry at C-6 was *S*. The stereochemistry at C-34 was assumed to be *S* because all acetogenins known so far, have C-34*S* stereochemistry (*79*)
absolute stereochemistry: 13*R*, 14*R*, 17*R*, 18*R*, 21*R*, 22*R*, 34*S* (?), 6*S* (?)

Bulladecinone

Annona bullata (barks) (*80*)

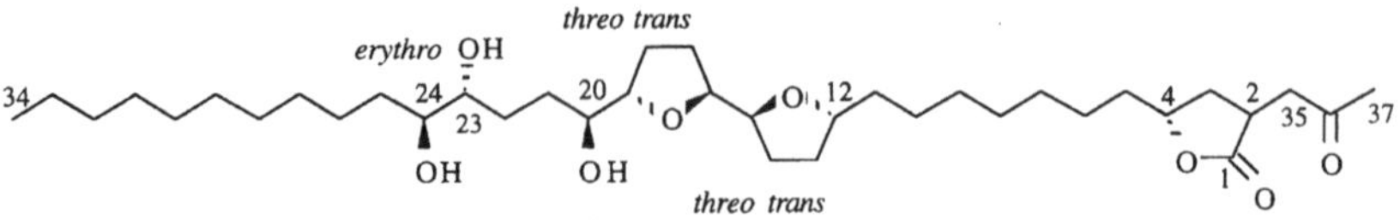

mixture of 2,4-*cis*- and 2,4-*trans*-isomers

$C_{37}H_{66}O_8$; M = 638
^{1}H nmr (500 MHz) (*80*), ^{13}C nmr (125 MHz) (*80*)
HRFABms (glycerol) (*80*), EIms (*80*)
triacetate: ^{1}H nmr (500 MHz) (*80*)
tri-TMSi and tri-TMSi$_{d9}$ derivatives: EIms (*80*)
acetonide: ^{1}H nmr (500 MHz) (*80*): comparative study of ^{1}H nmr of *threo* and *erythro* diols with their acetates and acetonides (*80*); (*S*)- and (*R*)-MTPA esters of acetonide: ^{1}H nmr (500 MHz) (*81*) → 20*S* (*81*)
absolute stereochemistry: 4*R*, 12*R*, 15*S*, 16*S*, 19*S*, 20*S*, 23*R*, 24*S*

Bullanin

Asimina triloba (stem bark) (*82*)

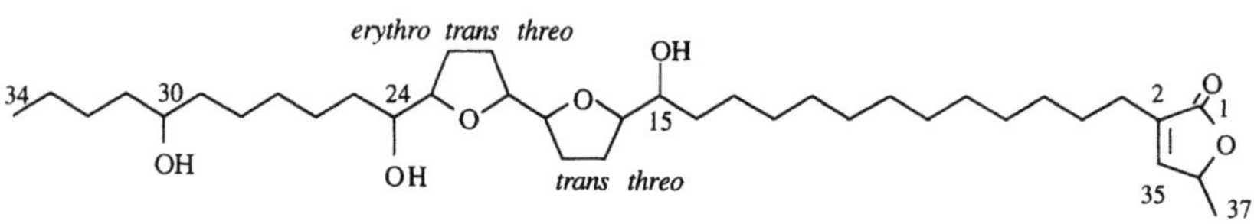

$C_{37}H_{66}O_7$; M = 622
$[\alpha]_D = +28°$ (c = 0.5, EtOH) (*82*)
IR (82); UV (*82*)
^{1}H nmr (500 MHz) (*82*); ^{13}C nmr (125 MHz) (*82*)
HRFABms (*82*), EIms (*82*)
triacetate: ^{1}H nmr (500 MHz) (*82*)
tri-TMSi, tri-TMSi$_{d9}$ derivatives: EIms (*82*)

The absolute stereochemistry was not determined; however the near identities of the pertinent nmr signals and biogenetic considerations suggested an absolute stereochemistry identical with that of bullatacin, 15*R*, 16*R*, 19*R*, 20*R*, 23*R*, 24*S*, 36*S* (*82*)

Bullatacin or **Rolliniastatin-2**

Bullatacinone = Isorolliniastatin-2

Annona bullata (bark) (*83*); *Annona cherimolia* (roots) (*17*); *Annona squamosa* (bark) (*84*); *Asimina triloba* (bark) (*72*); hemisynthesis (*46, 83*)

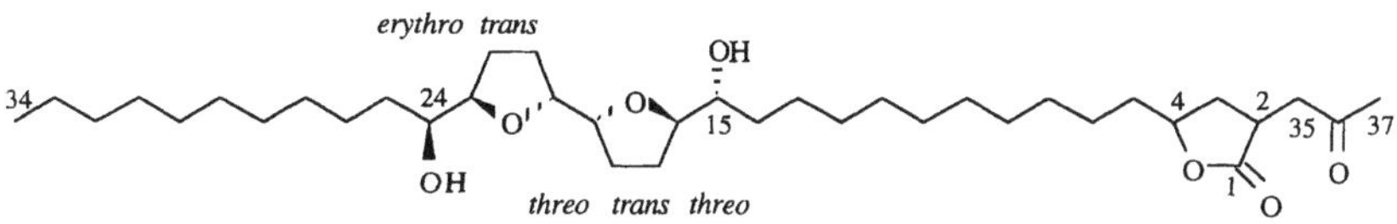

mixture of 2,4-*cis*- and 2,4-*trans*-isomers

$C_{37}H_{66}O_7$, M = 622
mp = 90.5–90.7 °C (*83*)
$[\alpha]_{589} = +12°$, $[\alpha]_{578} = +12.5°$, $[\alpha]_{546} = +14.5°$, $[\alpha]_{436} = +29.75°$, $[\alpha]_{365} = +51.25°$ (c = 0.4; $CHCl_3$) (*83*)
CD: negative Cotton effect → (*S*)-C-4 (*83*)
IR (*83*); UV (*83*)
^{1}H nmr (C_6D_6, 470 MHz) (*83*); ^{13}C nmr (50 MHz) (*83*)
EIms (*83*), CIms (*83*)
diacetate: ^{1}H nmr (470 MHz) (*83*); CIms (*83*), EIms (*83*)
di-TMSi and di-$TMSi_{d9}$ derivatives: EIms (*83*)
conversion of bullatacin to bullatacinone (KOH in *t*-BuOH) (*83*)
(*R*)- and (*S*)-bis-MTPA esters; C-15-mono(*R*)- and (*S*)-MTPA esters; C-24-mono(*R*)- and (*S*)-MTPA esters: ^{1}H nmr (500 MHz) (*46*); ^{19}F nmr (*46*) → 15*R*, 24*S* (*46*)
absolute stereochemistry: 4*S*, 15*R*, *24S*, 16*R*, 19*R*, 20*R*, 23*R*, 24*S*

Bullatalicin or **Cherimolin-1** (*4, 16, 85*), = Squamostatin-B* (*20*), = Annonin VIII (*10*), revised str. (*6, 11, 12*)

Annona bullata (*86*); *Annona cherimolia* (seeds) (wrong str.) (*4, 66*), (revised str.) (*16*); *Annona purpurea* (leaves) (*85*); *Annona squamosa* (*10, 20*); *Rollinia mucosa* (leaves) (*87*)

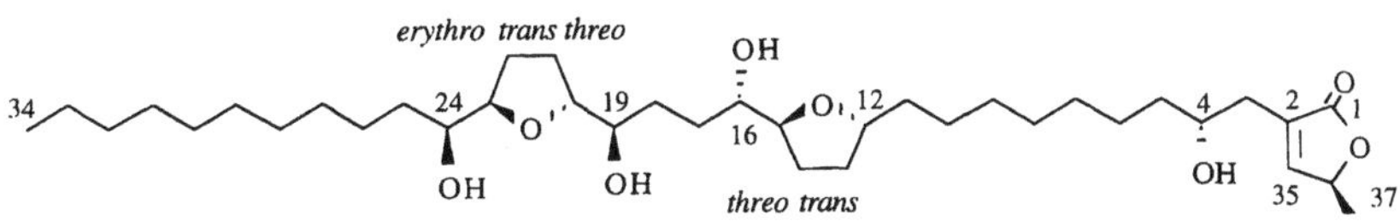

* Another acetogenin with a different structure has been named squamostatin-B (*13*)

$C_{37}H_{66}O_8$; M = 638
mp = 120–121 °C (*86*), mp = 116–117 °C (*66*), mp = 98–101 °C (EtOAc) (*20*)
$[\alpha]_D$ = + 64° (c = 0.3, MeOH) (*16*), $[\alpha]_D$ = + 13.25° (c = 0.4, EtOH) (*86*), $[\alpha]_D$ = + 10.5° (c = 0.10, MeOH) (*20*), $[\alpha]_D$ = + 85° (c = 0.88, MeOH) (*85*)
CD (*20, 86*): → 36*S*
IR (*16, 20, 86*); UV (*20, 85, 86*)
^{1}H nmr (500 MHz, C_6D_6) (*86*), (200 MHz) (*16, 20, 5*), (500 MHz, C_6D_6 and $CDCl_3$) (*11*); ^{13}C-nmr (50 MHz) (*16, 85*), (50 MHz, CD_3OD) (*86*), (125 MHz) (*11, 88*)
CIms (*4, 16, 85, 86*), HRCIms (*86*), EIms (*4, 20*), HRFABms (*20*)
tetraacetate (*4, 16, 86*): IR (*4, 16*); CIms (*86*); ^{1}H nmr (90 MHz) (*4*), (500 MHz) (*11, 86*), (C_6D_6 and $CDCl_3$) (*89*); ^{13}C nmr (50 MHz) (*16, 20, 86, 89*)
tetra-TMSi derivative: CIms (*86*), EIms (*86*)
tetramethyl: ^{1}H nmr (90 MHz) (*4*),
(*R*)-MTPA ester: ^{1}H nmr (500 MHz) (*20*); ^{13}C nmr (125 MHz) (*20*)
dihydrocherimolin-1: ^{1}H nmr (200 MHz) (*16*), CIms (*16*); ^{13}C nmr (50 MHz) (*16*)
Inversion of relative stereochemistry (*11, 12, 89*): 15,16-*threo*, 19,20-*threo*, 23,24-*erythro*, reassignments of nmr in (*11*).
16,19-formaldehyde acetal derivative: HRFABms, TMSi derivative (EIms), ^{1}H nmr (500 MHz), ^{13}C nmr (125 MHz) → 16,19 *cis* (*18*)
(*R*)- and (*S*)-MTPA esters of 16,19-formaldehyde acetal derivative: ^{1}H (500 MHz) → 24*S* (*18*). Configuration 4*R* deduced from nmr by comparison with literature data (*64, 46*)
absolute stereochemistry: 4*R*, 12*R*, 15*S*, 16*S*, 19*R*, 20*R*, 23*R*, 24*S*, 36*S*

12,15-*cis*-Bullatalicin

Annona bullata (bark) (*88*)

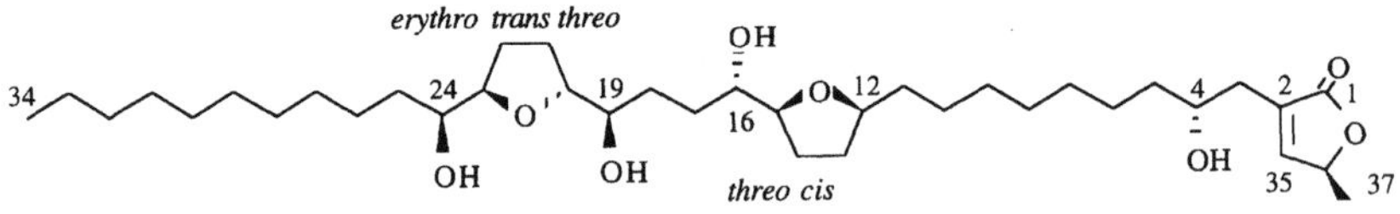

$C_{37}H_{66}O_8$; M = 638
IR (*88*)
^{1}H nmr (500 MHz) (*88*); ^{13}C nmr (125 MHz) (*88*)
HRFABsm (*88*)

TMSi derivative: EIsm (*88*)
16,19-formaldehyde acetal derivative (*88*)
(*R*)- and (*S*)-MTPA esters of 16,19-formaldehyde acetal derivative: ^{1}H nmr (500 MHz) → 4*R*, 24*S* (*88*)
absolute stereochemistry: 4*R*, 12*S*, 15*S*, 16*S*, 19*R*, 20*R*, 23*R*, 24*R*, 36*S*

The corresponding isoacetogenin resulting from translactonization has been detected in a mixture of 8 products as evidenced by a signal corresponding to H-15 in the ^{1}H nmr spectrum (*88*)

Bullatalicinone or **Isocherimolin-1**

Annona bullata (bark) (*90*); *Annona cherimolia* (*17*)

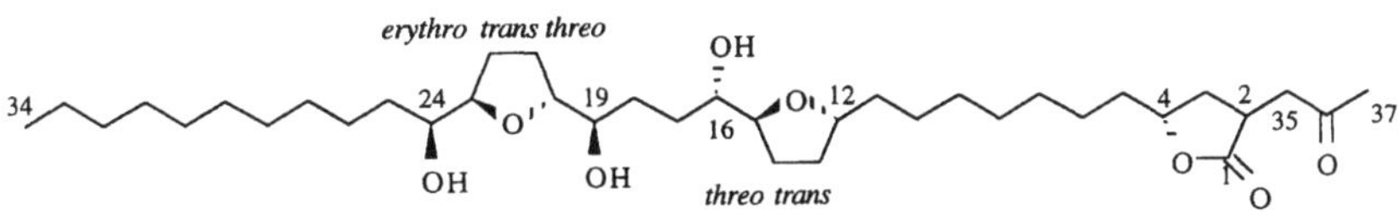

mixture of 2,4-*cis*- and 2,4-*trans*-isomers

$C_{37}H_{66}O_8$; M = 638
mp = 125–126° (*90*)
$[\alpha]_D = +23°$ (c = 0.4, $CHCl_3$) (*90*), $[\alpha]_D = +27°$ (c = 0.15, MeOH) (*17*)
CD (*90*)
IR (*90*); UV (*17, 90*)
^{1}H nmr (500 MHz, C_6D_6) (*90*), (500 MHz) (*11*), (200 MHz) (*17*); ^{13}C nmr (125 MHz, C_6D_6) (*90*), (125 MHz) (*11*), (50 MHz) (*17*)
HRCIms (*90*), CIms (*17, 90*), FAB-Li ms (*17*)
triacetate: ^{1}H nmr (500 MHz) (*11, 90*), CIms (*90*)
tri-TMSi derivative: HRCIms, EIms (*90*)
4*R* stereochemistry (positive Cotton effect) (*90*)
conversion of bullatalicin to bullatalicinone (KOH, BuOH) (*90*)
stereochemistry revised to 23,24 *erythro* and 15,16 *threo* (*11*)
16,19-formaldehyde acetal derivative: HRFABms, TMSi derivative (EIms), ^{1}H nmr (500 MHz), ^{13}C nmr (125 MHz) → 16,19 *cis* (*18*)
(*R*)- and (*S*)-MTPA esters of 16,19-formaldehyde acetal derivative: ^{1}H (500 MHz) → 24*S* (*18*)
absolute stereochemistry: (*2,4-cis*-bullatalicinone) 2*R*, 4*R*, 12*R*, 15*S*, 16*S*, 19*R*, 20*R*, 23*R*, 24*S*; (2,4-*trans*-bullatalicinone) 2*S*, 4*R*, 12*R*, 15*S*, 16*S*, 19*R*, 20*R*, 23*R*, 24*S*

Bullatanocin or **Cherimolin-2** (*12*) = Squamostatin-C (*20*), = Crassiflorin (*91*), = Annonin IV (wrong str.) (*10*), (revised struct.) (*11*), = Purpureacin-1* (*85*)

Annona bullata (bark) (*89*); *Annona cherimolia* (seeds) (*16*); *Annona crassiflora* (seeds) (*91*); *Annona glabra* (seeds) (*92*); *Annona purpurea* (leaves) (*85*); *Annona squamosa* (seeds) (*10, 20*)

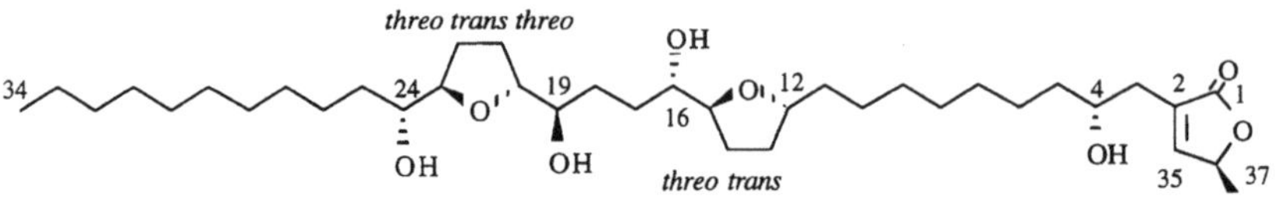

$C_{37}H_{66}O_8$; M = 638
mp = 95–97 °C (AcOEt) (*20*), mp = 107–109 °C (*10*), mp = 107.3–108.8 °C (*91*)
$[\alpha]_D = +14.4°$ (c = 0.55, $CHCl_3$) (*89*), $[\alpha]_D = +12°$ (c = 0.20, MeOH) (*20*), $[\alpha]_D = -3.3°$ (c = 0.12, MeOH) (*85*)*, $[\alpha]_D = +6°$ (c = 0.31, MeOH) (*16*)
CD (*20*): → absolute stereochemistry 36*S*
IR (*16, 20, 89, 91*); UV (*20, 85, 89*)
^{1}H nmr (500 MHz, C_6D_6 and $CDCl_3$) (*20, 89*), (500 MHz) (*91*), (200 MHz), (*10, 85*), (400 MHz) (*16*); ^{13}C nmr, (125 MHz) (*20, 88, 89, 91*), (50 MHz) (*10, 85*), (100 MHz) (*16*)
EIms (*10, 11, 16, 20, 85*), HRFABms (*20*), EIms, CIms (*10, 16, 85, 89*), HRCIms (*89*)
tetraacetate: ^{1}H nmr (500 MHz) (*20, 89*)
tetra-TMSi derivative: EIms (*11, 85, 89*); ^{13}C nmr (50 MHz) (*85*); tetra-$TMSI_{d9}$: EIms (*89*)
16,19-formaldehyde acetal derivative: HRFABms, TMSi derivative (EIms), ^{1}H nmr (500 MHz), ^{13}C nmr (125 MHz) → 16,19 *cis* (*18*)
relative stereochemistry 12,16-*trans* (*20*)
stereochemistry 4*R* deduced from nmr data by comparizing with literature data (*46, 64*)
tetra-(*R*)-MPTA ester: ^{1}H nmr (500 MHz) → 4*R*, 36*S* (*20*)
(*R*)- and (*S*)-MTPA esters of 16,19-formaldehyde acetal derivative: ^{1}H (500 MHz) → 24*R* (*18*), 36*S* is based on the usual ubiquitous 4*R*, 36*S* relationship (*18*)
absolute stereochemistry: 4*R*, 12*R*, 15*S*, 16*S*, 19*R*, 20*R*, 23*R*, 24*R*, 36*S*

---

* We have attributed to purpureacin-1 the same structure as bullatanocin after examination of the nmr (*85*); however the value of $[\alpha]_D = -3.3°$ is somewhat disquieting because a negative $[\alpha]_D$ is exceptional

12,15-*cis*-Bullatanocin

Annona bullata (bark) (*88*)

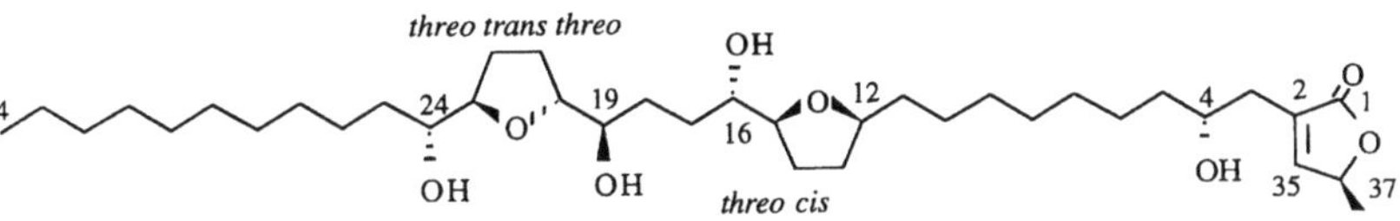

$C_{37}H_{66}O_8$; M = 638
IR (*88*)
^{1}H nmr (500 MHz) (*88*); ^{13}C nmr (125 MHz) (*88*)
HRFABms (*88*)
TMSi derivative: EIms (*88*)
16,19-formaldehyde acetal derivative (*88*)
(*R*)- and (*S*)-MTPA esters of 16,19-formaldehyde acetal derivative: ^{1}H nmr (500 MHz) → 4*R*, 24*S* (*88*)
absolute stereochemistry: 4*R*, 12*S*, 15*S*, 16*S*, 19*R*, 20*R*, 23*R*, 24*R*, 36*S*

Bullatanocinone

Annona bullata (bark) (*88*, *89*)

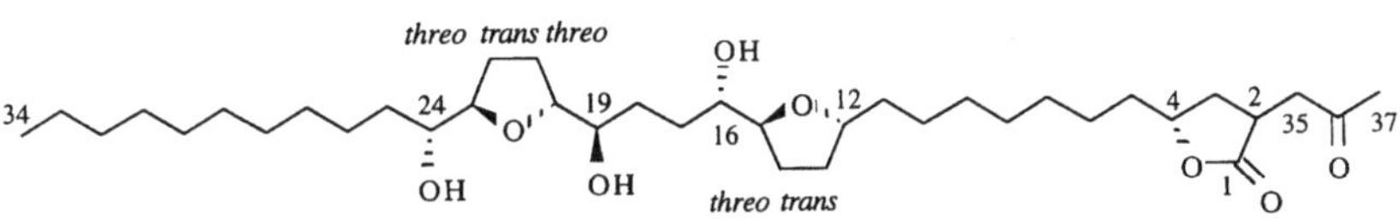

mixture of 2,4-*cis*- and 2,4-*trans*-isomers

$C_{37}H_{66}O_8$; M = 638
$[\alpha]_D = +21.5°$ (c = 0.20, $CHCl_3$) (*89*)
IR (*89*); UV (*89*)
^{1}H nmr (500 MHz, C_6D_6 and $CDCl_3$) (*89*); ^{13}C nmr, (125 MHz) (*89*)
CIms (*89*), HRCIms (*89*)
triacetate: ^{1}H nmr (500 MHz) (*89*), CIms (*89*)
tetra-TMSi: EIms (*89*, *11*); tetra-TMSi$_{d9}$: EIMS (*89*)
16,19-formaldehyde acetal derivative: HRFABms, TMSi derivative (EIms), ^{1}H nmr (500 MHz), ^{13}C nmr (125 MHz) → 16,19 *cis* (*18*)
(*R*)- and (*S*)-MTPA esters of 16,19-formaldehyde acetal derivative: ^{1}H (500 MHz) → 24*R* (*18*)

cis-Bullatanocinone and *trans*-bullatanocinone have been separated by HPLC on silica gel, eluent $CHCl_3$-MeOH (99:1) (*89*)

2,4-*cis*-Bullatanocinone

$[\alpha]_D = +30.1°$ (c = 0.20, $CHCl_3$) (*89*)
^1H nmr (500 MHz, C_6D_6 and $CDCl_3$) (*89*); ^{13}C nmr (125 MHz) (*89*)
16,19-formaldehyde acetal derivative → 24*R* (*18*)
absolute stereochemistry: 2*R*, 4*R*, 12*R*, 15*S*, 16*S*, 19*R*, 20*R*, 23*R*

2,4-*trans*-Bullatanocinone

$[\alpha]_D = +14.4°$ (c = 0.20, $CHCl_3$) (*89*)
^1H nmr (500 MHz, C_6D_6 and $CDCl_3$) (*89*); ^{13}C nmr (125 MHz) (*89*)
16,19-formaldehyde acetal derivative → 24*R* (*18*)
absolute stereochemistry: 2*S*, 4*R*, 12*R*, 15*S*, 16*S*, 19*R*, 20*R*, 23*R*

12,15-*cis*-Bullatanocinone

Annona bullata (bark) (*88*)

mixture of 2,4-*cis*- and 2,4-*trans*-isomers. The presence of 12,15-*cis*-bullatanocinone has been deduced by analysis of the ^1H nmr spectrum of a mixture of 8 products ! (*88*)

Bullatencin

Annona bullata (bark) (*93*)

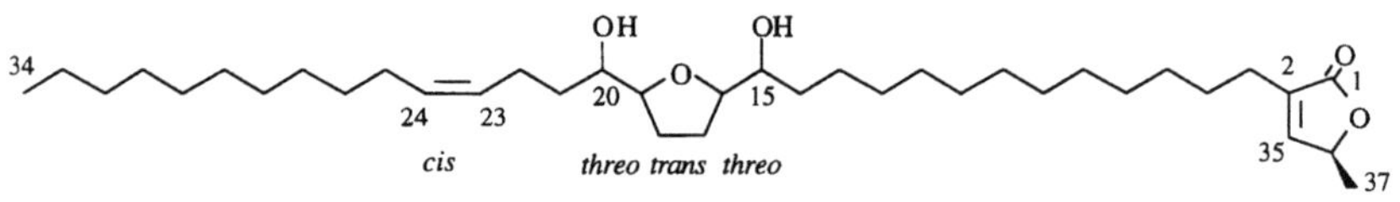

$C_{37}H_{66}O_5$; M = 590
$[\alpha]_D = +12.75°$ (c = 0.4, $CHCl_3$) (*93*)
^1H nmr (500 MHz) (*93*); ^{13}C nmr (125 MHz) (*93*)
CIms (*93*), HRCIms (*93*)
diacetate (*93*): ^1H nmr (500 MHz) (*93*)
di-TMSi, di-$TMSi_{d9}$ derivatives: EIms (*93*)

Bullatin

Asimina triloba (stem bark) (*82*)

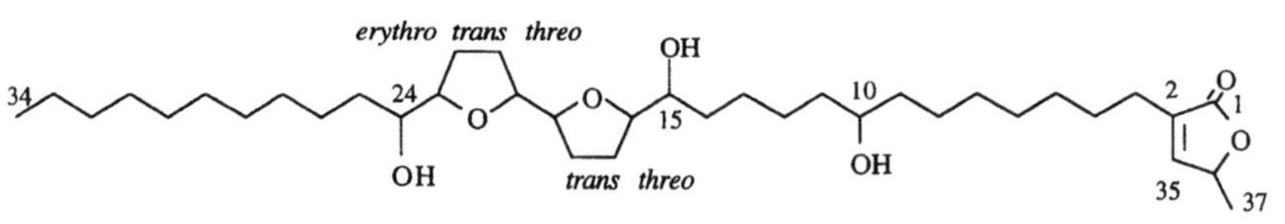

$C_{37}H_{66}O_7$; M = 622
$[\alpha]_D = +7.5°$ (c = 0.04, EtOH) (*82*)
IR (*82*); UV (*82*)
^1H nmr (500 MHz) (*82*); ^{13}C nmr (125 MHz) (*82*)
HRFABms (*82*), EIms (*82*)
triacetate: ^1H nmr (500 MHz) (*82*)
tri-TMSi, tri-$TMSi_{d9}$ derivatives: EIms (*82*)

The absolute stereochemistry was not determined; however the authors stated that the near identities of the pertinent nmr signals and biogenetic considerations suggest an absolute stereochemistry identical with that of bullatacin or rolliniastatin-2: 15*R*, 16*R*, 19*R*, 20*R*, 23*R*, 24*S*, 36*S* (*82*)

Cherimolin-1 or **Bullatalicin** = Annonin-VIII, = Squamostatin-B

Cherimolin-2 or **Bullatanocin** = Annonin-IV = Squamostatin-C = Crassiflorin = Purpureacin-1

Corepoxylone

Annona muricata (seeds) (*94*)

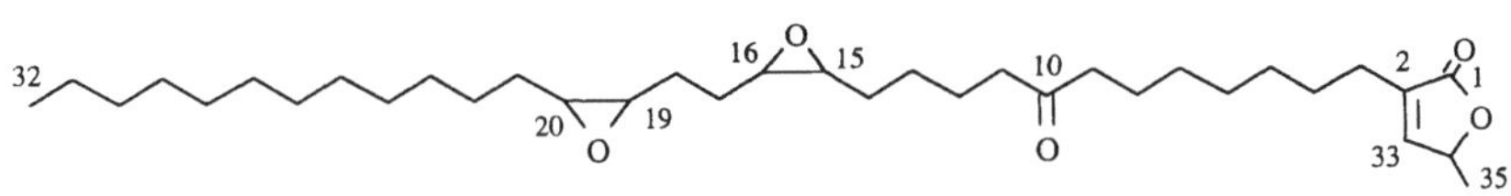

$C_{35}H_{60}O_5$, M = 560
$[\alpha]_D = +36.8°$ (c = 0.08, $CHCl_3$) (*94*)
IR (*94*); UV (*94*)
^1H nmr (200 MHz) (*94*); ^{13}C nmr (50 MHz) (*94*),
EIms (*94*), FAB-Li ms (*94*), L-SIms (*94*)
hemisynthesis of corossolone and its tetra-epimer from corepoxylone (m-CPBA; $HClO_4$) (*94*)

Coriacin

Annona coriacea (roots) (*95*)

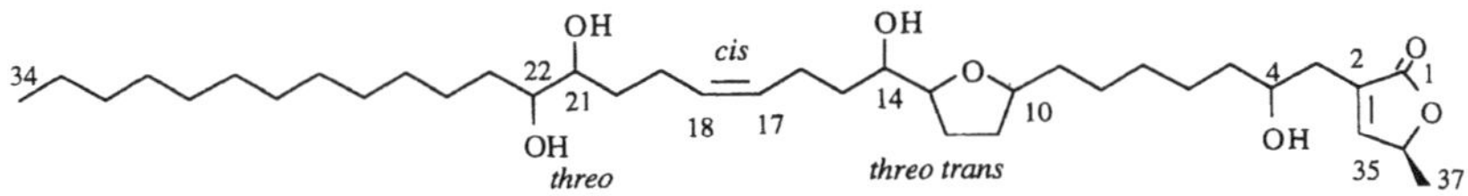

$C_{37}H_{66}O_7$; M = 622
mp = 49–50 °C (*95*)

$[\alpha]_D = +14.0°$ (c = 1.00, EtOH) (*95*)
IR (*95*); UV (*95*)
^{1}H nmr (400 MHz) (*95*); ^{13}C nmr (50 MHz) (*95*)
FAB-Li ms (*95*), EIms (*95*), CID B/E ms (*95*)
acetonide: CIms (*95*), EIms (*95*); ^{1}H nmr (200 MHz) (*95*); ^{13}C nmr (50 MHz) (*95*)
tetraacetate: CIms (*95*), EIms (*95*); ^{1}H nmr (200 MHz) (*95*); ^{13}C nmr (50 MHz) (*95*)
tetrahydrocoriacin: CIms (*95*), EIms (*95*), ^{1}H nmr (200 MHz) (*95*); ^{13}C nmr (50 MHz) (*95*)
conversion into gigantecin (m-CPBA): mp = 102–104 °C (*95*); $[\alpha]_D = +5°$ (c = 0.3, EtOH) (*95*) EIms, FAB-Li ms (*95*), ^{1}H nmr (200 MHz) (*95*), ^{13}C nmr (50 MHz) (*95*)

Coriadienin

Annona coriacea (roots) (*96*)

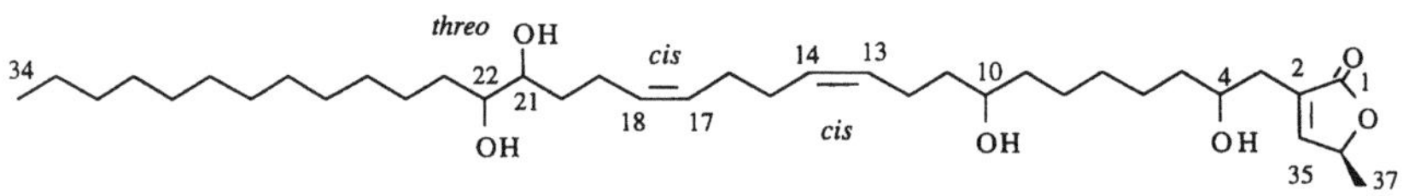

$C_{37}H_{66}O_6$; M = 606
mp = 58–60 °C (*96*)
$[\alpha]_D = +6.2°$ (c = 0.3, EtOH) (*96*)
IR (*96*); UV (*96*)
^{1}H nmr (400 MHz) (*96*); ^{13}C nmr (50 MHz) (*96*)
CIms (*96*), EIms (*96*), HRFAB ms (*96*), FAB-Li ms (*96*)
acetonide: CIms (*96*), ^{1}H nmr (400 MHz) (*96*)
diepoxide (*96*): CIms (*96*), EIms (*96*)

Corossolin = Howiicin-C

Annona muricata (seeds) (*38*); *Goniothalamus howii* (seeds) (*34*); synthesis (*97*)

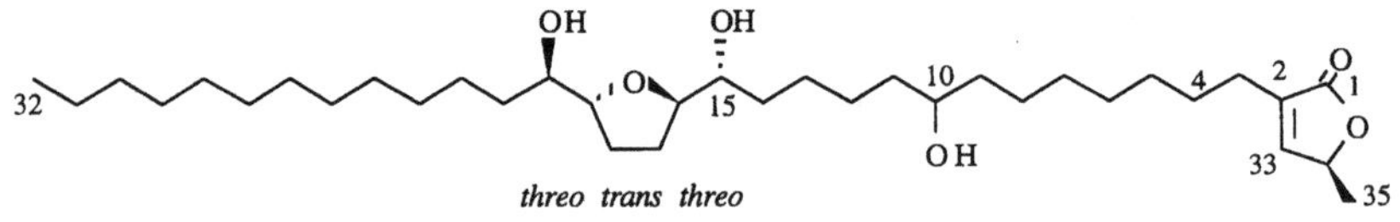

$C_{35}H_{64}O_6$; M = 580
mp = 45–50 °C; (*38*), mp (10*RS*-corossolin) = 66.5–67.5 °C (*97*), mp = 83–85 °C (*34*)

$[\alpha]_D = +19°$ (c = 0.2, MeOH) (*38*), $[\alpha]_D$ (10*RS*-corossolin) = +22.1° (c = 0.1, MeOH) (*97*), $[\alpha]_D = +24.02$ (c = 3.41, $CHCl_3$) (*34*)
IR (*34, 38, 97*); UV (*38, 97*)
^1H nmr (200 MHz) (*38*), (300 MHz) (*97*), (400 MHz) (*34*); ^{13}C nmr (50 MHz) (*38*); (125 MHz) (*97*), (100 MHz) (*34*)
CIms (*38*), FABms (*97*), EIms (*34*)
triacetate: ^1H nmr (200 MHz) (*34, 38*); ^{13}C nmr (50 MHz) (*38*); CIms (*38*), EIms (*34*)
synthesis (10*RS*) (*97*)
absolute stereochemistry: 15*R*, 16*R*, 19*R*, 20*R*, 34*S*

Synthesis of 16,19,20,34-epi-corossolin: (10ξ, 15*R*, 16*S*, 19*S*, 20*S*, 34*R*)-corossolin (*98*)

mp = 47–50 °C (*98*)
$[\alpha]_D = -23.7°$ (c = 0.057, MeOH) (*98*)
IR (*98*); UV (*98*)
^1H nmr (600 MHz) (*98*)
FABms (*98*)

Corossolone

Annona glabra (bark) (*36*); *Annona muricata* (seeds) (*38*); synthesis (*97*)

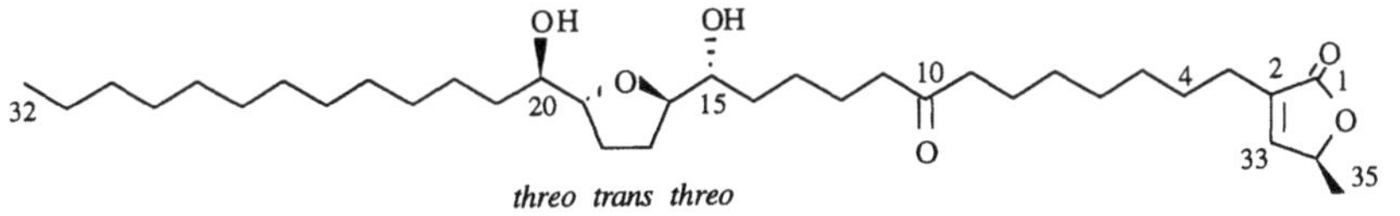

$C_{35}H_{62}O_6$; M = 578
mp = 55–57 °C (EtOAc) (*38*), mp = 79.5–80.5 °C (*97*), mp = 53–55 °C (*36*)
$[\alpha]_D = +15°$ (c = 0.13, MeOH) (*38*), $[\alpha]_D = +20.7°$ (c = 0.2 MeOH) (*97*), $[\alpha]_D = +10°$ (c = 0.18, MeOH) (*36*)
IR (*38, 36, 97*); UV (*36, 38, 97*)
^1H nmr (200 MHz) (*38*), (300 MHz) (*97*), (600 MHz) (*36*); ^{13}C nmr (50 MHz) (*38*); (125 MHz) (*36, 97*)
CIms (*38*), EIms (*36*), FABms (*36, 97*)
diacetyl-corossolone: ^1H nmr (200 MHz) (*38*); HRCIms (*38*)
dihydrocorossolone = corossolin (*38*)
synthesis (*97*)
absolute stereochemistry: 15*R*, 16*R*, 19*R*, 20*R*, 34*S*

Cyclogoniodenin-C

hemisynthesis (*75*)

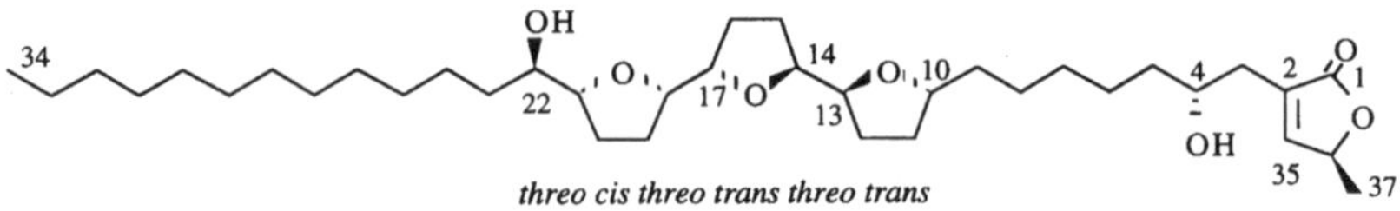

obtained by hemisynthesis from goniodenin (m-CPBA; toluenesulfonic acid) (*87*)

$C_{37}H_{64}O_7$; M = 620
$[\alpha]_D = +0.6°$ (c = 1.00, CH_2Cl_2) (*75*)
IR (*75*); UV (*75*)
^1H nmr (500 MHz) (*75*); ^{13}C nmr (125 MHz) (*75*)
HRCIms (*75*), HRFABms (*75*)
diacetate: ^1H nmr (500 MHz) (*75*)
di-TMSi derivative: EIms (*75*)
(*S*)- and (*R*)-MPTA esters: ^1H nmr (500 MHz) → 4*R*, 22*R*, 36*S* (*75*),
absolute stereochemistry: 4*R*, 10*R*, 13*R*, 14*R*, 17*R*, 18*S*, 21*R*, 22*R*, 36*S*.
This absolute configuration is, as regard the THF pattern (*75*), the mirror of that of goniocin (*99*)

Cyclogoniodenin-T

hemisynthesis (*75*)

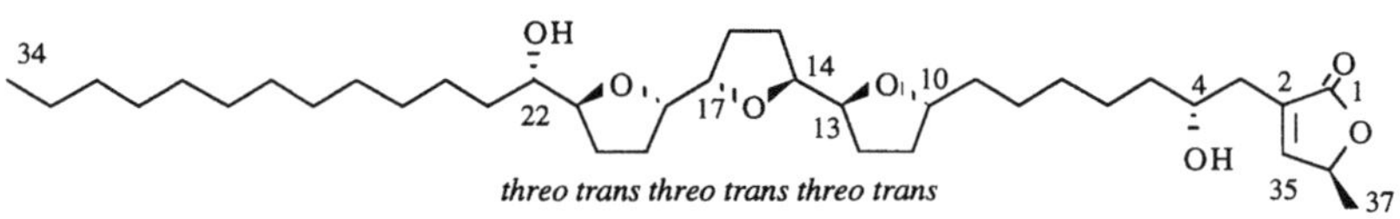

obtained by hemisynthesis from goniodenin (m-CPBA; toluenesulfonic acid) (*87*)

$C_{37}H_{64}O_7$; M = 620
$[\alpha]_D = +1.5°$ (c = 1.20, CH_2Cl_2) (*75*)
IR (*75*); UV (*75*)
^1H nmr (500 MHz) (*75*); ^{13}C nmr (125 MHz) (*75*)
HRCIms (*75*), HRFABms (*75*)
diacetate: ^1H nmr (500 MHz) (*75*)
di-TMSi derivative: EIms (*75*)
(*S*)- and (*R*)-MPTA esters: ^1H nmr (500 MHz) → 4*R*, 22*S*, 36*S* (*75*)
absolute stereochemistry: 4*R*, 10*R*, 13*S*, 14*S*, 17*S*, 18*S*, 21*S*, 22*S*, 36*S*

Cyclogonionenin-C

hemisynthesis (*100*)

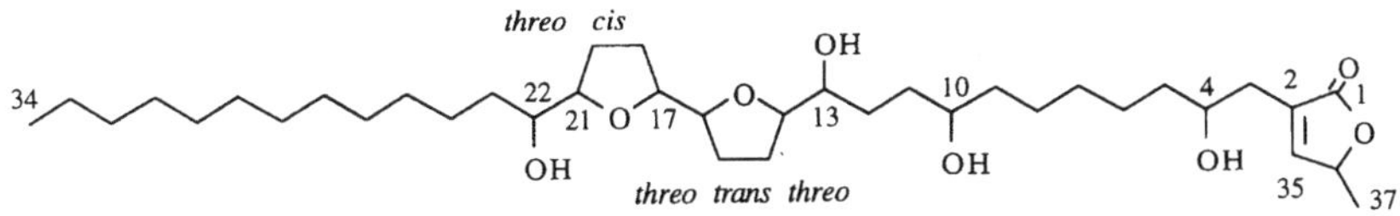

obtained by hemisynthesis from gonionenin (m-CPBA; $HClO_4$) (*100*)

$C_{37}H_{66}O_8$; M = 638
mp = 69–70 °C (*100*)
$[\alpha]_D = +3°$ (c = 0.33, MeOH) (*100*)
IR (*100*); UV (*100*)
^1H nmr (500 MHz) (*100*); ^{13}C nmr (125 MHz) (*100*)
HRFABms (glycerol) (*100*), CIms (*100*)
tetraacetate: ^1H nmr (500 MHz) (*100*)
tetra-TMSi derivative: EIms (*100*)

Cyclogonionenin-T

hemisynthesis (*100*)

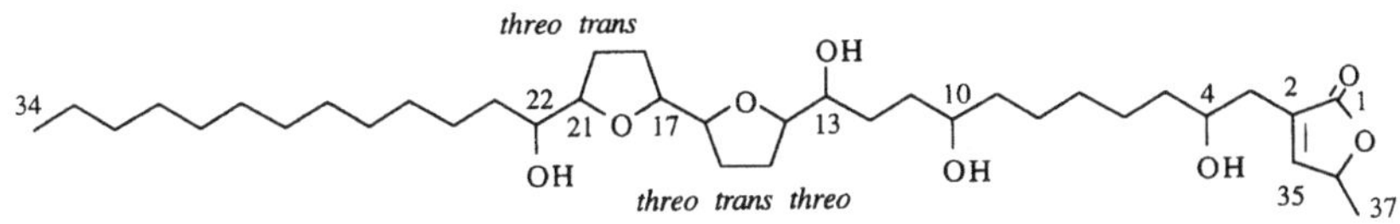

obtained by hemisynthesis from gonionenin (m-CPBA; $HClO_4$) (*100*)

$C_{37}H_{66}O_8$; M = 638
mp = 61–62 °C (*100*)
$[\alpha]_D = +18.3°$ (c = 0.46, MeOH) (*100*)
^1H nmr (500 MHz) (*100*); ^{13}C nmr (125 MHz) (*100*)
HRFABms (glycerol) (*100*), CIms (*100*)
tetraacetate: ^1H nmr (500 MHz) (*100*)
tetra-TMSi derivative: EIms (*100*)

Densicomacin-1 = 13,14-*erythro*-Densicomacin

Annona densicoma (stem bark) (*101*)

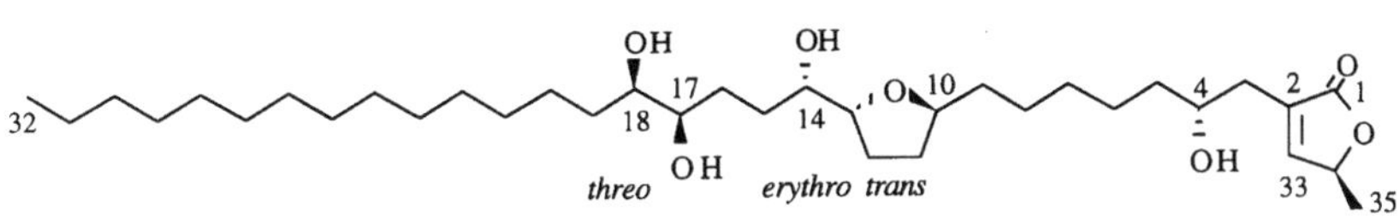

mixture of two stereoisomers 13,14-*erythro*-densicomacin and 13,14-*threo*-densicomacin (*101*)

$C_{35}H_{64}O_7$; M = 596
mp = 83–84 °C (*101*)
$[\alpha]_D^{25} = +26°$ (c = 0.05, MeOH) (*101*)
IR (*101*); UV (*101*)
^{1}H nmr (500 MHz) (*101*); ^{13}C nmr (125 MHz) (*101*)
HRCIms (*101*), EIms (*101*)
tetraacetate: ^{1}H nmr (500 MHz) (*101*)
TMSi derivative: CIms (*101*)
acetonide: ^{1}H nmr (500 MHz) (*101*), acetonide diacetate: ^{1}H nmr (500 MHz) (*101*)
sodium periodate → pentadecanoic acid (*101*)
isodensicomacin-1 et isodensicomacin-2 (mixture of 2,4-*cis*- and 2,4-*trans*-isomers) obtained from the mixture of densicomacin-1 and densicomacin-2: (KOH/*t*-BuOH): ^{1}H nmr (500 MHz) (*101*), CIms (*101*), EIms (*101*)
tetramesitoates: stereoisomers were resolvable by preparative TLC: ^{1}H nmr (C_6D_6 and $CDCl_3$, 500 MHz), study by comparison with THF pattern synthetic model (*101*)

Except for the ^{1}H nmr and ^{13}C nmr spectra of the mesitoates, all data were obtained from a mixture of 13,14-*erythro*- and 13,14-*threo*-densicomacin

Densicomacin-2 = 13,14-*threo*-Densicomacin, = Gigantetrocin-A (or B) = Howiicin-F

Annona densicoma (stem bark) (*101*); *Annona muricata* (seeds) (*174*)

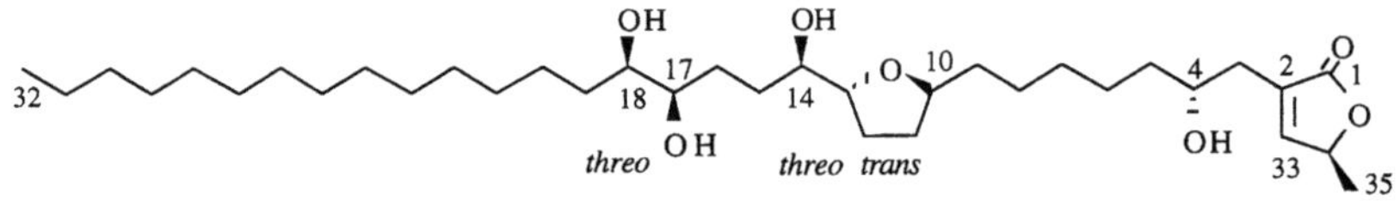

mixture of two stereoisomers 13,14-*erythro*-densicomacin and 13,14-*threo*-densicomacin (*101*)

$C_{35}H_{64}O_7$; M = 596
mp = 83–84 °C (*101*)
$[\alpha]_D{}^{25} = +26°$ (c = 0.05, MeOH) (*101*)
IR (*101*); UV (*101*)
^{1}H nmr (500 MHz) (*101*), ^{13}C nmr (125 MHz) (*101*)
HRCIms (*101*), EIms (*101*)
tetraacetate: ^{1}H nmr (500 MHz) (*101*)
TMSi derivative: CIms (*101*)

acetonide: ^1H nmr (500 MHz) (*101*), acetonide diacetate: ^1H nmr (500 MHz) (*101*)
sodium periodate → pentadecanoic acid (*101*)
isodensicomacin-1 et isodensicomacin-2 (mixture of 2,4-*cis*- and 2,4-*trans*-isomers) obtained from the mixture of densicomacin-1 and densicomacin-2: (KOH/*t*-BuOH): ^1H nmr (500 MHz) (*101*), CIms (*101*), EIms (*101*)
tetramesitoates: stereoisomers were resolvable by preparative TLC: ^1H nmr (C_6D_6 and $CDCl_3$, 500 MHz), study by comparison with THF pattern synthetic model (*101*)
absolute stereochemistry of 13,14-*threo*-densicomacin deduced from analysis of ^1H nmr of Mosher esters: 4*R*, 10*S*, 13*R*, 14*R*, 17*R*, 18*R*, 34*S* (*102*)

Except for the ^1H nmr and ^{13}C nmr spectrum of the mesitoates, all data were obtained from a mixture of densicomacin-1 (13,14-*erythro*) and densicomacin-2 (13,14-*threo*-densicomacin). The spectral data of densicomacin-2 are very close to those of gigantetrocin-A. Densicomacin-2 (13,14-*threo*-densicomacin) appears to differ from gigantetrocin-A only in the absolute configuration at C-10, C-13 and C-14 (10*R*, 13*S*, 14*S*)

4-Deoxycoriacin

Annona coriacea (roots) (*95*)

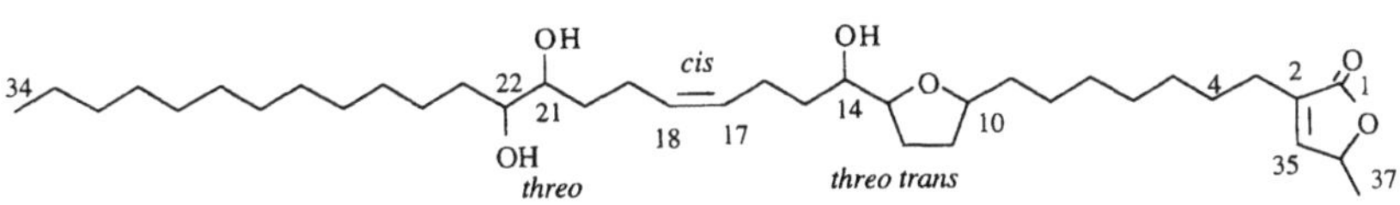

$C_{37}H_{66}O_6$; M = 606
$[\alpha]_D = +10°$ (c = 1.0, $CHCl_3$) (*95*)
IR (*95*); UV (*95*)
^1H nmr (400 MHz) (*95*); ^{13}C nmr (50 MHz) (*95*)
CIms (*95*), EIms (*95*)
acetonide: CIms (*95*), EIms (*95*); ^1H nmr (400 MHz) (*95*); ^{13}C nmr (50 MHz) (*95*) conversion into 4-deoxygigantecin (m-CPBA): EIms, ^1H nmr (200 MHz) (*95*)

4-Deoxygigantecin

Goniothalamus giganteus (bark) (*103*); hemisynthesis (*95*)

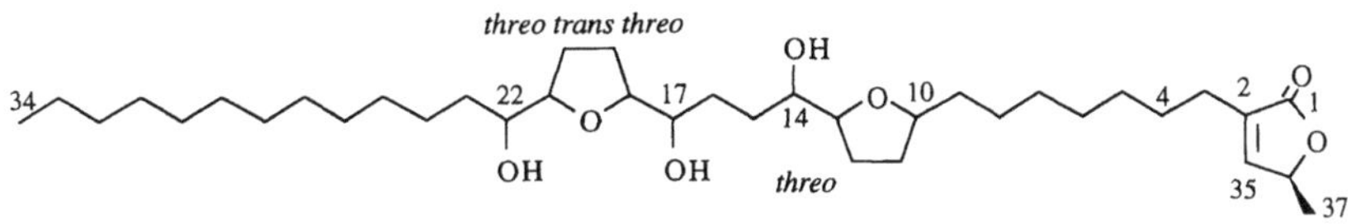

$C_{37}H_{66}O_7$; M = 622
mp = 97–99 °C (*103*)
$[\alpha]_D^{25}$ = + 15.5° (c = 0.2, MeOH) (*103*)
IR (*103*); UV (*103*)
^{1}H nmr (500 MHz) (*103*); ^{13}C nmr (125 MHz) (*103*)
CIms (*103*), FABms (*103*), HRFABms (*103*), EIms (*103*)
triacetate: ^{1}H nmr (500 MHz) (*103*), EIms (*103*)
tri-TMS and tri-TMSi$_{d9}$ derivatives: EIms (*103*)
hemisynthesis from 4-deoxycoriacin (m-CPBA) (*95*)

Desacetyluvaricin = Squamocin-L, = Neodesacetyluvaricin (*104*)

Annona atemoya (seeds) (*78*); *Annona bullata* (bark) (*89*); *Annona glabra* (seeds) (*67, 92*); *Annona muricata* (seeds) (*174*); *Annona squamosa* (seeds) (*14, 64, 104*); *Uvaria acuminata* (roots) (*105*)

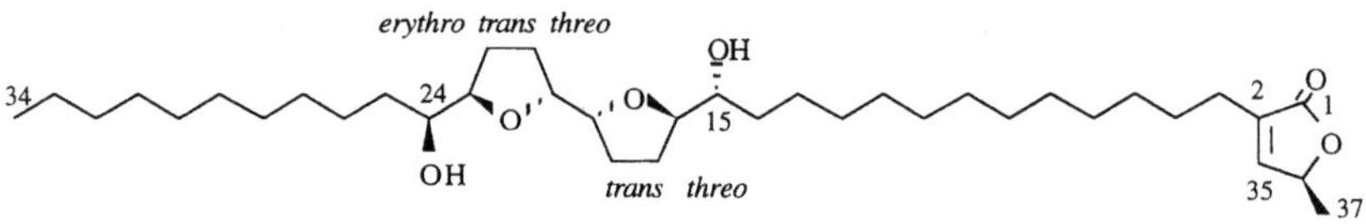

$C_{37}H_{66}O_6$; M = 606
mp = 63–65 °C (*105*), mp = 67.5–69 °C (*64*), mp = 65–67 °C (*14*), mp = 68.5–70.5 °C (*104*)
$[\alpha]_D$ = + 9.3° (c = 1.41, MeOH) (*105*), $[\alpha]_D$ = + 19.3 (c = 0.98, MeOH) (*64*), $[\alpha]_D$ = + 21.1 (c = 0.5, $CHCl_3$) (*89*), $[\alpha]_D$ = + 31.4 (c = 0.67, $CHCl_3$) (*104*)
IR (*64, 89, 105*), UV (*89*)
^{1}H-nmr (*105*), (500 MHz) (*14, 64, 89*); (300 MHz) (*104*); ^{13}C-nmr (*105*), (125 MHz) (*14, 64*), (75 MHz) (*104*)
EIms (*64, 104, 105*), CIms (*89*), FABms (*14, 64*)
diacetate: ^{1}H-nmr (500 MHz) (*14*), (300 MHz) (*104*), EIms (*104*)
di-(*R*)-MTPA ester: ^{1}H-nmr (500 MHz) → 15*R*, 24*R* (*64*),
(*S*)-36 stereochemistry by analogy with uvaricin (ozonolysis of uvaricin → (*S*)-lactic acid) (*105*)
absolute stereochemistry: 15*R*, 16*R*, 19*R*, 20*R*, 23*R*, 24*R*, 36*S*

Diepomuricanin-A = Epoxyrollin-B (*106*) = Diepomuricanin

Annona muricata (seeds) (*107, 108*), (stem bark) (*109*); *Annona reticulata* (seeds) (*110*); *Rollinia membranacea* (seeds) (*111*); *Rollinia ulei* (leaves) (*106*)

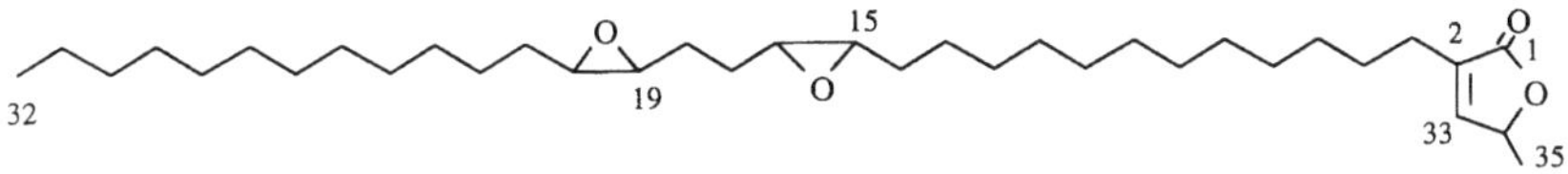

$C_{35}H_{62}O_4$; †M = 546
$[\alpha]_D = +13.5°$ (c = 0.2, MeOH) (*107*)
IR (*107*); UV (*107*)
^1H nmr (200 MHz) (*107*), (400 MHz) (*109*); ^{13}C nmr (50 MHz) (*107*, *109*), CIms (*110*), EIms (*109*, *110*), FABms (*109*), FAB-Li ms (*107*, *109*, *110*), HREIms (*108*)
transformation into solamin (*107*, *108*, *110*)
hemisynthesis from epomuricenin (m-CPBA) (*108*)
diepomuricanin (*106*, *107*, *109*, *110*), has to be named diepomuricanin-A

Diepomuricanin-B

Rollinia membranacea (seeds) (*111*)

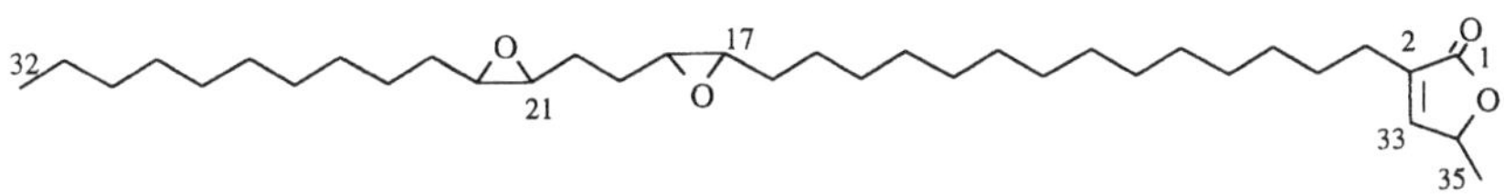

$C_{35}H_{62}O_4$; M = 546
$[\alpha]_D = +10°$ (c = 0.75, $CHCl_3$) (*111*)
IR (*111*); UV (*111*)
^1H nmr (200 MHz) (*111*); ^{13}C nmr (50 MHz) (*111*)
HRCIms (*111*), EIms (*111*), FAB-Li ms (*111*), CID-FAB-Li (*111*)

Dieporeticanin-1 = Epoxyrollin-A (*106*)

Annona reticulata (seeds) (*110*); *Rollinia membranacea* (seeds) (*111*); *Rollinia ulei* (leaves) (*106*)

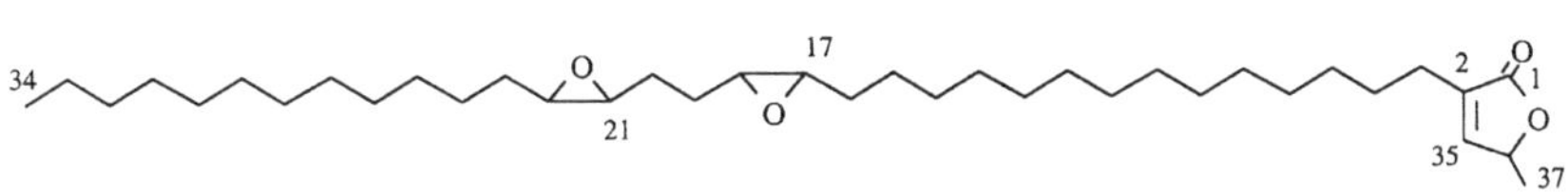

isolated as a mixture with dieporeticanin-2

$C_{37}H_{66}O_4$; M = 574
$[\alpha]_D$ (mixture) = + 12° (c = 1, $CHCl_3$) (*110*)
IR (*110*); UV (*110*)
^1H nmr (200 MHz); (*110*); ^{13}C nmr (50 MHz) (*110*)
FAB-Li ms (*110*), CIms (*110*)

mixture with dieporeticanin-2 + m-CPBA; $HClO_4$/acetone → hemisynthetic reticulatacin + a new mono-THF acetogenin (*110*)

Dieporeticanin-2

Annona reticulata (seeds) (*110*); *Rollinia membranacea* (seeds) (*111*)

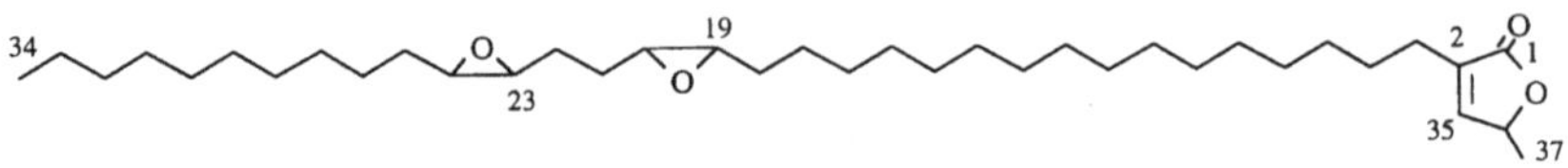

isolated as a mixture with dieporeticanin-1

$C_{37}H_{66}O_4$; M = 574
data and hemisynthesis: see dieporeticanin-1 (*110*)

Dieporeticenin

Annona reticulata (seeds) (*110*)

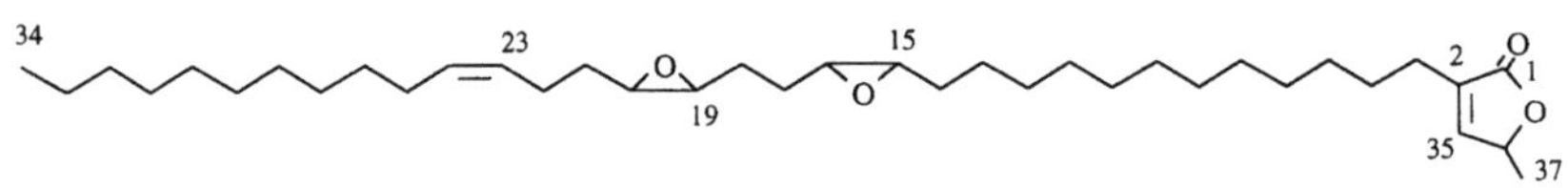

$C_{37}H_{64}O_4$; M = 572
$[\alpha]_D = +11°$ (c = 1, $CHCl_3$) (*110*)
IR (*110*); UV (*110*)
^1H nmr (200 MHz); (*110*); ^{13}C nmr (50 MHz) (*110*)
FAB-Li ms (*110*), CIms (*110*), EIms (*110*)
epoxidation → triepoxyrollin (= trieporeticanin) (*110, 165*)

Diepoxymontin*

Annona montana (seeds) (*112*)

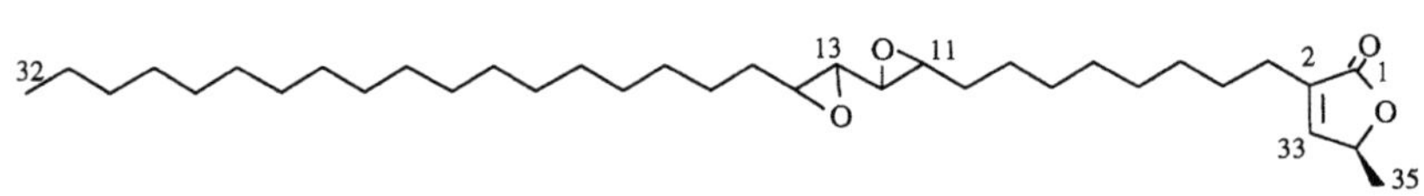

$C_{35}H_{62}O_4$; M = 546
IR (*112*)
^1H nmr (200 MHz) (*112*); ^{13}C nmr (50 MHz) (*112*)
HREIms (*112*), FABms tandem (*112*), CIms (*112*), EIms (*112*)

---

* Structure probably inexact as it is biogenetically improbable. It could be an analogue of diepomuricanin-A or diepomuricanin-B

Diepoxyrollin

Rollinia membranacea (seeds) (*111*)

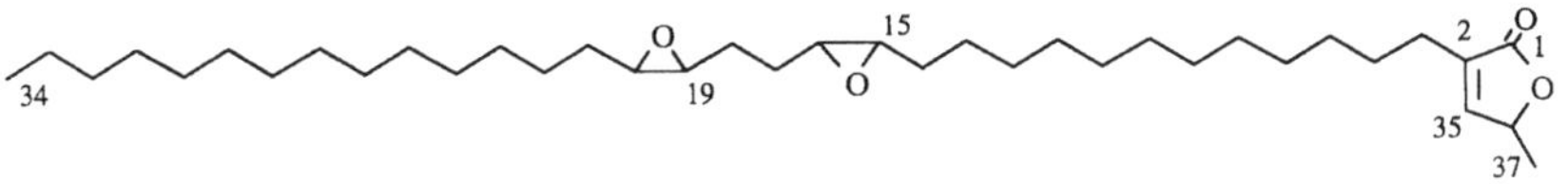

$C_{37}H_{66}O_4$; M = 574
$[\alpha]_D = +11°$ (c = 0.85, $CHCl_3$) (*111*)
IR (*111*); UV (*111*)
^1H nmr (200 MHz), (*111*); ^{13}C nmr (50 MHz) (*111*)
HRCIms (*111*), FAB-Li ms (*111*), CID-FAB-Li ms (*111*), EIms (*111*)

Epomuricenin-A = Epoxymurin-A

Annona muricata (seeds) (*108*), (stem bark) (*109*)

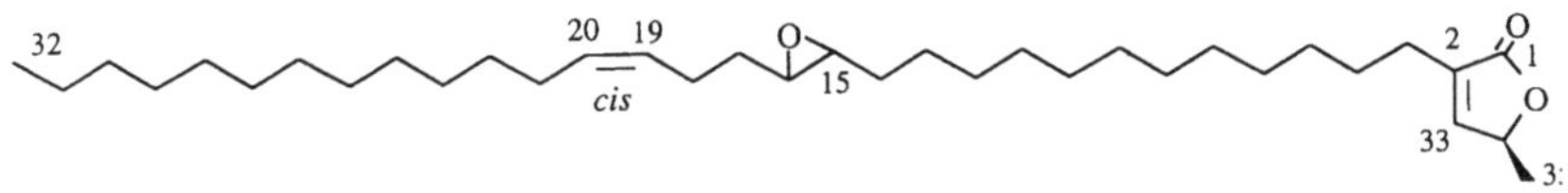

isolated as a mixture with epomuricenin-B (*108*)
isolated as a mixture with epoxymurin-B (*109*)

$C_{35}H_{62}O_3$; M = 530
$[\alpha]_D$ (mixture) = + 18° (c = 0.22, MeOH) (*108*)
IR (*108*); UV (*108*)
^1H nmr (200 MHz) (*108*), (400 MHz) (*109*); ^{13}C nmr (50 MHz) (*108*, *109*)
HREIms (*108*), EIms (*108*, *109*), FAB-Li ms (*108*, *109*), FAB-Li ms/ms (*109*)
epoxidation → diepomuricanin-A and a C-13,14-C-17,18-bis-epoxy isomer: FAB-Li ms (B/E linked scan spectrum) (*108*)
cyclization of mixture of diepomuricanin-A and its isomer → solamin + mono-THF isomer (*108*)

Epomuricenin-B

Annona muricata (seeds) (*108*)

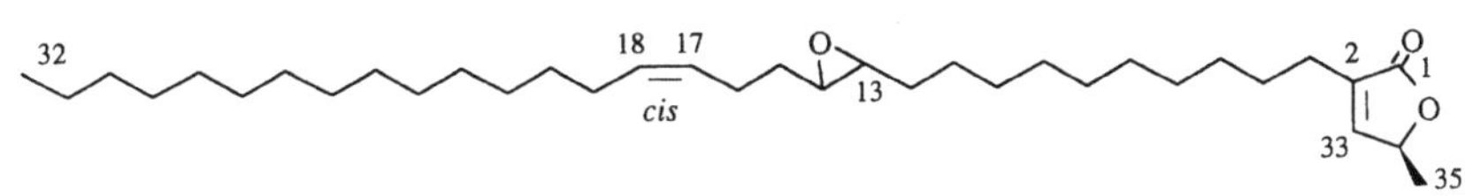

isolated as a mixture with epomuricenin-A (*108*)

$C_{35}H_{62}O_3$; M = 530
data: see epomuricenin-A (*108*)

Epoxymurin B

Annona muricata (stem bark) (*109*)

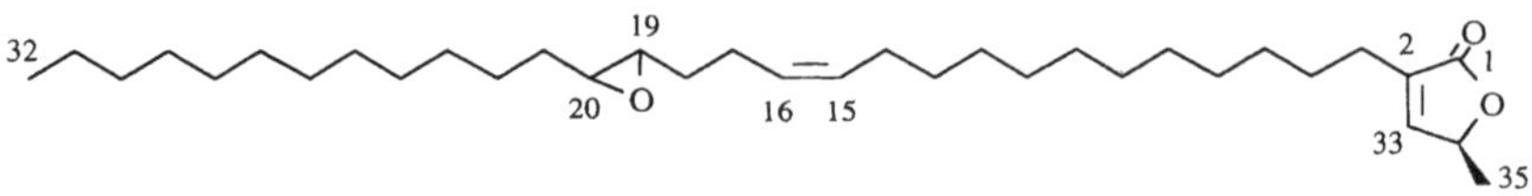

isolated as a mixture with epomuricenin-A (= epoxymurin-A) (*109*)

$C_{35}H_{62}O_3$; M = 530
^1H nmr (400 MHz) (*109*); ^{13}C nmr (200 MHz) (*109*)
EIms (*109*), FAB-Li ms (*109*), FAB-Li ms/ms (*109*)

Epoxyrollin-A = Dieporeticanin-1

Rollinia ulei (leaves) (*106*)

Has been assigned the formula $C_{38}H_{70}O_3$; M = 574 (*106*). Such a formula is improbable for an acetogenin since all of them possess 35 or 37 carbons; the correct formula is probably $C_{37}H_{66}O_4$; M = 574, and corresponds to dieporeticanin-1 (*110, 111*). Spectral data are in accord with this hypothesis.

Epoxyrollin-B = Diepomuricanin-A

Rollinia ulei (leaves) (*106*)

Has been assigned the formula $C_{36}H_{66}O_3$; M = 546 (*106*). Such a formula is improbable for an acetogenin since all of them possess 35 or 37 carbons; the correct formula is probably $C_{35}H_{62}O_4$; M = 546 and corresponds to diepomuricanin-A (*107, 108, 109, 110, 111*). Spectral data are in accord with this hypothesis.

Giganenin*

Goniothalamus giganteus (bark) (*103*)

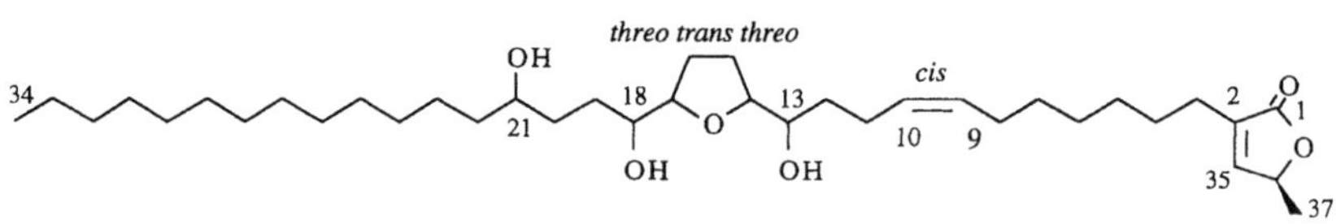

* Note added in proof: the revised structure places the double bond between C-21, 22 and hydroxyl at C-10 (*245*)

$C_{37}H_{66}O_6$; M = 606
F: 60–62 °C (*103*)
$[\alpha]_D^{25} = +21.4°$ (c = 0.23, MeOH) (*103*)
IR (*103*); UV (*103*)
^1H nmr (500 MHz) (*103*); ^{13}C nmr (125 MHz) (*103*)
CIms (*103*), FABms (*103*), HRFABms (*103*), EIms (*103*)
triacetate: ^1H nmr (500 MHz) (*103*), EIms (*103*)
tri-TMSi and tri-$TMSi_{d9}$ derivatives: EIms (*103*)

Giganin

Goniothalamus giganteus (bark) (*29, 113*)

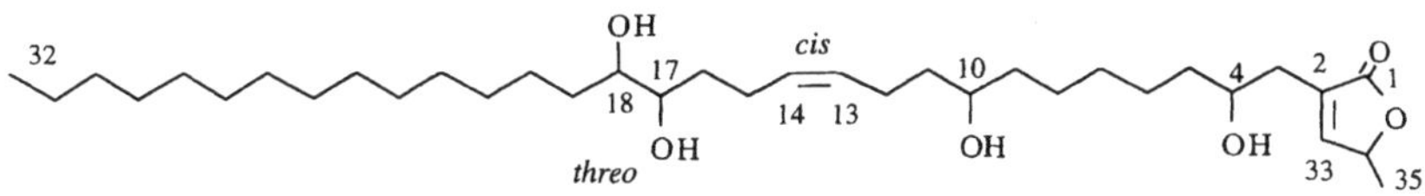

$C_{35}H_{64}O_6$; M = 580
$[\alpha]_D^{25} = +22.8°$ (c = 0.35, $CHCl_3$) (*113*)
IR (*113*); UV (*113*)
^1H nmr (500 MHz) (*113*); ^{13}C nmr (125 MHz) (*113*)
FABms (*113*), HRFABms (*113*), CIms (*113*)
tetra-TMSi derivative: EIms (*113*)
acetonide: ^1H nmr (500 MHz) (*113*)

Gigantecin

Annona coriacea (seeds) (*114*); *Goniothalamus giganteus* (stem bark) (*115*); hemisynthesis (*95, 100*)

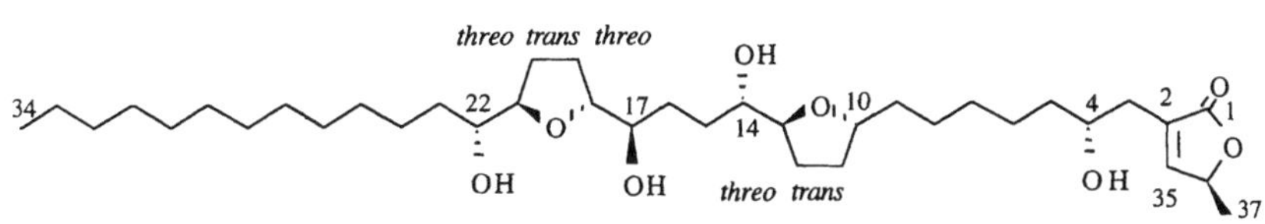

$C_{37}H_{66}O_8$; M = 638
mp = 96–98 °C (*115*), mp = 108–109 °C (*114*), mp = 109–110 °C (*100*), mp = 102–104 °C (*95*)
$[\alpha]_D = +3.15°$ (c = 0.025, MeOH) (*115*), $[\alpha]_D = +15.5°$ (c = 0.22, $CHCl_3$) (*114*), $[\alpha]_D = +5.3°$ (c = 1.8, MeOH) (*100*), $[\alpha]_D = +5°$ (c = 0.3, EtOH) (*95*)
IR (*100, 114, 115*); UV (*100, 114, 115*)

^{1}H nmr (*115*), (500 MHz) (*100*, *114*), (200 MHz) (*95*); ^{13}C nmr (*115*), (125 MHz) (*100*, 114), (50 MHz) (*95*)
CIms (*115*), EIms (*95*), HREIms (fragments) (*114*), FABms (*114*); HRFABms (glycerol) (*100*, *114*), FAB-Li ms (*95*)
tetraacetate: ^{1}H nmr (*114*, *115*)
TMSi, $TMSi_{d9}$ derivatives: EIms (*115*); TMSi derivative: Elms (*100*)
2,35-dihydro-TMSi derivative: EIms (*115*)
hemisynthesis from gigantetronenin (*100*)
hemisynthesis from coriacin (m-CPBA) (*95*)
RX → relative stereochemistry (*114*)
(*R*)- and (*S*)-MTPA esters: ^{1}H nmr (500 MHz) (*114*), ^{19}F nmr → 4*R* (*114*)
absolute stereochemistry: 4*R*, 10*R*, 13*S*, 14*S*, 17*R*, 18*R*, 21*R*, 36*S*

Gigantetrocin-A = gigantetrocin (*116*), = densicomacin-2 (*101*), = howiicin-F (*117*)

Annona cherimolia (seeds) (*22*); *Annona densicoma* (stem bark) (*101*); *Annona muricata* (seeds) (*40*, *118*), (leaves) (*31*); *Annona senegalensis* (seeds) (*22*); *Asimina longifolia* (leaves and twigs) (*42*); *Goniothalamus giganteus* (stem bark) (*116*); *Goniothalamus howii* (seeds) (*117*)*; *Xylopia aromatica* (bark) (*167*)

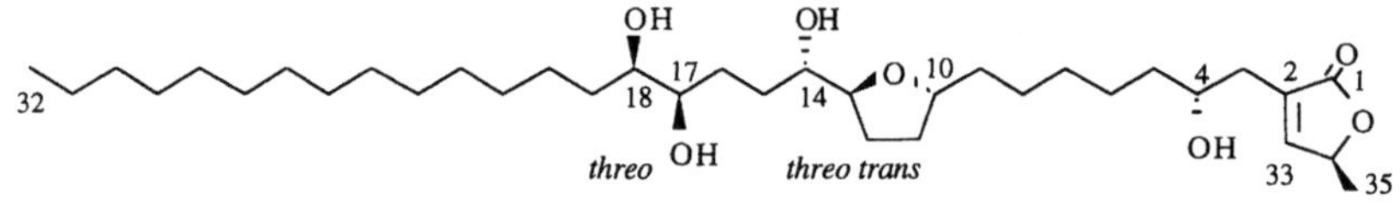

$C_{35}H_{64}O_7$; M = 596
mp = 80–81 °C (*116*), mp = 93–94 °C (*119*), mp = 85–87 °C (*117*)
$[\alpha]_D = +10.3°$ (*116*), $[\alpha]_D = +14.3°$ (c = 0.45, $CHCl_3$) (*119*), $[\alpha]_D = +8.82$ (c = 8.84, $CHCl_3$) (*117*)
IR (*22*, *116*, *117*, *119*); UV (*116*)
^{1}H nmr (500 MHz) (*116*, *119*), (400 MHz) (*118*), (600 MHz) (*117*); ^{13}C nmr (100 MHz) (*117*, *118*), (125 MHz) (*116*)
HRCIms (*116*), FABms (*119*), FAB-LI ms (*22*), EIms (*116*, *117*, *118*)
tetraacetate: ^{1}H nmr (500 MHz) (*116*), (200 MHz) (*117*); EIms (*116*, *117*)
tetra-TMSi derivative: CIms (*116*), EIms (*116*)
acetonide: ^{1}H nmr (500 MHz) (*116*, *119*); CIms (*116*)
14,17-formaldehyde acetal derivative: HRFABms, TMSi derivative (EIms), ^{1}H nmr (500 MHz), ^{13}C nmr (125 MHz) (*18*)

* Data (*117*) obtained from a mixture howiicin-F and howiicin-G

17,18-formaldehyde acetal derivative: HRFABms, TMSi derivative (EIms), ^{1}H nmr (500 MHz), ^{13}C nmr (125 MHz) → C-17,18-*threo* (*18*)
tetra-(*R*)- and (*S*)-MTPA esters: ^{1}H nmr (500 MHz) (*119*) → 4*R*, 14*S*, (*119*)
18-(*R*)- and (*S*)-mono-MTPA esters: ^{1}H nmr (500 MHz) (*119*) → 18*R*
(*R*)- and (*S*)-MTPA esters of 17,18-formaldehyde acetal derivative: ^{1}H nmr (500 MHz) → 14*S* (*18*)
absolute stereochemistry: 4*R*, 10*R*, 13*S*, 14*S*, 17*R*, 18*R*, 36*S*

gigantetrocin (*116*) should now be named gigantetrocin-A (*119*)
Spectral data of gigantetrocin-A are very close to those published for densicomacin-2. Densicomacin-2 (13,14-*threo*-densicomacin) would differ from gigantetrocin-A only in the absolute stereochemistry at C-10, C-13 and C-14: 10*R*, 13*S*, 14*S* (*102*)

Gigantetrocin-B = Howiicin-G ? (*117*)

Annona muricata (seeds) (*119*, *174*); *Asimina longifolia* (leaves and twigs) (*42*); *Goniothalamus howii* (seeds) (*117*)*

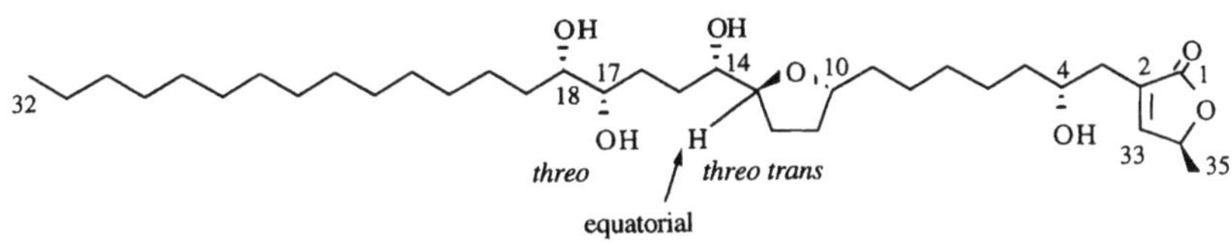

$C_{35}H_{64}O_7$; M = 596
mp = 91–92 °C (*119*), mp = 85–87 °C (*117*)
IR (*117*)
$[\alpha]_D = +4.1°$ (c = 0.41, $CHCl_3$) (*119*), $[\alpha]_D = +8.82$ (c = 8.84, $CHCl_3$) (*117*)
IR (*119*); UV (*119*)
^{1}H nmr (500 MHz) (*119*), (600 MHz) (*117*); ^{13}C nmr (125 MHz) (*119*), (100 MHz) (*117*)
HRFABms (*119*), EIms (*117*, *119*)
tetraacetate: ^{1}H nmr (200 MHz) (*117*); EIms (*117*)
TMSi derivative: HREIms (*119*)
acetonide: ^{1}H nmr (500 MHz) (*119*)
tetra-(*R*)- and (*S*)-MTPA esters: ^{1}H nmr (500 MHz) (*119*) → 4*R*, 14*S*, (*119*)
18-(*R*)- and (*S*)-mono-MTPA esters: ^{1}H nmr (500 MHz) (*119*) → 18*S*
absolute stereochemistry: 4*R*, 10*S*, 13*S*, 14*S*, 17*S*, 18*S*, 36*S*

Gigantetrocinone

Asimina longifolia (leaves and twigs) (*42*); *Asimina triloba* (stem bark) (*60*)

* Data (*117*) obtained from a mixture howiicin-F and howiicin-G

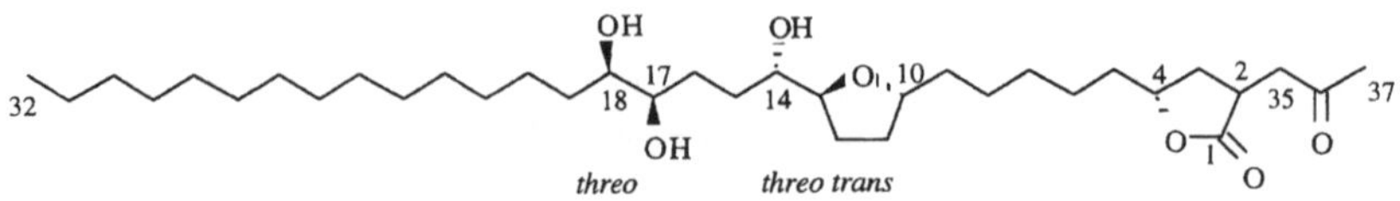

mixture of 2,4-*cis* and 2,4-*trans*-isomers

$C_{35}H_{64}O_7$; M = 596
mp = 91–92 °C (*60*)
$[\alpha]_D = +10°$ (c = 0.2, $CHCl_3$) (*60*)
IR (*60*); UV (*60*)
^1H nmr (500 MHz) (*60*); ^{13}C nmr (125 MHz) (*60*)
FABms (glycerol) (*60*), CIms (*60*), EIms (*60*)
triacetate: ^1H nmr (500 MHz) (*60*)
tri-TMSi derivative: EIms (*60*)
acetonide: ^1H nmr (500 MHz) (*60*)

Gigantetronenin

Annona muricata (leaves) (*32*); *Goniothalamus giganteus* (bark) (*29*); *Annona coriacea* (*96*); *Xylopia aromatica* (bark) (*167*)

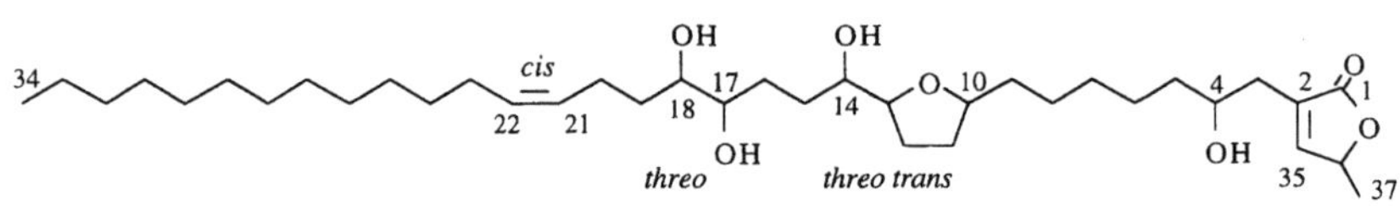

$C_{37}H_{66}O_7$; M = 622
mp = 57–59 °C (*29*)
$[\alpha]_D = +10°$ (c = 0.2, $CHCl_3$) (*29*)
IR (*29*); UV (*29*)
^1H nmr (500 MHz) (*29*); ^{13}C nmr (125 MHz) (*29*)
HRFABms (*29*), CIms (*29*), FABms (*29*)
tetraacetate: CIms, ^1H nmr (*29*)
tetra-TMSi, tetra-$TMSi_{d9}$: EIms (*29*)
acetonide: FABms, ^1H nmr (500 MHz) (*29*)
21/22-epoxy-gigantetronenin (*100*)
hemisynthesis of 18/21-*trans*-gigantecin and 18/21-*cis*-gigantecin from gigantetronenin (*100*)

Gigantriocin = Howiicin-D (*117*)

Goniothalamus giganteus (stem bark) (*116*); *Goniothalamus howii* (seeds) (*117*)

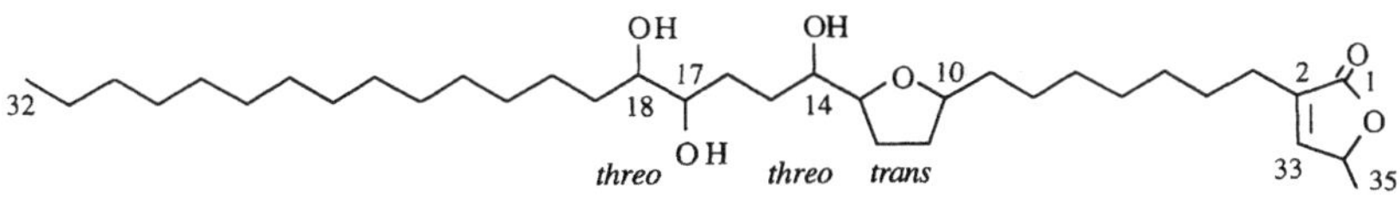

$C_{35}H_{64}O_6$; M = 580
mp = 69–71 °C (*116*), mp = 95–96 °C (*117*)
$[\alpha]_D = +18°$ ($CHCl_3$) (*116*), $[\alpha]_D = +25.07$ (c = 2.49, $CHCl_3$) (*117*)
IR (*117*)
^1H nmr (500 MHz) (*116*), (200 MHz) (*117*); ^{13}C nmr (125 MHz) (*116*), (100 MHz) (*117*)
HRCIms (*116*), EIms (*117*)
triacetate: ^1H nmr (500 MHz) (*116*), (200 MHz) (*117*), EIms (*117*)
TMSi derivative: EIms (*116, 117*), $TMSi_{d9}$ derivative: EIms (*116*)
acetonide: CIms (*116*)

Gigantrionenin

Goniothalamus giganteus (bark) (*29*)

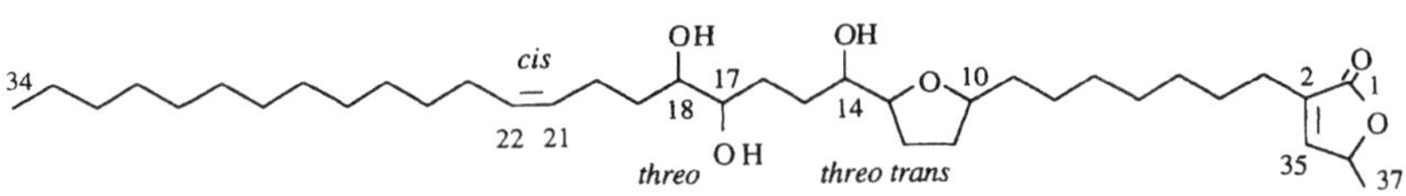

$C_{37}H_{66}O_6$; M = 606
mp = 55–57 °C (*29*)
$[\alpha]_D = +17°$ (c = 0.2, $CHCl_3$) (*29*)
IR (*29*); UV (*29*)
^1H nmr (500 MHz) (*29*); ^{13}C nmr (125 MHz) (*29*)
CIms (*29*), FABms (*29*), HRFABms (*29*)
triacetate: CIms (*29*); ^1H nmr (500 MHz) (*29*)
tri-TMSi, tri-$TMSi_{d9}$ derivatives: EIms (*29*)
acetonide: FABms, ^1H nmr (500 MHz) (*29*)

Glaucanisin

Annona glauca (seeds) (*120*)

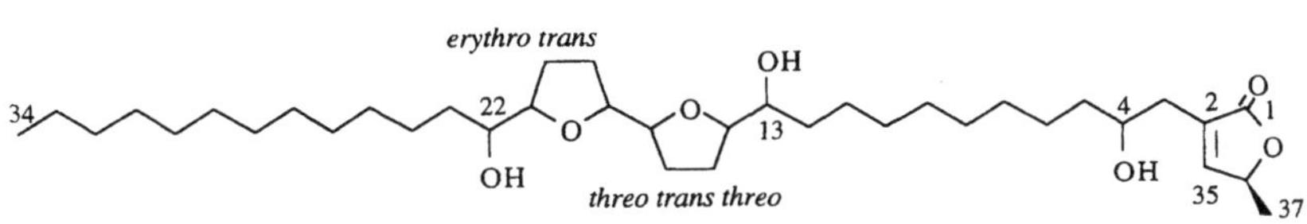

$C_{37}H_{66}O_7$; M = 622
$[\alpha]_D = +13°$ (c = 0.57, MeOH) (*120*)
IR (*120*)
^{1}H nmr (200 MHz) (*120*); ^{13}C nmr (50 MHz) (*120*)
CIms (*120*), EIms (*120*)
triacetate: ^{1}H nmr (200 MHz) (*120*), CIms (*120*)

Goniocin

Goniothalamus giganteus (bark) (*99*)

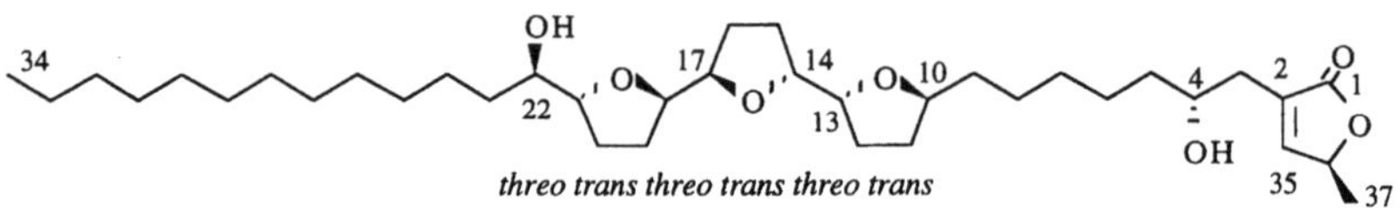

$C_{37}H_{64}O_7$; M = 620
^{1}H nmr (500 MHz) (*99*); ^{13}C nmr (125 MHz) (*99*)
HRFABms (*99*)
bis-TMSi: EIms (*99*)
(*S*)- and (*R*)-MPTA esters: ^{1}H nmr (500 MHz) → 22*R*, 4*R* (*99*)
absolute stereochemistry: 4*R*, 10*S*, 13*R*, 14*R*, 17*R*, 18*R*, 21*R*, 22*R*, 36*S*

Goniodenin

Goniothalamus giganteus (barks) (*75*)

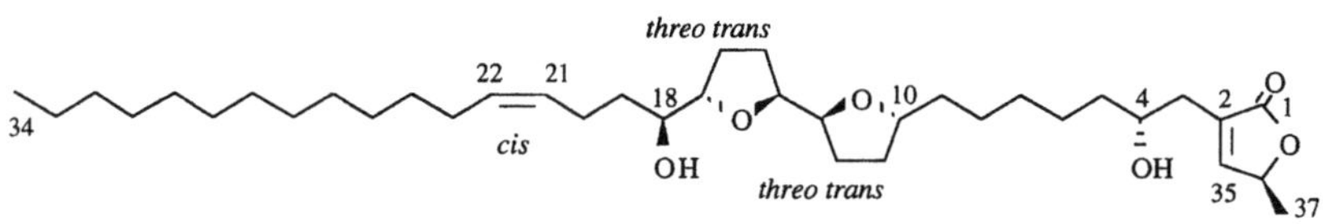

$C_{37}H_{64}O_6$; M = 604
$[\alpha]_D = +0.5°$ (c = 1.10, CH_2Cl_2) (*75*) (0.5 in text and 5.0 in experimental part)
IR (*75*); UV (*75*)
^{1}H nmr (500 MHz) (*75*); ^{13}C nmr (125 MHz) (*75*)
HRFABms (*75*), HRCIms (*75*)
diacetate: ^{1}H nmr (500 MHz) (*75*)
di-TMSi derivative: EIms (*75*)
hemisynthesis of cyclogoniodenins T and C (type D acetogenins) (m-CPBA; tosylic acid) (*87*)
(*S*)- and (*R*)-MPTA esters: ^{1}H nmr (500 MHz) → 4*R*, 18*S*, 36*S* (*75*)
absolute stereochemistry: 4*R*, 10*R*, 13*S*, 14*S*, 17*S*, 18*S*, 36*S*

Gonionenin

Goniothalamus giganteus (bark) (*100*)

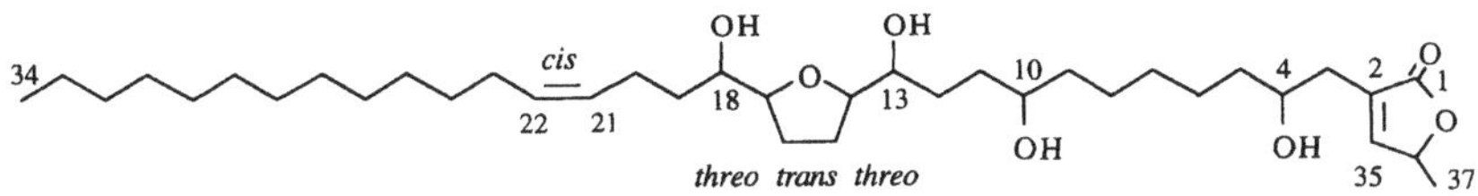

$C_{37}H_{66}O_7$; M = 622
mp = 87–88 °C (*100*)
$[\alpha]_D = +19.5°$ (c = 0.22, MeOH) (*100*)
IR (*100*); UV (*100*)
^{1}H nmr (500 MHz) (*100*); ^{13}C nmr (125 MHz) (*100*)
HRFABms (glycerol) (*100*), CIms (*100*)
tetraacetate: ^{1}H nmr (500 MHz) (*100*)
TMSi derivative: CIms (*100*)
gonionenin + *m*-CPBA → epoxides of gonionenin: ^{1}H nmr (500 MHz) (*100*); + $HClO_4$ → cyclogonionenin T and cyclogonionenin C (*100*)

Goniothalamicin

Annona densicoma (stem bark) (*101*); *Annona montana* (seeds) (*26*); *Annona muricata* (seeds) (*40*), (leaves) (*31*); *Asimina longifolia* (leaves and twigs) (*42*); *Asimina parviflora* (twigs) (*43*); *Goniothalamus giganteus* (stem bark) (*44*); *Goniothalamus howii* (seeds) (*34*)

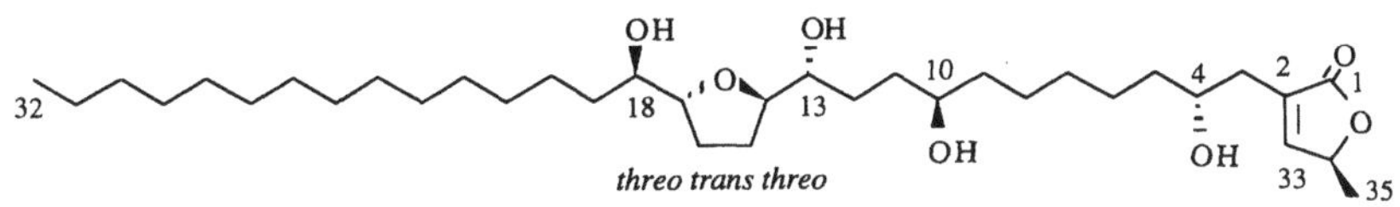

$C_{35}H_{64}O_7$; M = 596
mp = 86–88 °C (*44*); F = 91–92 °C (*101*)
$[\alpha]_D = +1.6°$ (*44*), $[\alpha]_D = +10.4$ (c = 0.08, MeOH) (*101*)
IR (*44, 101*); UV (*44, 101*)
^{1}H nmr (470 MHz) (*44*), (500 MHz) (*101*); ^{13}C nmr (100 MHz) (*44*), (125 MHz) (*101*)
HRCIms (*44*), FABms (*101*)
tetraacetate, tetraacetate$_{d3}$: CIms (*44*), EIms (*44*), tetraacetate: (*101*), CIms (*101*)
TMSi, TMSi$_{d9}$ derivatives: CIms (*44*), EIms (*44*), TMSi: EIms (*101*)
2,33-dihydrogoniothalamicin: EIms (*44*)
10,13-formaldehyde acetal derivative: HRFABms, TMSi derivative (EIms), ^{1}H nmr (500 MHz), ^{13}C nmr (125 MHz) → 10,13-*cis* (*18*)

(*R*)- and (*S*)-MTPA esters of 10,13-formaldehyde acetal derivative: ^{1}H nmr (500 MHz) → 18*R* (*18*)
absolute stereochemistry: 4*R*, 10*R*, 13*R*, 14*R*, 17*R*, 18*R*, 36*S*

Goniothalamicinone

Asimina longifolia. (leaves and twigs) (*42*)

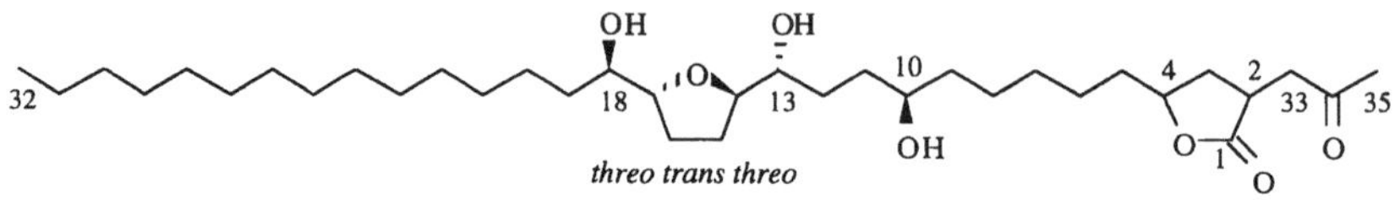

mixture of 2,4-*cis*- and 2,4-*trans*-isomers

$C_{35}H_{64}O_7$; M = 596
mp = 98 °C (*42*)
$[\alpha]_D = +22.9°$ (c = 1, CH_2Cl_2) (*42*)
IR (*42*); UV (*42*)
^{1}H nmr (500 MHz) (*42*); ^{13}C nmr (125 MHz) (*42*)
CIms (*42*), HRCIms (*42*), FABms (*42*), EIms (*42*)
tri-TMSi derivative: EIms (*42*), HRCIms (*42*)
(*R*)- and (*S*)-MTPA esters: ^{1}H nmr (500 MHz) (*42*), → 10*R*, 13*R*, 18*R* (*42*)
absolute stereochemistry: 4*R*, 10*R*, 13*R*, 14*R*, 17*R*, 18*R*

8-Hydroxyannonacin

Annona densicoma (stem bark) (*101*)

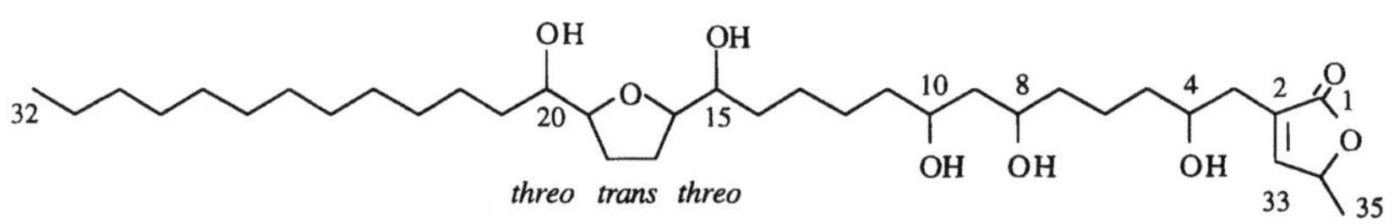

$C_{35}H_{64}O_8$; M = 612
$[\alpha]_D = +6.1°$ (c = 0.12, MeOH) (*101*)
IR (*101*); UV (*101*)
^{1}H nmr (500 MHz) (*101*)
FABms (*101*), EIms (*101*)
TMSi derivative: EIms (*101*)
pentaacetate: ^{1}H nmr (500 MHz) (*101*), EIms (*101*)

30-Hydroxybullatacin

Annona bullata (bark) (*81*)

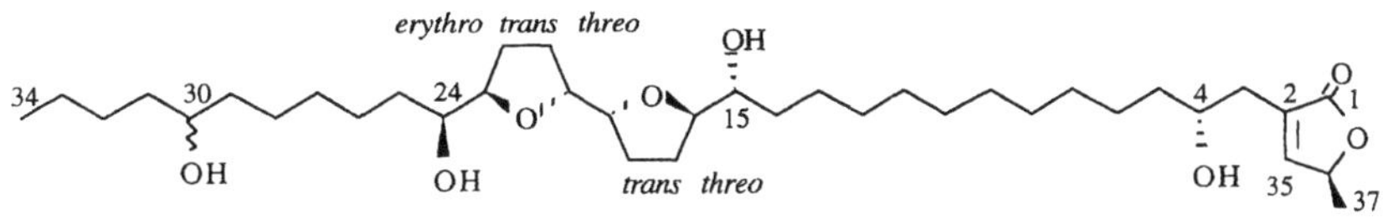

mixture of 30*R*- and 30*S*-hydroxybullatacin

$C_{37}H_{66}O_8$; M = 638
$[\alpha]_D = +14°$ (c = 0.50, $CHCl_3$) (*81*)
IR (*81*)
^1H nmr (500 MHz) (*81*); ^{13}C nmr (125 MHz) (*81*)
HRFABms (glycerol) (*81*)
tetra-TMSi derivative: EIms (*81*)
(*S*)- and (*R*)-MTPA esters: ^1H nmr (500 MHz) (*81*) → 30*S* and 30*R*, 15*R*, 24*S* (*81*)
The absolute configuration of C-36 was assigned as *S* based on the C-4*R*/C-36S relationship found in all 4-OH α,β-unsaturated γ-lactone acetogenins (*81*)
absolute stereochemistry: 4*R*, 15*R*, 16*R*, 19*R*, 20*R*, 23*R*, 24*S*, 30*R* (and 30*S*), 36*S*

31-Hydroxybullatacin

Annona bullata (bark) (*81*)

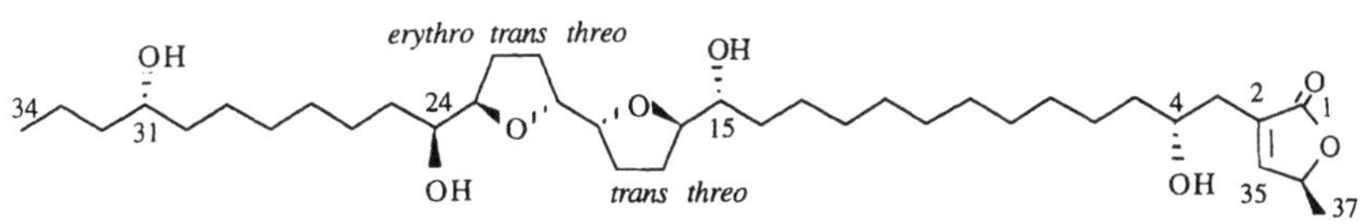

$C_{37}H_{66}O_8$; M = 638
$[\alpha]_D = +19°$ (c = 0.08, $CHCl_3$) (*81*)
IR (*81*)
^1H nmr (500 MHz) (*81*); ^{13}C nmr (125 MHz) (*81*)
HRFABms (glycerol) (*81*)
tetra-TMSi derivative: EIms (*81*)
(*S*)- and (*R*)-MTPA esters: ^1H nmr (500 MHz) (*81*) → 31*R*, 15*R*, 24*S*, (*81*)
The absolute configuration of C-36 was assigned as *S* based on the C-4*R*/-C-36*S* relationship found in all 4-OH α,β-unsaturated γ-lactone acetogenins (*81*)
absolute stereochemistry: 4*R*, 15*R*, 16*R*, 19*R*, 20*R*, 23*R*, 24*S*, 31*R*, 36*S*

32-Hydroxybullatacin

Annona bullata (bark) (*81*)

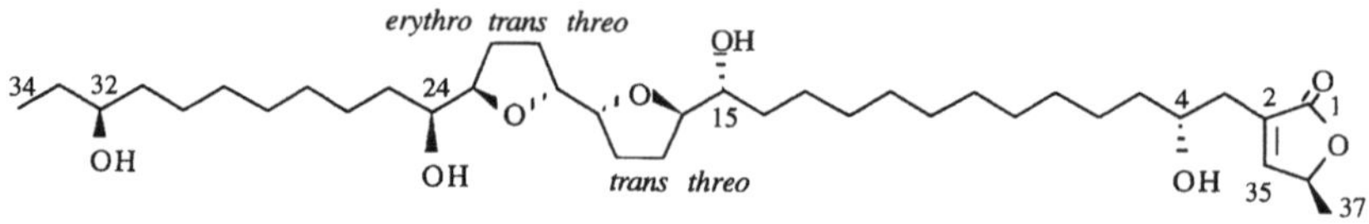

$C_{37}H_{66}O_8$; M = 638
$[\alpha]_D = +18°$ (c = 0.10, $CHCl_3$) (*81*)
IR (*81*)
^1H nmr (500 MHz) (*81*); ^{13}C nmr (125 MHz) (*81*)
HRFABms (glycerol) (*81*)
tetra-TMSi derivative: EIms (*81*)
(*S*)- and (*R*)-MTPA esters: ^1H nmr (500 MHz) (*81*) → 32*R*, 15*R*, 24*S* (*81*)
The absolute configuration of C-36 was assigned to *S* based on the C-4*R*/-C-36*S* relationship found in all 4-OH α,β-unsaturated γ-lactone acetogenins (*81*)
absolute stereochemistry: 4*R*, 15*R*, 16*R*, 19*R*, 20*R*, 23*R*, 24*S*, 32*R*, 36*S*

10-Hydroxybullatacinone

Annona bullata (barks) (*121*)

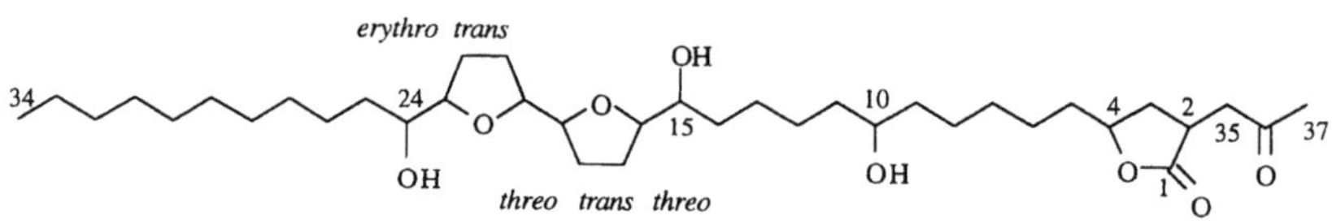

mixture of 2,4-*cis*- and 2,4-*trans*-isomers

$C_{37}H_{66}O_8$; M = 638
IR (*121*)
^1H nmr (500 MHz) (*121*), ^{13}C nmr (125 MHz) (*121*)
HRFABms (glycerol) (*121*)
tri-acetate: ^1H nmr (500 MHz) (*121*)
tri-TMSi derivative: EIms (*121*)

12-Hydroxybullatacinone

Annona bullata (barks) (*121*)

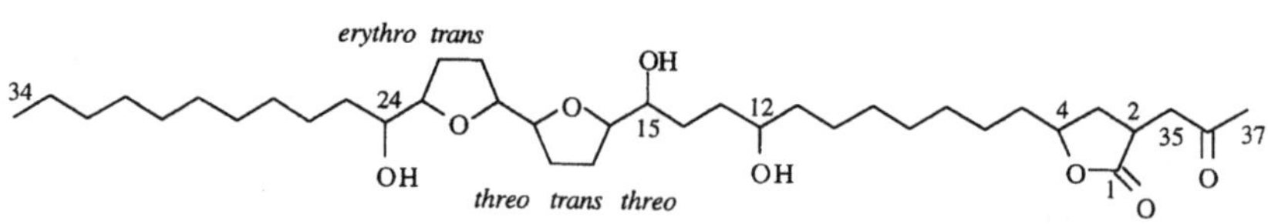

mixture of 2,4-*cis*- and 2,4-*trans*-isomers

$C_{37}H_{66}O_8$; M = 638
IR (*121*)
HRFABms (glycerol) (*121*)
^{1}H nmr (500 MHz) (*121*); ^{13}C nmr (125 MHz) (*121*)
tri-acetate: ^{1}H nmr (500 MHz) (*121*)
tri-TMSi derivative: EIms (*121*)

28-Hydroxybullatacinone

Annona bullata (barks) (*81*)

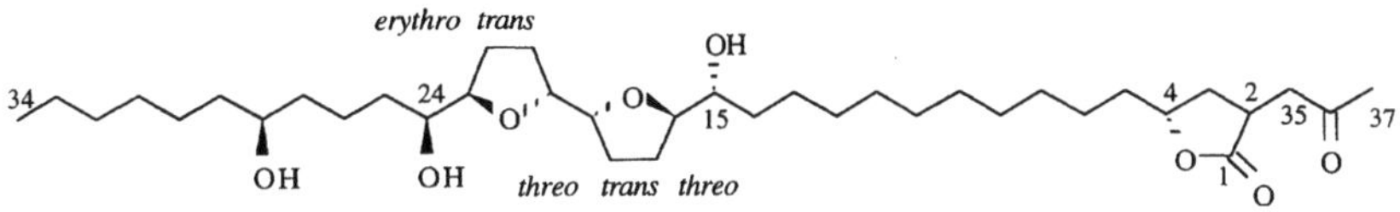

mixture of 2,4-*cis*- and 2,4-*trans*-isomers

$C_{37}H_{66}O_8$; M = 638
$[\alpha]_D = +18°$ (c = 0.10, $CHCl_3$) (*81*)
IR (*81*)
^{1}H nmr (500 MHz) (*81*); ^{13}C nmr (125MHz) (*81*)
HRFABms (glycerol) (*81*)
tetra-TMSi derivative: EIms (*81*)
24,28-formaldehyde derivative (of mixture): ^{1}H nmr (500 MHz) (*81*); (*S*)- and (*R*)-MTPA esters: ^{1}H nmr (500 MHz) (*81*) → 15*R* (*81*)
(*S*)- and (*R*)-MTPA esters: ^{1}H nmr (500 MHz) (*81*) → 28*S* (*81*)
absolute stereochemistry: 4*R*, 15*R*, 16*R*, 19*R*, 20*R*, 23*R*, 24*S*, 28*S*

The 2,4-*cis*- and 2,4-*trans*-isomers were separated by HPLC (*81*)

29-Hydroxybullatacinone

Annona bullata (barks) (*121*)

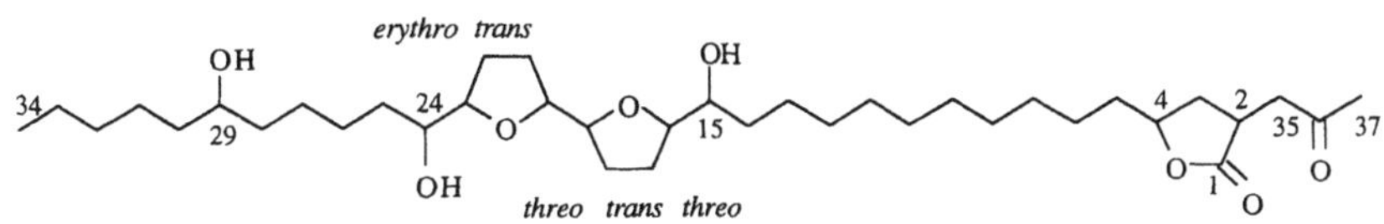

mixture of 2,4-*cis* and 2,4-*trans*-isomers

$C_{37}H_{66}O_8$; M = 638
IR (*121*)

^1H nmr (500 MHz) (*121*); ^{13}C nmr (125 MHz) (*121*)
HRFABms (glycerol) (*121*)
tri-acetate: ^1H nmr (500 MHz) (*121*)
tri-TMSi derivative: EIms (*121*)

Allocation of a hydroxyl group to C-28, C-29, C-30, C-31 or C-32 was based on the shift of the terminal methyl (C-34) (*121*)

30-Hydroxybullatacinone

Annona bullata (bark) (*122*)

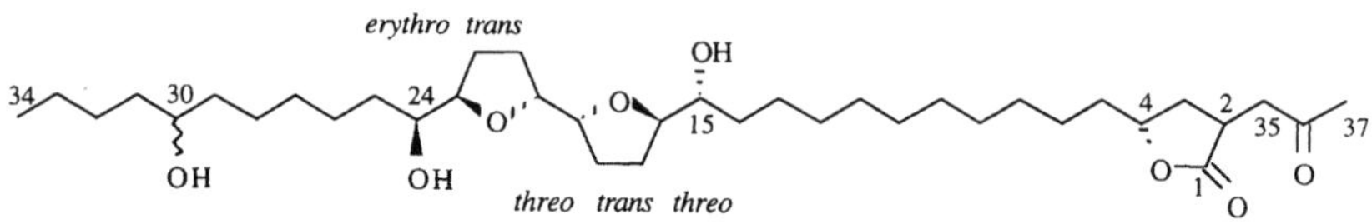

mixture of 2,4-*cis*- and 2,4-*trans*-isomers
mixture of 30*S* and 30*R* (*81*)

$C_{37}H_{67}O_8$; M = 638
IR (*122*)
^1H nmr (500 MHz) (*122*); ^{13}C nmr (125 MHz) (*122*)
FABms (*122*), HRFABms (glycerol) (*122*)
tri-acetate: ^1H nmr (500 MHz) (*122*)
tri-TMSi derivative: EIms (*122*)
tri-(*S*)- and (*R*)-MTPA esters: ^1H nmr (500 MHz) (*81*) → 30*R* and 30*S*, 15*R*, 24*S* (*81*)
absolute stereochemistry: 15*R*, 16*R*, 19*R*, 20*R*, 23*R*, 24*S*, 30*R* and 30*S*

Allocation of a hydroxyl group to C-28, C-29, C-30, C-31 or C-32 was based on the shift of the terminal methyl (C-34) (*121*)

31-Hydroxybullatacinone

Annona bullata (bark) (*122*)

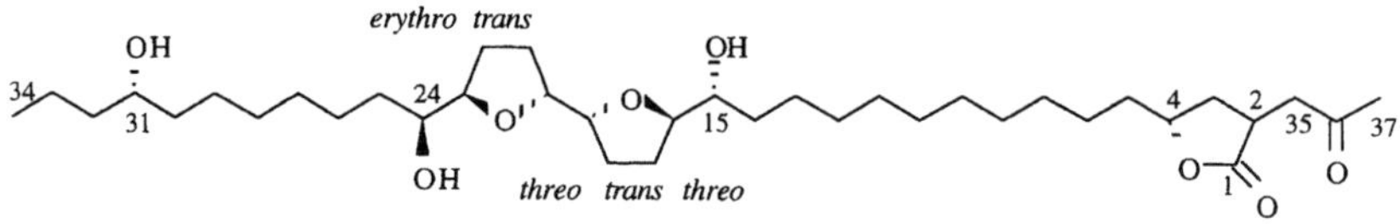

mixture of 2,4-*cis*- and 2,4-*trans*-isomers

$C_{37}H_{66}O_8$; M = 638
IR (*122*)

^{1}H nmr (500 MHz) (*122*); ^{13}C nmr (125 MHz) (*122*)
FABms (*122*), HRFABms (glycerol) (*122*)
tri-acetate: ^{1}H nmr (500 MHz) (*122*)
tri-TMSi derivative: EIms (*122*)
tri-(*S*)- and (*R*)-MTPA esters: ^{1}H nmr (500 MHz) (*81*) → 31*R*, 15*R*, 24*S* (*81*)
absolute stereochemistry: 15*R*, 16*R*, 19*R*, 20*R*, 23*R*, 24*S*, 31*R*

Allocation of a hydroxyl group to C-28, C-29, C-30, C-31 or C-32 was based on the shift of the terminal methyl (C-34) (*121*)

32-Hydroxybullatacinone

Annona bullata (bark) (*122*)

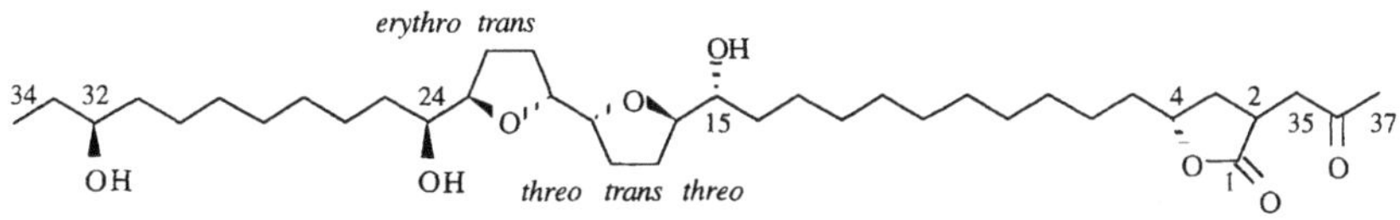

mixture of 2,4-*cis*- and 2,4-*trans*-isomers

$C_{37}H_{66}O_8$; M = 638
IR (*122*)
^{1}H nmr (500 MHz) (*122*); ^{13}C nmr (125 MHz) (*122*); ^{1}H nmr study allowing to locate hydroxy group at C-28, C-29, C-30, C-31 or C-32 according to terminal methyl chemical shift (*121*)
FABms (*122*), HRFABms (glycerol) (*122*)
tri-acetate (*122*): ^{1}H nmr (500 MHz) (*122*)
tri-TMSi derivative: EIms (*122*)
tri-(*S*)- and (*R*)-MTPA esters: ^{1}H nmr (500 MHz) (*81*) → 32*R*, 15*R*, 24*S* (*81*)
absolute stereochemistry: 15*R*, 16*R*, 19*R*, 20*R*, 23*R*, 24*S*, 32*R*

Isoannonacin

Annona densicoma (stem bark) (*7*), *Annona muricata* (seeds) (*40*), (leaves) (*59*), *Asimina longifolia* (leaves and twigs) (*42*), *Asimina triloba* (stem bark) (*60*)

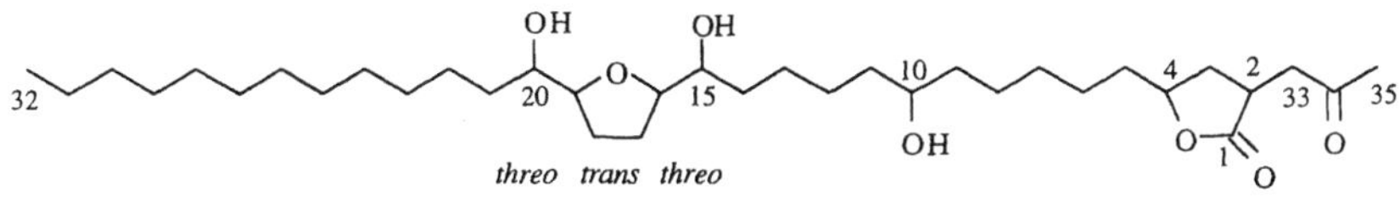

mixture of 2,4-*cis*- and 2,4-*trans*- isomers

$C_{35}H_{64}O_7$; M = 596
mp = 96–98 °C (7), mp = 95–96 °C (*60*)

$[\alpha]_D = +24.8°$ (c = 0.12, MeOH) (7), $[\alpha]_D = +20°$ (c = 0.2, $CHCl_3$) *(60)*
IR (7, *60*); UV (7, *60*)
^1H nmr (470 MHz) (7), (500 MHz) *(60)*; ^{13}C nmr (67.5 MHz) (7), (125 MHz) *(60)*
HRFABms *(60)*, HRCIms (7), EIms *(60)*
triacetate: ^1H nmr (470 MHz) (7), (500 MHz) *(60)*,
TMSi derivative: HREIms (7), EIms *(60)*
hemisynthesis from annonacin (KOH/*t*-BuOH) (7)

Isoannonacinone

Annona densicoma (stem bark) (7); *Annona muricata* (seeds) *(40)*

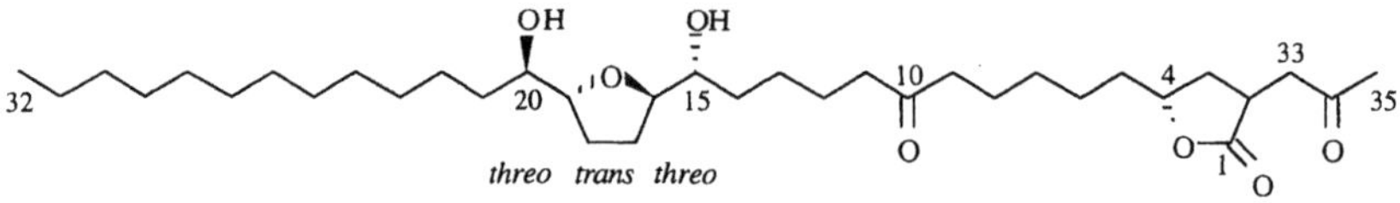

mixture of 2,4-*cis*- and 2,4-*trans*- isomers

$C_{35}H_{62}O_7$; M = 594
mp = 103 °C (7)
$[\alpha]_D = +19.8°$ (c = 0.05, MeOH) (7)
IR (7, *60*); UV (7, *60*)
^1H nmr (470 MHz) (7), (500 MHz) *(40, 60)*; ^{13}C nmr (125 MHz) *(40)*
FABms *(60)*, CIms (7), EIms *(60)*, HRCIms (7)
acetate: ^1H nmr (470 MHz) (7)
TMSi derivative: HREIms (7)
hemisynthesis from annonacin-10-one (KOH/*t*-BuOH) (7)
(*R*)- and (*S*)-bis-MTPA esters: ^1H nmr (500 MHz) *(46)*; ^{19}F nmr *(46)*; → 15*R*, 20*R* *(46)*
absolute stereochemistry: 4*R*, 15*R*, 16*R*, 19*R*, 20*R*

Isoannoreticuin

Annona reticulata (leaves) *(62)*

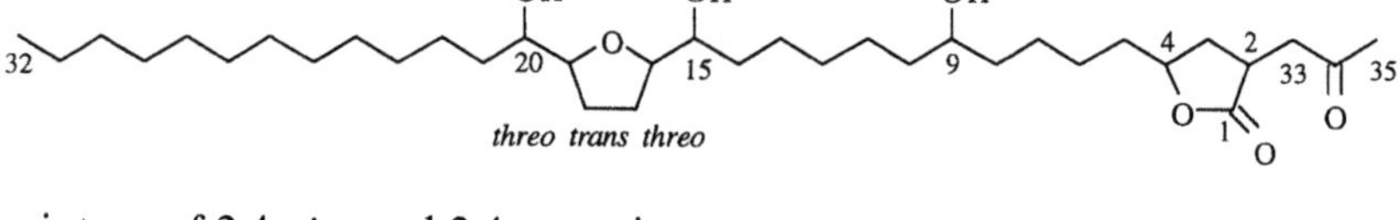

mixture of 2,4-*cis*- and 2,4-*trans*-isomers

$C_{35}H_{64}O_7$; M = 596
$[\alpha]_D = +9.7$ (c = 0.33, $CHCl_3$) *(62)*

IR (*62*); UV (*62*)
^{1}H nmr (200 MHz) (*62*); ^{13}C nmr (50 MHz) (*62*)
HRCIms (*62*), EIms (*62*)
tri-TMSi: EIms (*62*)

Isocherimolin-1 or **Bullatalicinone**

Isodesacetyluvaricin = 4-Deoxyasimicin (*93*) = Squamocin-M (*64*)

Annona atemoya (seeds) (*78*); *Annona bullata* (bark) (*93*); *Annona squamosa* (seeds) (*64*); *Uvaria grandiflora* (seeds) (*123*); *Uvaria hookeri* (root bark) (*137*); *Uvaria narum* (root bark) (*124*)

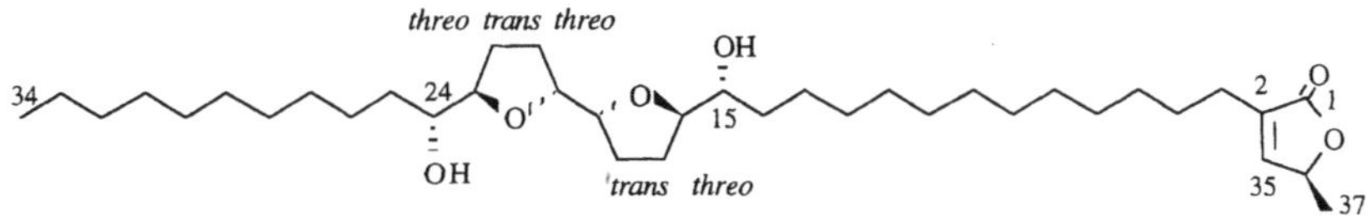

named 4-deoxyasimicin and considered as new by authors (*93*).

$C_{37}H_{66}O_6$; M = 606
mp = < 30 °C (*124*)
$[\alpha]_D$ = + 17.82° (c = 0.0059, $CHCl_3$) (*93*), $[\alpha]_D$ = + 26° (c = 0.55, MeOH) (*64*)
CD (*93*)
IR (*93, 124*); UV (*93*)
^{1}H nmr (300 MHz) (*124*), (500 MHz) (*64, 93*); ^{13}C nmr (75 MHz) (*124*), (125 MHz) (*64, 93*) FABms (*93, 124*), CIms, (*124*), EIms (*64, 93, 124*), HRFABms (glycerol) (*93*), HRFABms (nitrobenzylalcohol) (*64*)
diacetate: ^{1}H nmr (300 MHz) (*124*), (500 MHz) (*93*); EIms (*64, 93, 124*)
di-TMSi derivative: EIms (*124*)
di-(*R*)-MTPA ester: ^{1}H nmr (500 MHz) (*64*) → 15*R*, 24*R* (*64*)
absolute stereochemistry: 15*R*, 16*R*, 19*R*, 20*R*, 23*R*, 24*R*

Isomolvizarin-1

Annona cherimolia (roots) (*17*)

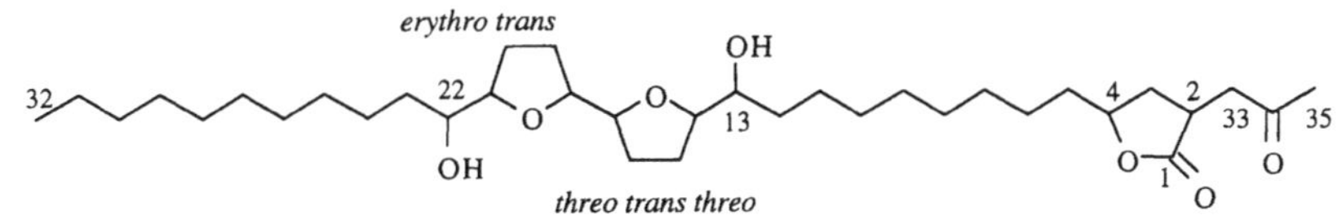

mixture of 2,4-*cis*- and 2,4-*trans*-isomers

$C_{35}H_{62}O_7$; M = 594
$[\alpha]_D = +24°$ (c = 0.32, MeOH) (*17*)
IR (*17*); UV (*17*)
^{1}H nmr (200 MHz) (*17*); ^{13}C nmr (50 MHz) (*17*)
CIms (*17*); EIms (*17*), FAB-Li ms (*17*)
artifact of extraction (*8, 9*)

Isomolvizarin-2

Annona cherimolia (roots) (*17*)

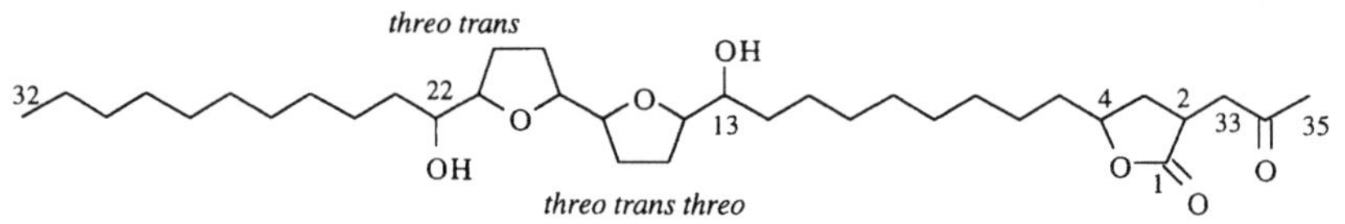

mixture of 2,4-*cis*- and 2,4-*trans*-isomers

$C_{35}H_{62}O_7$; M = 594
$[\alpha]_D = +33°$ (c = 0.19, MeOH) (mixture with a small quantity of isomolvizarin-1) (*17*)
IR (*17*); UV (*17*)
^{1}H nmr (200 MHz) (*17*); ^{13}C nmr (50 MHz) (*17*)
CIms (*17*); EIms (*17*), FAB-Li ms (*17*)

artifact of extraction (*8, 9*)

Isomurisolin

hemisynthesis (*125*)

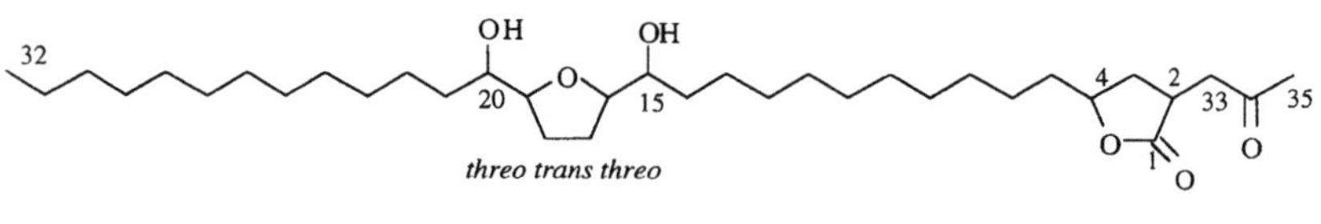

mixture of 2,4-*cis*- and 2,4-*trans*-isomers
obtained from murisolin (KOH/t-BuOH) (*125*)

$C_{35}H_{64}O_6$; M = 580
IR (*125*)
^{1}H nmr (400 MHz) (*125*); ^{13}C nmr (100 MHz) (*125*)
CIms (*125*)

Isoneoannonacinone

Annona muricata (seeds) (*126*)

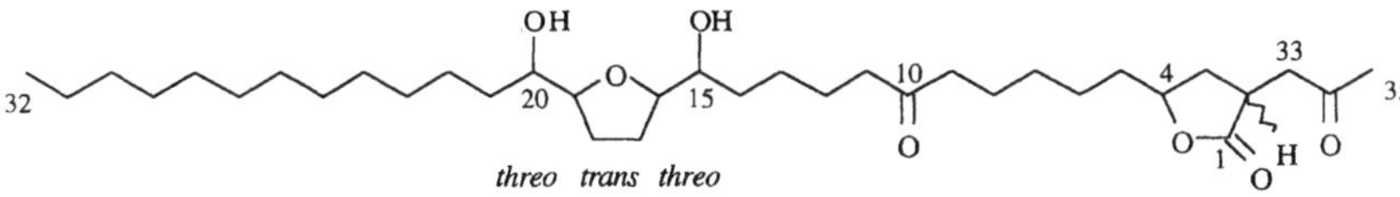

The argument that this acetogenin is a diastereoisomer of isoannonacinone at the level of the tetrahydrofuran pattern is not very reliable and is based only on the difference in $[\alpha]_D$ (*126*).

$C_{35}H_{62}O_7$; M = 594
mp = 107–109 °C (*126*)
$[\alpha]_D = +9.2°$ (c = 0.27, $CHCl_3$) (*126*), $[\alpha]_D = +9.4°$ (c = 1.87, MeOH) (*126*)
IR (*126*)
^1H nmr (400 MHz) (*126*); ^{13}C nmr (100 MHz) (*126*)
EIms (*126*)
acetate: ^1H nmr (400 MHz) (*126*); EIms (*126*)

Isorolliniastatin-1

hemisynthesis (*9*)

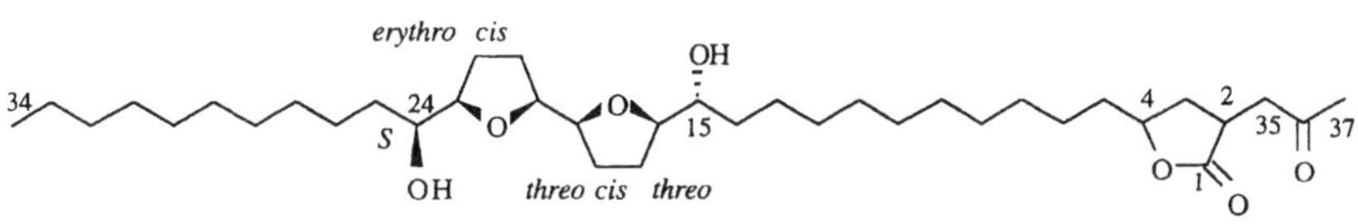

mixture of 2,4-*cis*- and 2,4-*trans*-isomers

$C_{37}H_{66}O_7$; M = 622
^1H nmr (200 MHz) (*9*)
hemisynthesis from rolliniastatin-1 (diethylamine/CH_2Cl_2; alkaloids base or salts/MeOH; heating of a methanolic solution) (*9*)

Isorollinicin

Rollinia papillionella (roots) incorrect structure (*2, 3*)

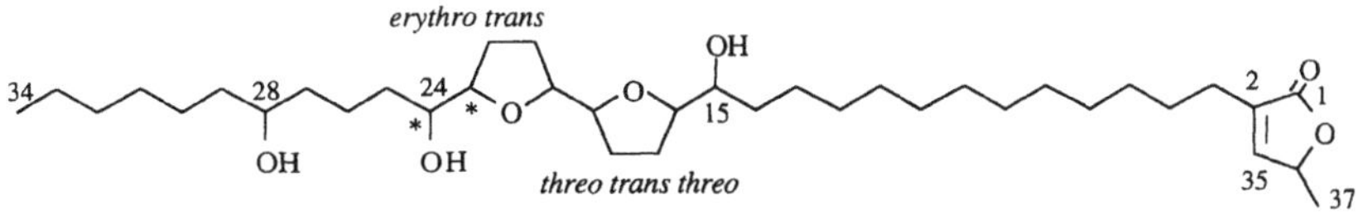

would have to be a diastereoisomer of C-23 or C-24 of squamocin (*12, 5*)

$C_{37}H_{66}O_7$; M = 622
mp = 66–68 °C (*3*)
IR (*3*); UV (*3*)
^{1}H nmr (360 MHz) (*3*); ^{13}C nmr (22.5 MHz) (*3*)
HRCIms (*3*), CIms (*3*)

Isosylvaticin

hemisynthesis (*111*)

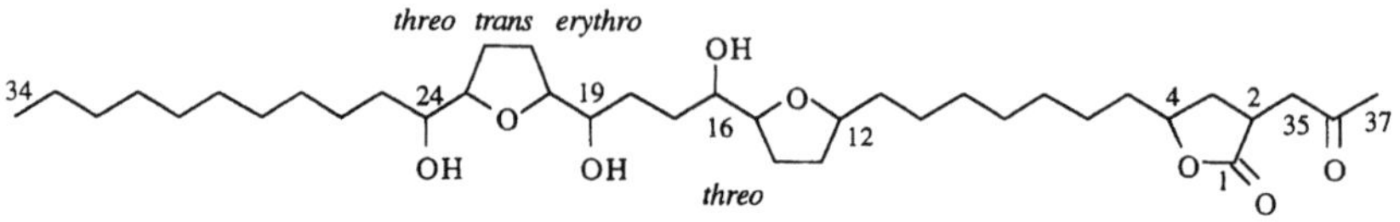

mixture of 2,4-*cis*- and 2,4-*trans*-isomers
obtained from sylvaticin (diethylamine/CH_2Cl_2) (*21*, *111*)

$C_{37}H_{66}O_8$; M = 638
$[\alpha]_D$ = + 23° (c = 0.30, MeOH) (*111*)
IR (*111*); UV (*111*)
^{1}H nmr (200 MHz) (*111*); ^{13}C nmr (50 MHz) (*111*)
FAB-Li ms (*111*)

Itrabin

Annona cherimolia (seeds) (*16*, *127*)

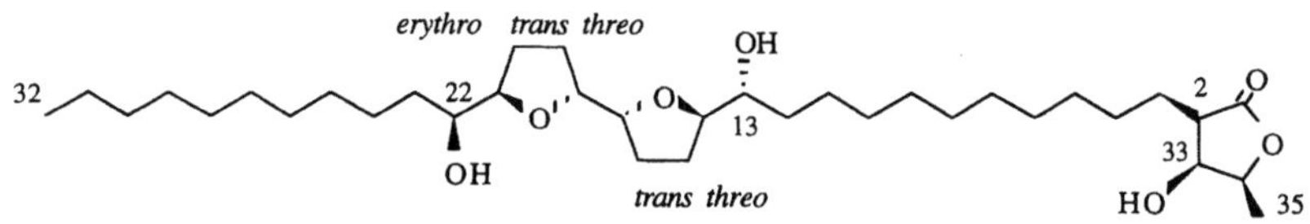

$C_{35}H_{64}O_7$; M = 596
mp = 38–40 °C (*127*)
$[\alpha]_D$ = + 21° (c = 0.2, MeOH) (*127*)
IR (*127*)
^{1}H nmr (200 MHz) (*127*); ^{13}C nmr (50 MHz) (*127*)
CIms (*127*)
triacetate: ^{1}H nmr (200 MHz) (*127*), CIms (*127*)
dehydration (tosyl derivative heated in DMSO) (*127*)
absolute stereochemistry 34*S* deduced from that of uvaricin (*127*)

Jetein

Annona cherimolia (seeds) (*16*, *127*)

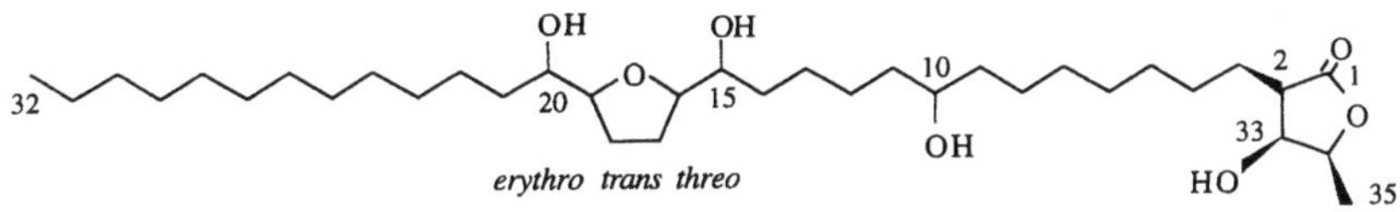

$C_{35}H_{66}O_7$; M = 598
mp = 58–60 °C (*127*)
$[\alpha]_D = +9°$ (c = 0.03, MeOH) (*127*)
IR (*127*)
^1H nmr (200 MHz) (*127*); ^{13}C nmr (50 MHz) (*127*)
CIms (*127*)
absolute stereochemistry 34*S* deduced from that of uvaricin (*127*)

Laherradurin

Annona cherimolia (seeds) (*66*), revised structure (*16*, *127*)

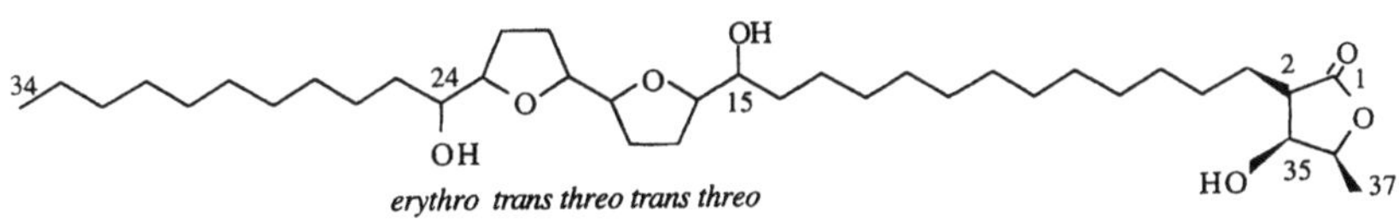

$C_{37}H_{68}O_7$; M = 624
$[\alpha]_D = +21°$ (c = 0.09, MeOH) (*16*)
IR (*16*)
^1H nmr (200 MHz) (*16*); ^{13}C nmr (50 MHz) (*16*)
CIms (*16*), EIms (*16*)
dehydration (tosyl derivative heated in DMSO) (*16*)

Longicin

Asimina longifolia (leaves and twigs) (*42*)

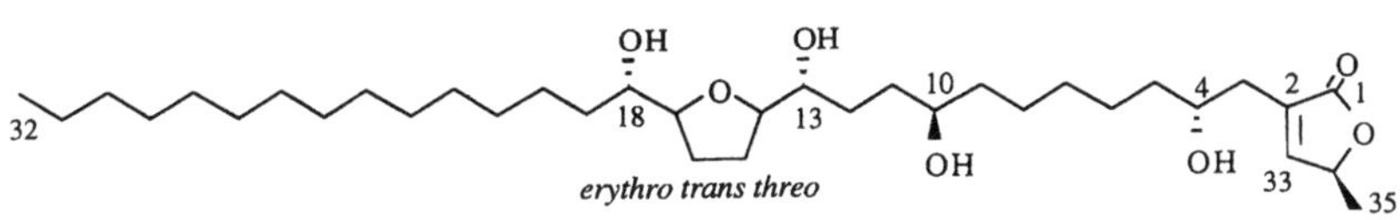

$C_{35}H_{64}O_7$; M = 596
mp = 83 °C (*42*)
$[\alpha]_D = -13.0°$ (c = 1, CH_2Cl_2) (*42*)
IR (*42*); UV (*42*)

^{1}H nmr (500 MHz) (*42*); ^{13}C nmr (125 MHz) (*42*)
CIms (*42*), HRCIms (*42*), EIms (*42*)
tetraacetate: ^{1}H nmr (500 MHz) (*42*)
tetra-TMSi derivative: EIms (*42*)
longicin ($EtOH/Na_2CO_3$ → mixture of 2,4-*cis*- and 2,4-*trans*-longicinone (*42*): ^{1}H nmr (500 MHz) (*42*); diacetyl derivatives: ^{1}H nmr (500 MHz) (*42*); (*R*)- and (*S*)-MTPA esters: ^{1}H nmr (500 MHz) (*42*) → 10*R*, 34*S* (*42*)
(*R*)- and (*S*)-MTPA esters: ^{1}H nmr (500 MHz) (*42*), → 4*R*, 13*R*, 18*S*, 34*S* (*42*)
absolute stereochemistry: 4*R*, 13*R*, 14*R*, 17*R*, 18*S*, 34*S*

Membranacin

Rollinia membranacea (seeds) (*128*)

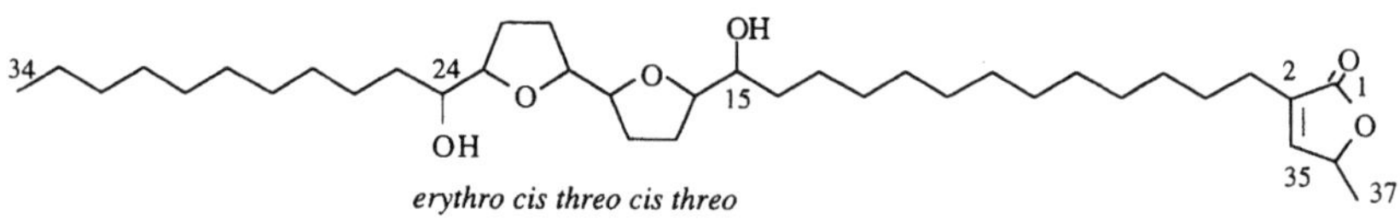

$C_{37}H_{66}O_6$; M = 606
$[\alpha]_D = +22.5°$ (c = 0.2, MeOH) (*128*)
IR (*128*)
^{1}H nmr (200 MHz) (*128*); ^{13}C nmr (50 MHz) (*128*)
CIms (*128*)
diacetate: ^{1}H nmr (200 MHz) (*128*), CIms (*128*)

Molvizarin

Annona atemoya (seeds) (*129*); *Annona cherimolia* (*130*), (seeds) (*16*); *Annona reticulata* (barks) (*131*); *Annona senegalensis* (seeds) (*68*, *69*); *Asimina parviflora* (twigs) (*43*)

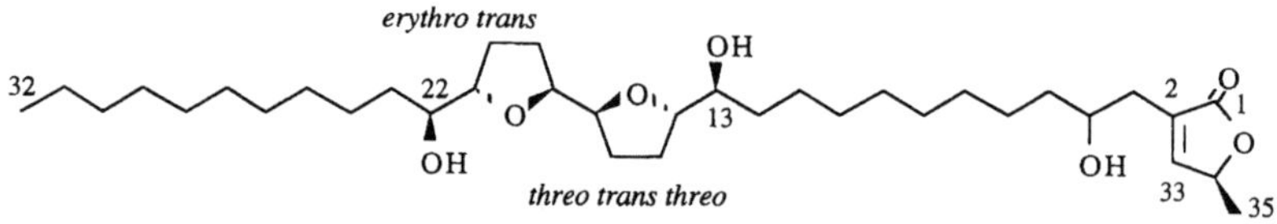

$C_{35}H_{62}O_7$; M = 594
mp = 36–38 °C (*130*)
$[\alpha]_D = +10°$ (c = 0.13, MeOH) (*130*), $[\alpha]_D = +10°$ (c = 0.20, $CHCl_3$) (*68*)
IR (*130*); UV (*130*)
^{1}H nmr (200 MHz) (*130*), (400 MHz) (*131*); ^{13}C nmr (50 MHz) (*130*)
CIms (*130*, *131*), FAB-Li ms (*131*), EIms (*131*)

triacetate: IR (*130*), ^{1}H nmr (200 MHz) (*130*); HRClms (*130*), Clms (*130*)
TMSi derivative (*131*)

Montanacin

Annona montana (seeds) (*26*)

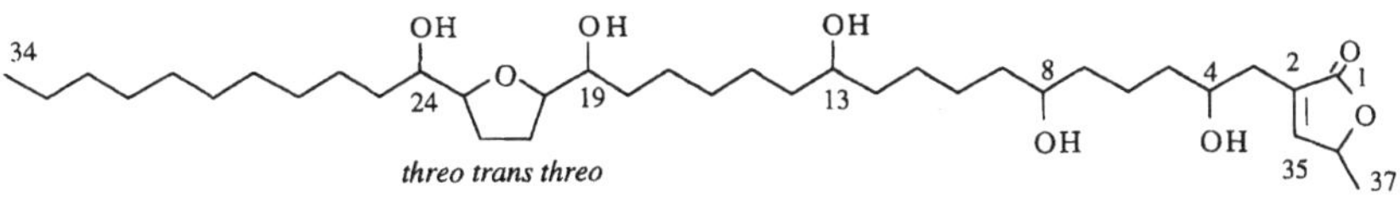

$C_{37}H_{68}O_8$; M = 640
mp = 53–55 °C (*26*)
$[\alpha]_D$ = + 15° (c = 1.1, MeOH) (*26*)
IR (*26*); UV (*26*)
[1]H nmr (300 MHz) (*26*), ^{13}C nmr (75 MHz) (*26*)
CIms (*26*)
pentaacetate: CIms (*26*); ^{1}H nmr (300 MHz) (*26*)

Motrilin = Annonin-III (*132*), = Squamocin-C (*64*)

Annona cherimolia (seeds) (*16, 130*); *Annona squamosa* (seeds) (*64, 132*); *Asimina triloba* (stem bark) (*82*)

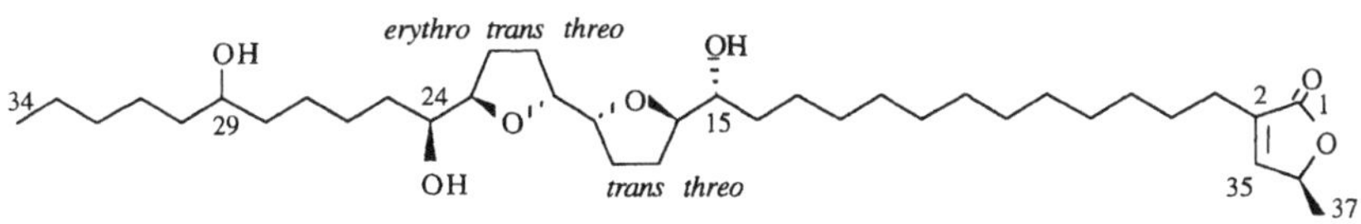

$C_{37}H_{66}O_7$; M = 622
mp = 50–51 °C (*64, 130*)
$[\alpha]_D$ = + 10.8° (c = 0.13, MeOH) (*130*), $[\alpha]_D$ = + 19.5° (c = 0.92, MeOH) (*64*)
IR (*130*); UV (*64, 130*)
[1]H nmr (200 MHz) (*130*), (500 MHz) (*64*); ^{13}C nmr (50 MHz) (*130, 132*), (125 MHz) (*64*)
CIms (*130*), EIms (*64*), HRFABms (nitrobenzylalcohol) (*64*)
triacetate: IR (*130*); ^{1}H nmr (200 MHz) (*130*), (500 MHz) (*64*); HRCIms (*130*), CIms (*130*)
dihydromotrilin: [1]H nmr (200 MHz) (*130*), HRCIms (*130*), CIms (*130*)
tri-(*R*)-MTPA ester: [1]H nmr (500 MHz) (*64*), proposed absolute stereochemistry because the chemical shifts resembled those of the (*R*)-MPTA of squamocin, as 15*R*, 16*R*, 19*R*, 20*R*, 23*R*, 24*S* (*64*)
advanced Mosher method: 15*R*, 16*R*, 19*R*, 20*R*, 23*R*, 24*S*, 29*S* (*19*)

In the ^{1}H nmr of the (*R*)- and (*S*)-MTPA esters the chemical shifts of the terminal methyl groups showed a small but diagnostic difference: C-29-*S* (*19*)

Mucocin

Rollinia mucosa (leaves) (*133*)

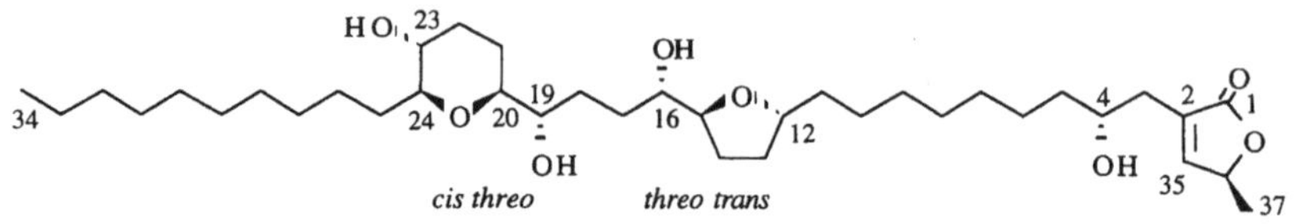

$C_{37}H_{66}O_8$; M = 638
mp = 57–58 °C (*133*)
$[\alpha]_D = -10.8°$ (*133*)
IR (*133*); UV (*133*)
^{1}H nmr (500 MHz) (*133*); ^{13}C nmr (75 MHz) (*133*)
EIms (*133*), HRFABms (*133*)
tetra-TMSi derivative: EIms (*133*)
formaldehyde acetal derivative: ^{1}H nmr (500 MHz) (*133*)
(R)- and (S)-MTPA esters of formaldehyde acetal derivative: ^{1}H nmr (500 MHz) (*133*), → 23*R*, 4*R*, 36*S* (*133*)
absolute stereochemistry: 4*R*, 12*R*, 15*S*, 16*S*, 19*S*, 20*S*, 23*R*, 24*S*, 36*S*

Muricatacin

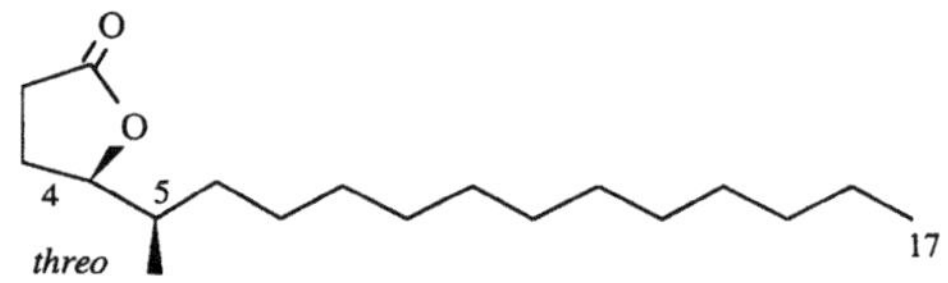

results from oxidation of annonacin: see annonacin

Muricatalin

Annona muricata (seeds) (*134*)

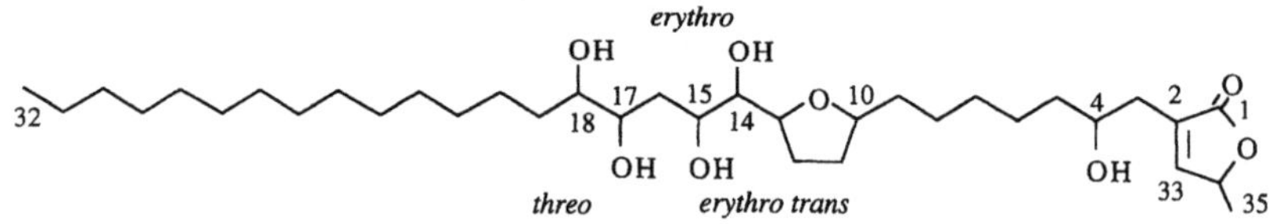

$C_{35}H_{64}O_8$; M = 612
mp = 143–144 °C (*134*)
$[\alpha]_D$ = 8.8° (c = 0.06, MeOH) (*134*)
IR (*134*); UV (*134*)
^{1}H nmr (500 MHz) (*134*); ^{13}C nmr (125 MHz) (*134*)
HREIms (*134*), FABms (*134*)
pentaacetate: ^{1}H nmr (500 MHz) (*134*); EIms (*134*)
tetra-TMSi derivative: HREIms (*134*)
pentamesitoate: ^{1}H nmr (C_6D_6, 500 MHz) (*134*)

Muricatetrocin-A = Howiicin-E (*117*)

Annona muricata (seeds) (*119*), (leaves) (*31*); *Asimina longifolia* (leaves and twigs) (*42*); *Goniothalamus howii* (seeds) (*117*)

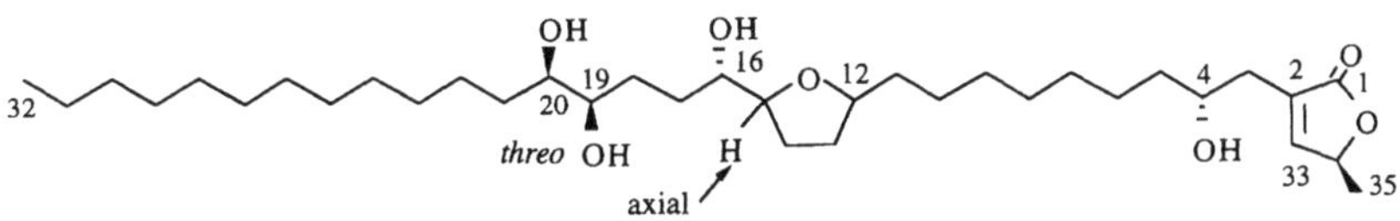

$C_{35}H_{64}O_7$; M = 596
mp = 102 °C (*119*), mp = 102–104 °C (*117*)
$[\alpha]_D$ = +10.3° (c = 0.15, $CHCl_3$) (*119*), $[\alpha]_D$ = +15.79° (c = 0.8, $CHCl_3$) (*117*)
IR (*117*, *119*); UV (*119*)
^{1}H nmr (500 MHz) (*119*), (200 MHz) (*117*); ^{13}C nmr (125 MHz) (*119*), (100 MHz) (*117*)
HRFABms (*119*), EIms (*117*, *119*)
tetraacetate: ^{1}H nmr (200 MHz) (*117*); EIms (*117*)
TMSi and $TMSi_{d9}$ derivatives: HREIms (*119*), EIms (*119*)
acetonide: ^{1}H nmr (500 MHz) (*119*)

The only differences in the nmr spectra of muricatetrocin-A and muricatetrocin-B are the chemical shifts, of C-15 and H-15: δ 82.01 and 3.70 for muricatetrocin-A, δ 81.73 and 3.78 for muricatetrocin-B. It has been suggested that the THF of muricatetrocin-A has the cis configuration (*119*)
tetra-(*R*)- and (*S*)-MTPA-esters: ^{1}H nmr (500 MHz) (*119*) → 4*R*, 16*S*, 19*R*, 20*R*

Muricatetrocin-B

Annona muricata (seeds) (*119*), (leaves) (*31*); *Asimina longifolia* (leaves and twigs) (*42*); *Rollinia mucosa* (leaves) (*87*)

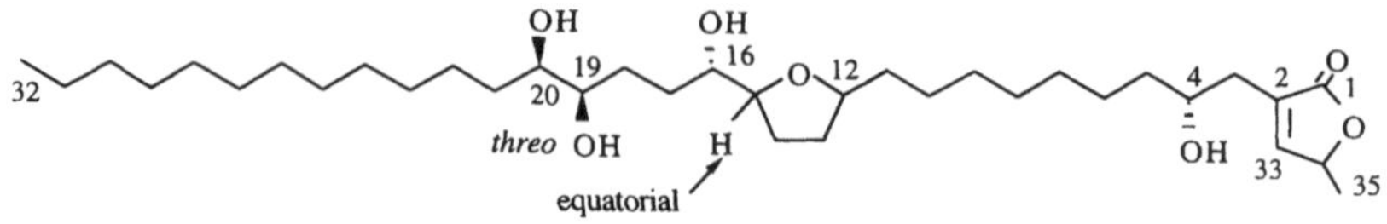

$C_{35}H_{64}O_7$; M = 596
mp = 89–90 °C (*119*)
$[\alpha]_D = +15°$ (c = 0.43, $CHCl_3$) (*119*)
IR (*119*); UV (*119*)
^1H nmr (500 MHz) (*119*); ^{13}C nmr (125 MHz) (*119*)
HRFABms (*119*), EIms (*119*)
TMSi and $TMSi_{d9}$ derivatives: HREIms (*119*), EIms (*119*)
acetonide: ^1H nmr (500 MHz) (*119*)
tetra-(*R*)- and (*S*)-MTPA-esters: ^1H nmr (500 MHz) (*119*) → 4*R*, 16*S*, 19*R*, 20*R*

The only differences in the nmr spectra of muricatetrocin-A and muricatetrocin-B are the chemical shifts of C-15 and H-15: δ 82.01 and 3.70 for muricatetrocin-A, δ 81.73 and 3.78 for muricatetrocin-B. It has been suggested that the THF of muricatetrocin-B has the *trans* configuration (*119*)

Muricatin-A

Annona muricata (seeds and stem barks) (*118*)

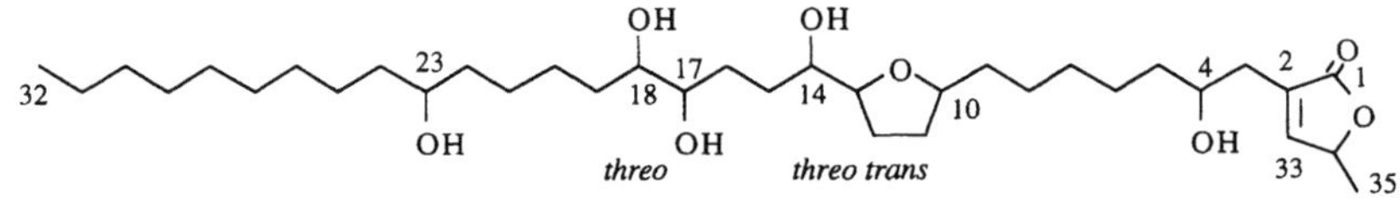

$C_{35}H_{64}O_8$; M = 612
mp = 83 °C (*118*)
$[\alpha]_D = +7.10°$ (c = 0.63, MeOH) (*118*)
IR (*118*)
^1H nmr (400 MHz) (*118*); ^{13}C nmr (100 MHz) (*118*)*
EIms (*118*), FABms Li (*118*)

---

* Some assignments in the ^{13}C nmr spectrum are doubtful

Muricatin-B

Annona muricata (seeds and stem barks) (*118*)

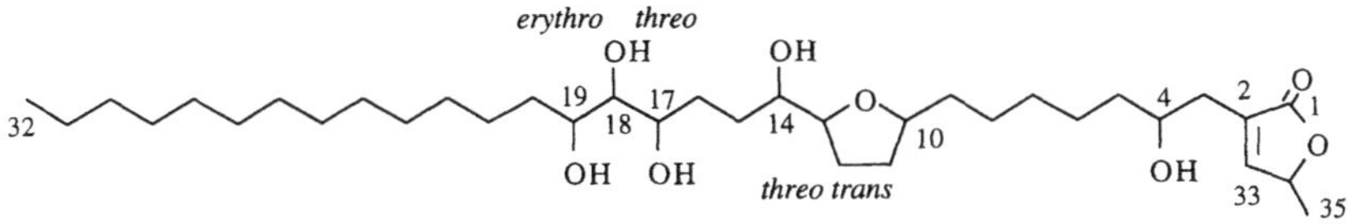

$C_{35}H_{64}O_8$; M = 612
mp = 128 °C (*118*)
$[\alpha]_D = +11.43°$ (c = 0.18, MeOH) (*118*)
IR (*118*)
^{1}H nmr (400 MHz) (*118*), ^{13}C nmr (100 MHz) (*118*)†
EIms (*118*), FAB-Lims (*118*)

Muricatin-C

Annona muricata (seeds, stem barks) (*118*)

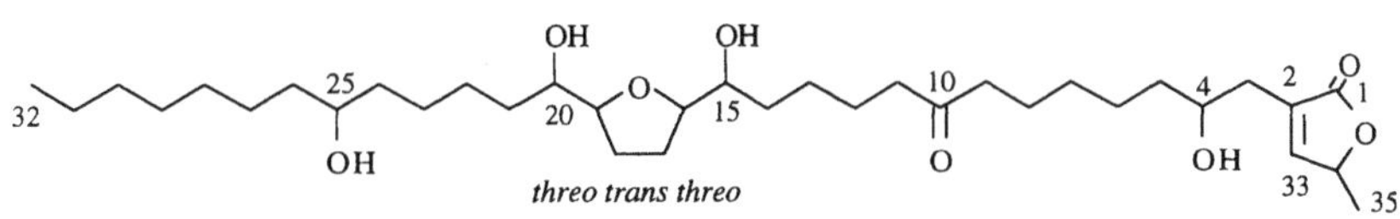

$C_{35}H_{62}O_8$; M = 610
mp = 61 °C (*118*)
$[\alpha]_D = +15.40°$ (c = 0.49, MeOH) (*118*)
IR (*118*)
^{1}H nmr (400 MHz) (*118*). ^{13}C nmr (100 MHz) (*118*)†
EIms (*118*), FABms Li (*118*)

Muricatocin-A

Annona muricata (leaves) (*59*)

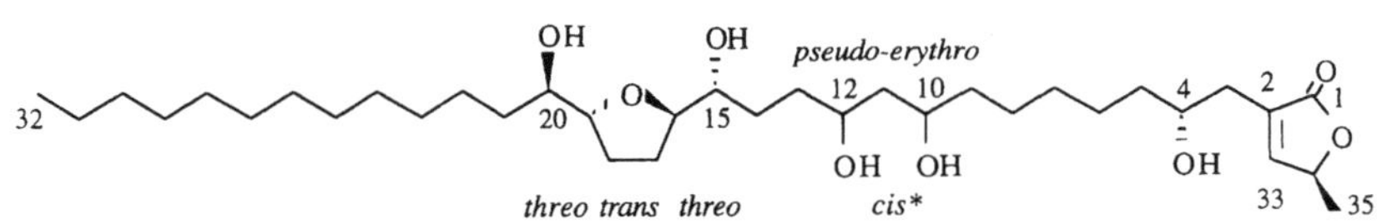

---

† Some assignments in the ^{13}C nmr spectrum are doubtful

* The *cis* configuration is suggested by the ^{1}H nmr spectrum of the acetonide, by analogy with vicinal diols (59)

$C_{35}H_{64}O_8$; M = 612
$[\alpha]_D = +21.8°$ (c = 0.001, EtOH) (*59*)
IR (*59*); UV (*59*)
^{1}H nmr (500 MHz) (*59*); ^{13}C nmr (125 MHz) (*59*)
HRFABms (*59*), CIms (*59*)
pentaacetate: ^{1}H nmr (500 MHz) (*59*)
penta-TMSi derivative: EIms (*59*)
acetonide: ^{1}H nmr (500 MHz) (*59*)
(*R*)- and (*S*)-penta-MTPA esters: ^{1}H nmr (500 MHz) (*59*) → 4*R*, 15*R*, 20*R*
absolute configuration: 4*R*, 15*R*, 16*R*, 19*R*, 20*R*, 34*S*

Muricatocin-B

Annona muricata (leaves) (*59*)

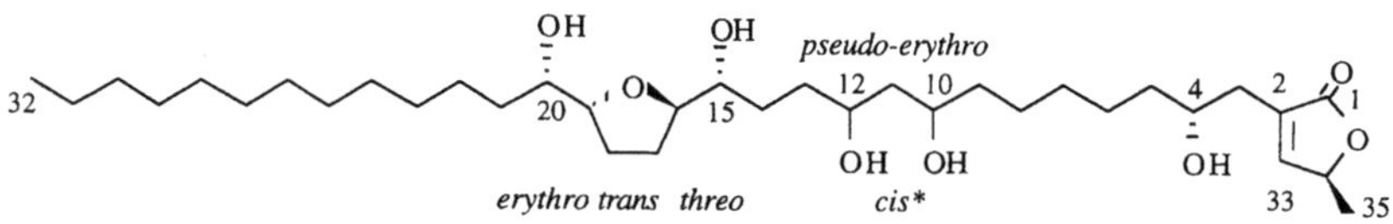

$C_{35}H_{64}O_8$; M = 612
$[\alpha]_D = +62.5°$ (c = 0.001) (*59*)
IR (*59*); UV (*59*)
^{1}H nmr (500 MHz) (*59*); ^{13}C nmr (125 MHz) (*59*)
HRFABms (*59*), CIms (*59*)
pentaacetate: ^{1}H nmr (500 MHz) (*59*)
penta-TMSi derivative: EIms (*59*)
acetonide: ^{1}H nmr (500 MHz) (*59*)
(*R*)- and (*S*)-penta-MTPA esters: ^{1}H nmr (500 MHz) (*59*) → 4*R*, 15*R*, 20*S*
absolute configuration: 4*R*, 15*R*, 16*R*, 19*R*, 20*S*, 34*S*

Muricatocin-C

Annona muricata (leaves) (*32*)

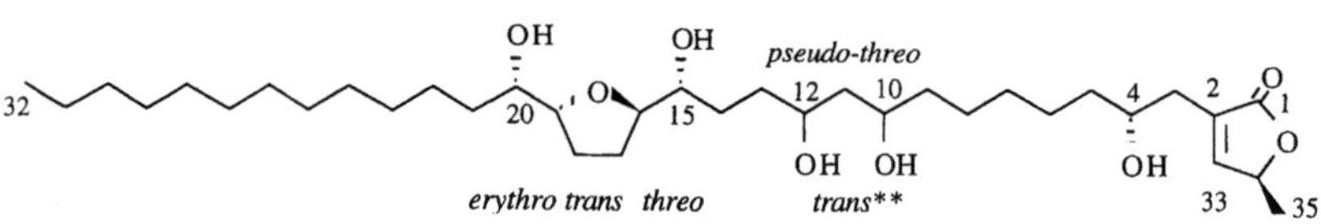

* The *cis* configuration is suggested by the ^{1}H nmr spectrum of the acetonide, by analogy with vicinal diols (*59*)

** The suggested *trans* relative (*pseudo-threo*) stereochemistry C-10/C-12 is based on the analogy of the ^{1}H nmr spectral data with those of the C-10/C-11 diol of annomuricin-C (*32*)

$C_{35}H_{64}O_8$; M = 612
$[\alpha]_D = +32.5°$ (c = 0.001) (*32*)
IR (*32*); UV (*32*)
^1H nmr (500 MHz) (*32*); ^{13}C nmr (125 MHz) (*32*)
HRFABms (*32*), CIms (*32*), EIms (*32*)
pentaacetate: ^1H nmr (500 MHz) (*32*)
penta-TMSi derivative: EIms (*32*)
acetonide: ^1H nmr (500 MHz) (*32*)
(*R*)- and (*S*)-penta-MTPA esters: ^1H nmr (500 MHz) (*32*) → 4*R*, 15*R*, 20*S*
absolute configuration: 4*R*, 15*R*, 16*R*, 19*R*, 20*S*, 34*S*

Murihexocin-A

Annona muricata (leaves) (*135*)

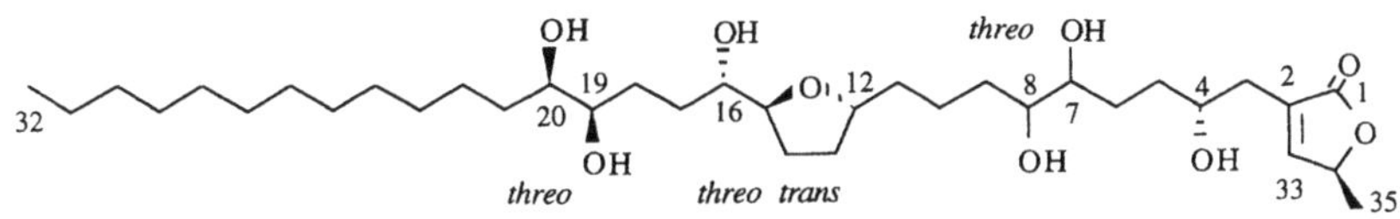

$C_{35}H_{64}O_9$; M = 628
^1H nmr (500 MHz) (*135*); ^{13}C nmr (125 MHz) (*135*)
HRFABms (*135*)
hexaacetate: ^1H nmr (500 MHz) (*135*); ^{13}C nmr (125 MHz) (*135*)
hexa-TMSi derivative: ms
bis-acetonide: ^1H nmr (500 MHz): relative stereochemistry 7,8 *threo*, 19,20 *threo*, absolute stereochemistry 19*R*, 20*R* (*135*)
(*R*)- and (*S*)-hexa-MTPA esters: ^1H nmr (500 MHz) (*135*); ^{13}C nmr (125 MHz) → 4*R*, 12*R*, 16*S* (*135*)
absolute stereochemistry: 4*R*, 12*R*, 15*S*, 16*S*, 19*R*, 20*R*, 34*S*

The magnitudes of the $\Delta\delta_{s-r}$ values for H-33 and H-34 are 0.20 and 0.06 which shows that C-34 has the usual *S* configuration (*45*)

Murihexocin-B

Annona muricata (leaves) (*135*)

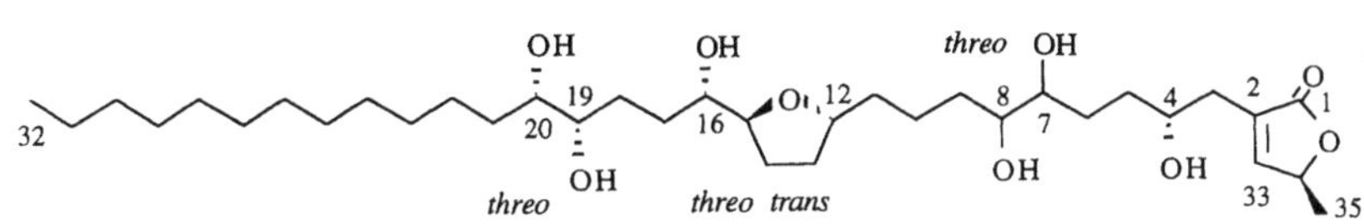

$C_{35}H_{64}O_9$; M = 628
^1H nmr (500 MHz) (*135*); ^{13}C nmr (125 MHz) (*135*)

HRFABms (*135*)
hexaacetate: ^{1}H nmr (500 MHz) (*135*); ^{13}C nmr (125 MHz) (*135*)
hexa-TMSi derivative: ms
bis-acetonide: ^{1}H nmr (500 MHz): relative stereochemistry 7,8 *threo*, 19,20 *threo*, absolute stereochemistry 19*S*, 20*S* (*135*)
(*R*)- and (*S*)-hexa-MTPA esters: ^{1}H nmr (500 MHz) (*135*); ^{13}C nmr (125 MHz): → 4*R*, 12*R*, 16*S*, (*135*) → C-34 *S* (*45*)
absolute stereochemistry: 4*R*, 12*R*, 15*S*, 16*S*, 19*S*, 20*S*, 34*S*

Murisolin = Howiicin-B (*34*)

Annona muricata (seeds) (*41, 125*); *Asimina triloba* (seeds) (*136*); *Goniothalamus howii* (seeds) (*34*)

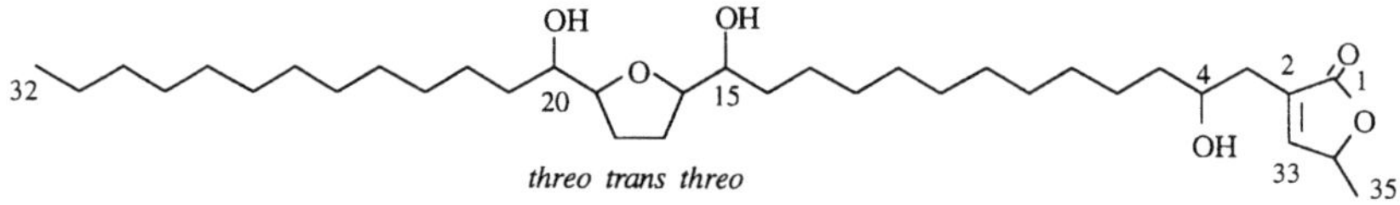

$C_{35}H_{64}O_6$; M = 580
mp = 72–73 °C (*136*), mp = 72–75 °C (*34*), mp = 70–72 °C (*41*)
$[\alpha]_D = +14.8°$ (c = 0.1, MeOH) (*125*), $[\alpha]_D = +16°$ (c = 0.1, $CHCl_3$) (*136*), $[\alpha]_D = +20.44°$ (c = 5.92, $CHCl_3$) (*34*), $[\alpha]_D = +19.05°$ (c = 0.84, $CHCl_3$) (*41*)
IR (*34, 41, 125*); UV (*125, 136*)
^{1}H nmr (400 MHz) (*34, 41, 125*); ^{13}C nmr (100 MHz) (*34, 41, 125*)
HRCIms (*125*), CIms (*125*), EIms (*34, 41*)
triacetate: ^{1}H nmr (400 MHz) (*125*), (200 MHz) (*34*); ^{13}C nmr (100 MHz) (*125*), EIms (*34*)
hemisynthesis of isomurisolin from murisolin (KOH/*t*-BuOH) (*125*): IR (*125*), CIms (*125*), ^{1}H nmr (400 MHz) (*125*); ^{13}C nmr (100 MHz) (*125*),

Murisolin-A

Asimina triloba (seeds) (*136*)

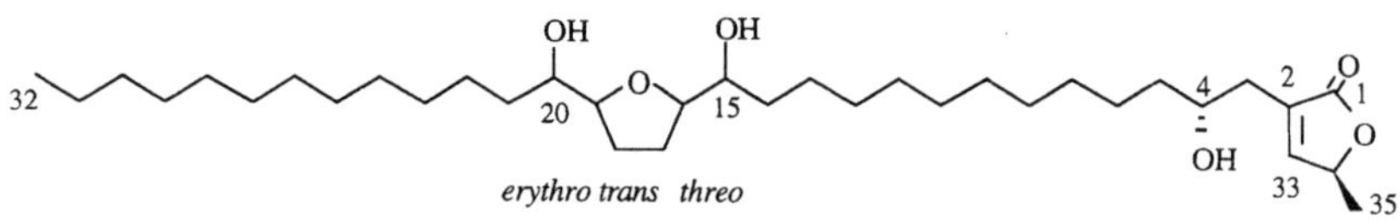

$C_{35}H_{64}O_6$; M = 580
mp = 83–84 °C (*136*)
$[\alpha]_D = +17°$ (c = 0.1, $CHCl_3$) (*136*)

IR (*136*); UV (*136*)
^{1}H nmr (500 MHz) (*136*); ^{13}C nmr (125 MHz) (*136*)
HRCIms (*136*), CIms (*136*)
tri-TMSi: EIms (*136*)
triacetate: EIms, ^{13}C nmr (125 MHz), ^{1}H nmr (500 MHz) (*136*)
(*R*)- and (*S*)-tri-MTPA esters: ^{1}H nmr (500 MHz) (*136*)
absolute stereochemistry: 4*R*, 34*S*, but ambiguities remain for 15, 16, 19, 20 (*R* or *S*) (*136*)

16,19-*cis*-Murisolin

Asimina triloba (seeds) (*136*)

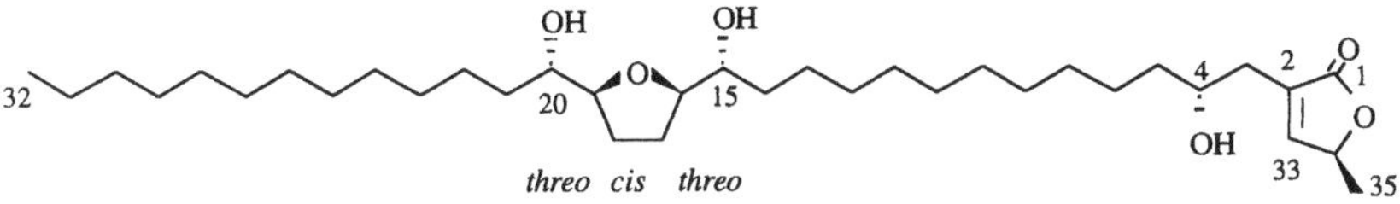

$C_{35}H_{64}O_6$; M = 580
mp = 67–68 °C (*136*)
$[\alpha]_D = +11°$ (c = 0.1, CH_2Cl_2) (*136*)
IR (*136*); UV (*136*)
^{1}H nmr (500 MHz) (*136*); ^{13}C nmr (125 MHz) (*136*)
HRCIms (*136*)
tri-TMSi: EIms (*136*)
triacetate: EIms, ^{13}C nmr (125 MHz), ^{1}H nmr (500 MHz) (*136*)
(*R*)- and (*S*)-tri-MTPA esters: ^{1}H nmr (500 MHz) (*136*) → 4*R*, 15*R*, 20*S*
absolute stereochemistry: 4*R*, 15*R*, 16*R*, 19*S*, 20*S*, 34*S*

Narumicin-I

Uvaria hookeri (root bark) (*137*); *Uvaria narum* (root bark) (*124*, *137*)

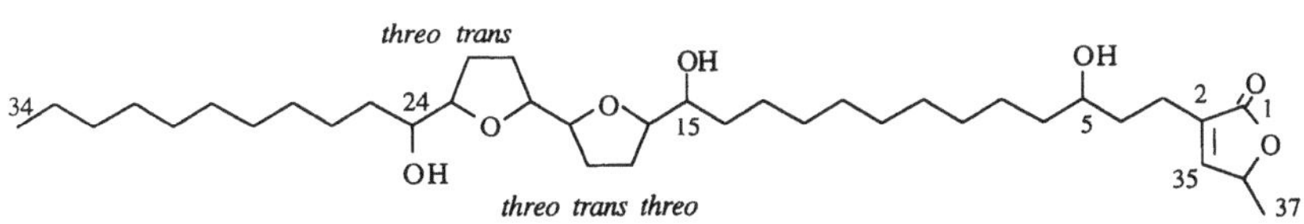

isolated as a mixture with narumicin-II

$C_{37}H_{66}O_7$; M = 622,
IR (*124*)
^{1}H nmr (300 MHz) (*124*, *137*); ^{13}C nmr (75 MHz) (*124*), (60 MHz) (*137*)
FABms, (*124*), EIms (*124*)

tri-TMSi derivative: EIms (*124*)
triacetate: ^{1}H nmr (300 MHz) (*124*); ^{13}C nmr (75 MHz) (*124*); EIms (*124*)

Narumicin-II

Uvaria hookeri (root bark) (*137*); *Uvaria narum* (root bark) (*124, 137*)

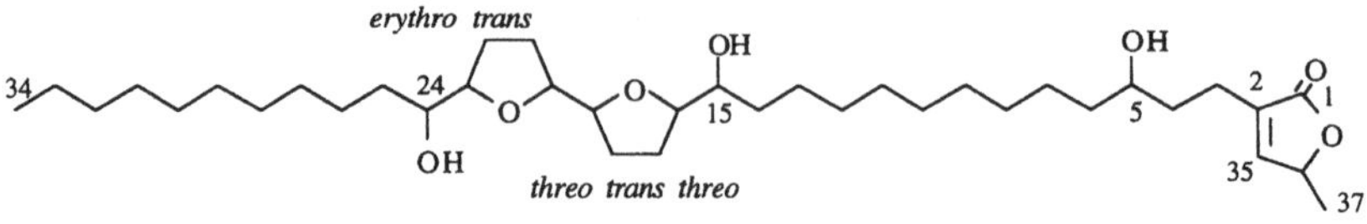

isolated as a mixture with narumicin-I

$C_{37}H_{66}O_7$; M = 622
data: see narumicin-I (mixture)

Neoannonacinone

Annona muricata (seeds) (*126*)

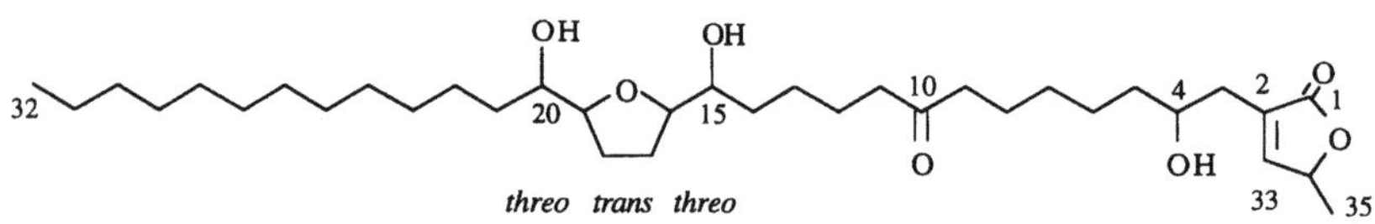

$C_{35}H_{62}O_7$; M = 594
mp = 72.5–75 °C (*126*)
$[\alpha]_D = +17.3°$ (c = 1.15, $CHCl_3$), $[\alpha]_D = +20°$ (c = 1.55, MeOH) (*126*)
IR (*126*)
^{1}H nmr (300 MHz) (*126*), ^{13}C nmr (75 MHz) (*126*)
EIms (*126*)
triacetate: ^{1}H nmr (300 MHz) (*126*), EIms (*126*)

The argument that this acetogenin should be a diastereoisomer of annonacinone at the level of the tetrahydrofuran pattern is insubstantial. The only point made is the difference of $[\alpha]_D$.

Neoannonin = Squamocin-J (*64*), = Neoannonin-B (*104*), = Atemoyacin-B (*129*)

Annona atemoya (seeds) (*78, 129*); *Annona squamosa* (seeds), (*64, 104, 138*)

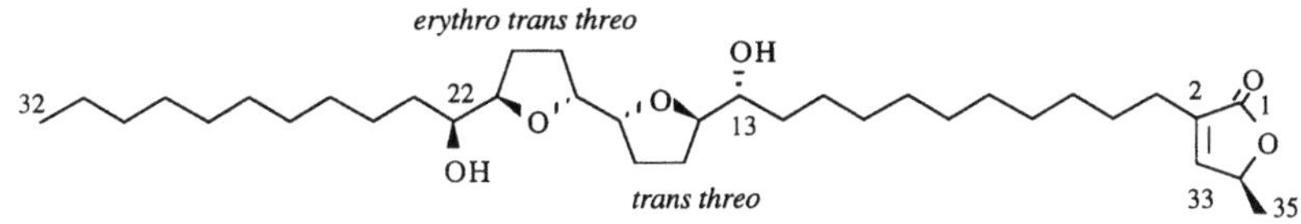

$C_{35}H_{62}O_6$; M = 578
mp = 85–86.5 °C (*64*), mp = 62–64 °C (*104*)
$[\alpha]_D$ = + 18.6° (c = 0.42, MeOH) (*64*), $[\alpha]_D$ = + 20° (c = 0.25, MeOH) (*78*), $[\alpha]_D$ = + 16.2° (c = 0.8, $CHCl_3$) (*104*)
CD (*64*)
IR (*64, 78, 104, 138*); UV (*64, 78*)
^{1}H nmr (500 MHz) (*64, 138*), (200 MHz) (*78*), (300 MHz) (*104*); ^{13}C nmr (125 MHz) (*64, 138*), (75 MHz) (*104*), (50 MHz) (*78*)
HRFABms (nitrobenzylalcohol) (*64*), FABms (*138*), CIms (*78, 138*), EIms (*64, 78, 104, 138*)
diacetate: IR (*64, 138*); ^{1}H nmr (500 MHz) (*64, 138*), (300 MHz) (*104*); CIms (*138*), EIms (*104*)
oxidation → α-(ω-formyldecyl)-α,β-angelica lactone + undecanal (*138*)
proposed relative stereochemistry *threo-trans-threo-trans-erythro* (*12*)
di-(*R*)-MTPA ester: ^{1}H nmr (500 MHz) (*64*) → absolute stereochemistry 13*R*, 14*R*, 17*R*, 18*R*, 21*R*, 22*S* although it could not be established which of the two isomers, neonnonin (squamocin-J) and squamocin-I, has the 13*R*, 14*R*, 21*R*, 22*S* or 13*S*, 14*R*, 21*R*, 22*R* absolute stereochemistry (*64*)
absolute stereochemistry 34*S* determined by negative Cotton effect (*64*)

Neoisoannonacinone

Annona muricata (seeds) (*126*)

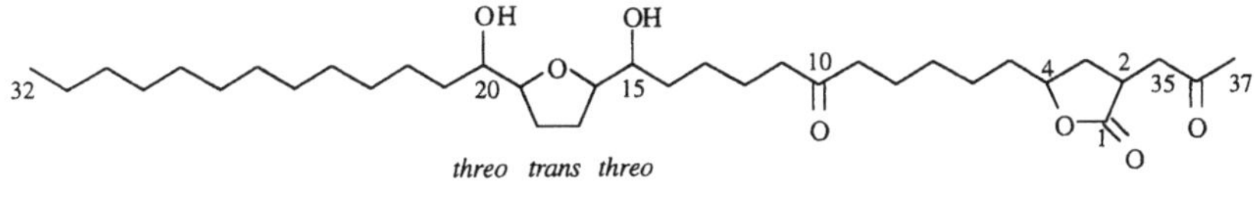

$C_{35}H_{62}O_7$; M = 594
mp = 106–108 °C (*126*)
$[\alpha]_D$ = + 29° (c = 0.59, $CHCl_3$) = + 33.3° (c = 0.4, MeOH) (*126*)
IR (*126*)
^{1}H nmr (300 MHz) (*126*); ^{13}C nmr (75 MHz) (*126*)
EIms (*126*)
acetyl derivative: ^{1}H nmr (300 MHz) (*126*), EIms (*126*)

The argument that this acetogenin should be a diastereoisomer of isoannonacinone at the level of the tetrahydrofuran pattern is insubstantial! The only proof points made are the differences of $[\alpha]_D$ and IR

Otivarin = revised structure of dihydrocherimolin (*4*)

Annona cherimolia (seeds) (*4, 16, 127*)

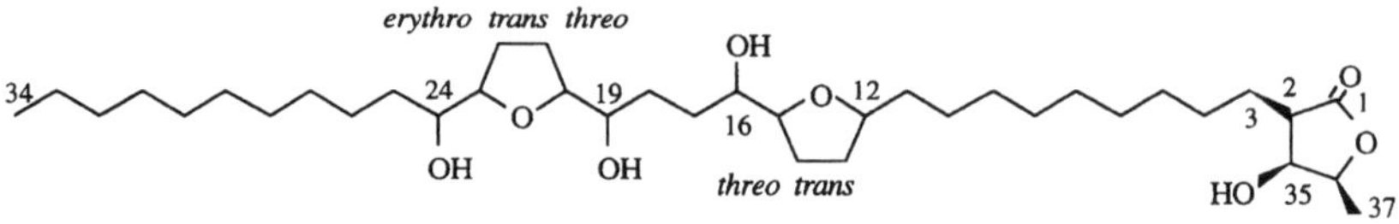

$C_{37}H_{68}O_8$; M = 640
$[\alpha]_D = +13°$ (c = 0.15, MeOH) (*16*)
IR (*16*)
^{1}H nmr (400 MHz) (*4, 16*); ^{13}C nmr (100 MHz) (*16*)
CIms (*16*)
tetraacetate: IR (*4*), ^{1}H nmr (400 MHz) (*4*)

Panalicin

Uvaria narum (root bark) (*139*)

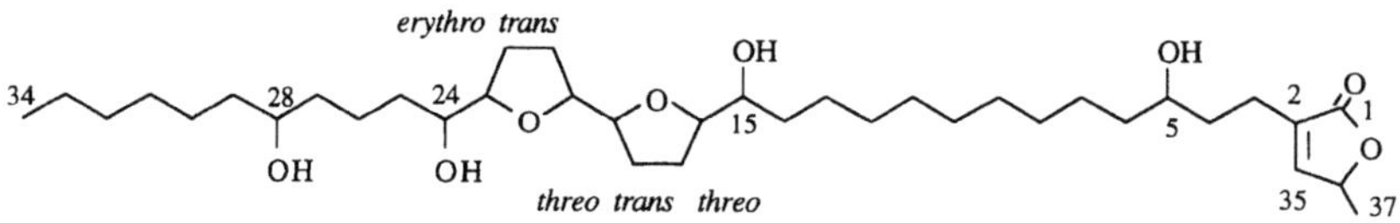

$C_{37}H_{66}O_8$; M = 638
IR (*139*)
CIms (*139*), FABms (*139*), EIms (*139*)
^{1}H nmr (200 MHz) (*139*); ^{13}C nmr (50 MHz) (*139*)
TMSi derivative: EIms (*139*)

Parvifloracin

Asimina parviflora (twigs) (*43*)

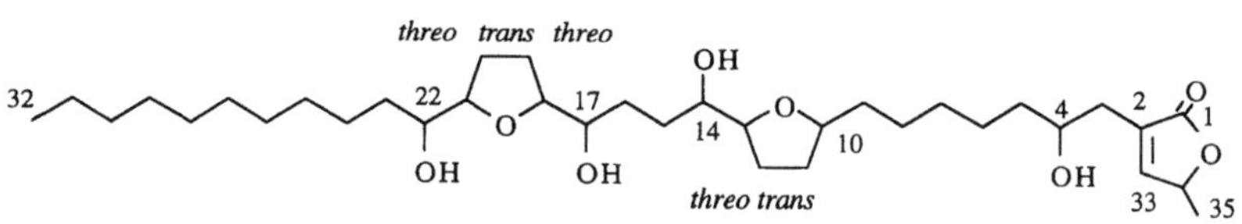

$C_{35}H_{62}O_8$; M = 610
$[\alpha]_D$ = + 18.75° (c = 0.08, EtOH) (*43*)
IR (*43*); UV (*43*)
^{1}H nmr (500 MHz) (*43*); ^{13}C nmr (125 MHz) (*43*)
FABms (*43*), CIms (*43*), EIms (*43*)
tetraacetate: ^{1}H nmr (200 MHz) (*43*); EIms (*43*)
tetra-TMSi derivative, tetra-TMSi$_{d9}$ derivative: EIms (*43*)

Parviflorin or **Squamocin-E** (*64*) = Atemoyacin-A (*77*)

Annona atemoya (seeds) (*77*) *Annona bullata* (bark) (*79*); *Annona squamosa* (seeds) (*64*); *Asimina parviflora* (twigs) (*43*); *Asimina triloba* (seeds) (*73*)

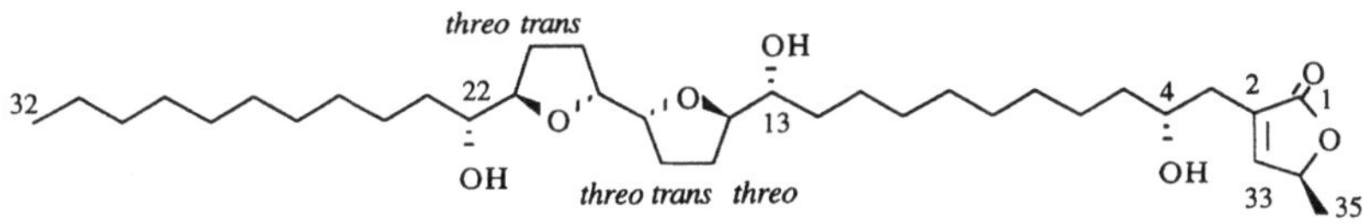

$C_{35}H_{62}O_7$; M = 594
mp = 48–50 °C (*64*)
$[\alpha]_D$ = + 18.33° (c = 0.06, EtOH) (*43*), $[\alpha]_D$ = + 20.9° (c = 0.25, MeOH) (*64*)
IR (*43, 64*); UV (*43*)
^{1}H nmr (500 MHz) (*43, 64*); ^{13}C nmr (125 MHz) (*43, 64*),
HRCIms (*43, 79*), CIms (*43*), EIms (*43, 64*), HRFABms (*64*)
triacetate: ^{1}H nmr (500 MHz) (*43, 64*); CIms (*43*), EIms (*43*)
tri-TMSi derivative, tri-TMSi$_{d9}$ derivative: CIms (*43*), EIms (*43*)
(*R*)- and (*S*)-tri-(*R*)-MTPA esters: ^{1}H nmr (500 MHz) (*79*), → 13*R*, 22*R* (*64*) → 34*S*, 4*R* (*79*)
absolute stereochemistry: 4*R*, 13*R*, 14*R*, 17*R*, 18*R*, 21*R*, 22*R*, 34*S*

Plagionicin-A

Polyalthia plagioneura (seeds) (*140*)

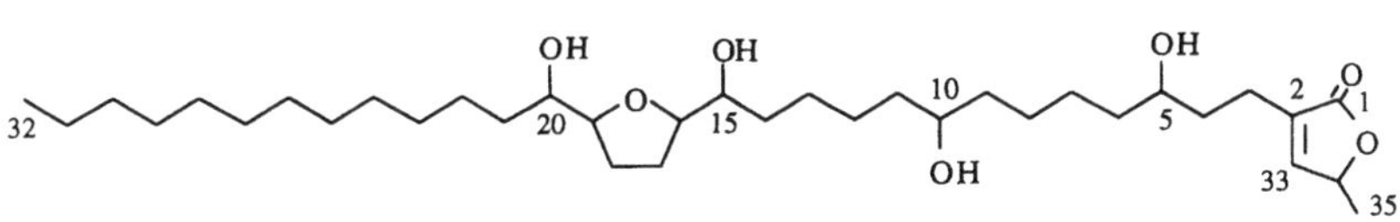

$C_{35}H_{64}O_7$; M = 596
mp = 73–75 °C (*140*)

$[\alpha]_D = +16.24°$ (c = 1.13, $CHCl_3$) (*140*)
IR (*140*)
^{1}H nmr (400 MHz) (*140*); ^{13}C nmr (75 MHz) (*140*)
EIms (*140*)
tetraacetate: ^{1}H nmr (400 MHz) (*140*); EIms (*140*)

Purpureacin-2

Annona purpurea (leaves) (*85*)

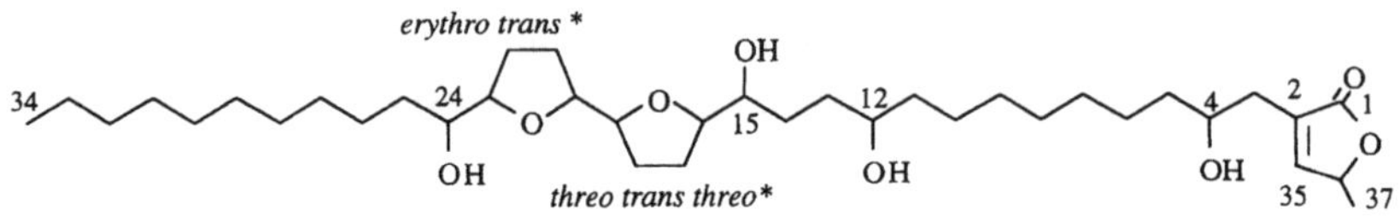

$C_{37}H_{66}O_8$; M = 638
$[\alpha]_D = +6.5°$ (c = 0.17, MeOH) (*85*)
IR (*85*); UV (*85*)
^{1}H nmr (200 MHz) (*85*); ^{13}C nmr (50 MHz) (*85*)
CIms (*85*), EIms (*85*)
TMSi derivative: EIms (*85*)

Reticulacinone

Annona reticulata (barks) (*131*)

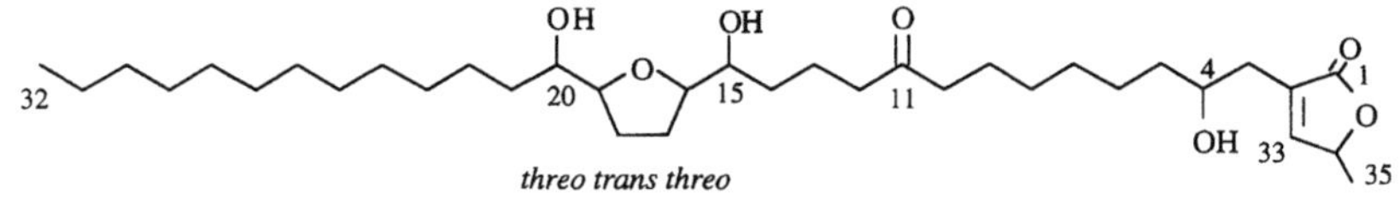

$C_{35}H_{62}O_7$; M = 594
^{1}H nmr (400 MHz) (*131*); ^{13}C nmr (50 MHz) (*131*)
CIms (*131*), FAB-Li ms (*131*), EIms (*131*)

Reticulatacin = Uvariamicin-II (*93, 137, 141*)

Annona reticulata (bark) (*142*), (seeds) (*143*); *Annona bullata* (bark) (mixture with uvariamicin-I, -III and -IV) (*93*); *Uvaria hookeri* (mixture with uvariamicin-I and -III) (root bark) (*137*); *Uvaria narum* (mixture with uvariamicin-I and -III) (*141*), (root bark) (*137*); synthesis (*144, 155*)

---

* We have proposed the relative stereochemistry shown

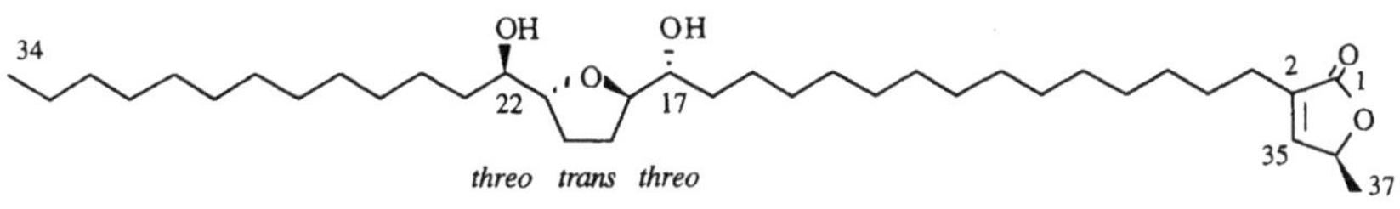

$C_{37}H_{68}O_5$; M = 592
mp = 80–80.5 °C (*142*), mp = 79–81 °C (*144*), mp = 80–81 °C (*155*)
$[\alpha]_D$ = + 26° (c = 0.5, MeOH) (*142*), $[\alpha]_D$ = + 24° (c = 0.05, MeOH) (*144*), $[\alpha]_D$ = +25.6° (c = 0.5, $CHCl_3$) (*155*)
CD (*142*)
IR (*142*); UV (*142*)
^1H nmr (500 MHz) (*142*); ^{13}C nmr (125 MHz) (*142*)
CIms (*142*), HRCIms (*142*), EIms (*142*)
diacetate: CIms (*142*), EIms (*142*), ^1H nmr (500 MHz) (*142*); ^{13}C nmr (125 MHz) (*142*)
tri-TMSi: EIms (*142*)
(*R*)- and (*S*)-bis-MTPA esters: ^1H nmr (500 MHz) (*46*), ^{19}F nmr (*46*) → 17*R*, 22*R* (*46*)
total synthesis (*144*, *155*)
synthesis of 17,18-diepi-reticulatacin (*144*): mp = 79–81 °C (*144*); $[\alpha]_D$ = + 24° (c = 0.05, MeOH) (*144*)
absolute stereochemistry: 17*R*, 18*R*, 21*R*, 22*R*, 35*S*

Reticulatain-1 = Neoreticulatacin-A (*104*)

Annona atemoya (seeds) (*63*); *Annona reticulata* (seeds) (*143*); *Annona squamosa* (seeds) (*104*)

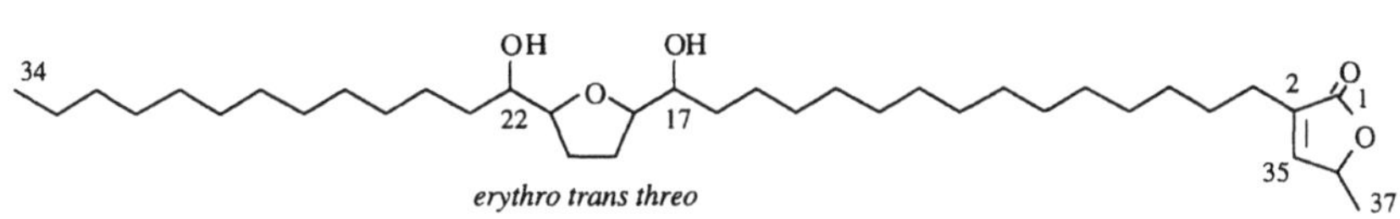

$C_{37}H_{68}O_5$; M = 592
$[\alpha]_D$ = + 22° (c = 1, $CHCl_3$) (*143*), $[\alpha]_D$ = + 23.61° (c = 0.61, $CHCl_3$) (*104*)
IR (*104*, *143*); UV (*143*)
^1H nmr (200 MHz) (*143*), (300 MHz) (*104*); ^{13}C nmr (50 MHz) (*143*), (75 MHz) (*104*)
CIms (*143*); EIms; (*104*, *143*); LSIms, LSIms (CID *B*/*E* + Li) (*143*)
diacetate: ^1H nmr (300 MHz) (*104*), EIms (*104*)

Reticulatain-2

Annona reticulata (seeds) (*143*)

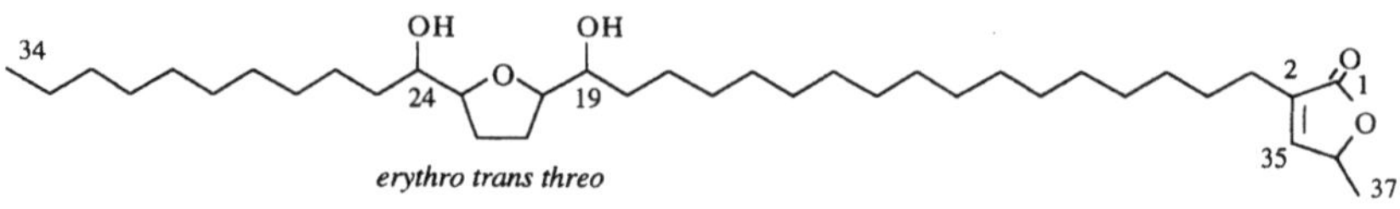

$C_{37}H_{68}O_5$; M = 592
$[\alpha]_D = +28$ (c = 1, $CHCl_3$) (*143*)
IR (*143*); UV (*143*)
^{1}H nmr (200 MHz) (*143*); ^{13}C nmr (50 MHz) (*143*)
CIms (*143*), EIms (*143*), LSIms, LSIms (CID *B*/*E* + Li) (*143*)

Reticulatamol

Annona reticulata (seeds) (*145*); synthesis (*145*)

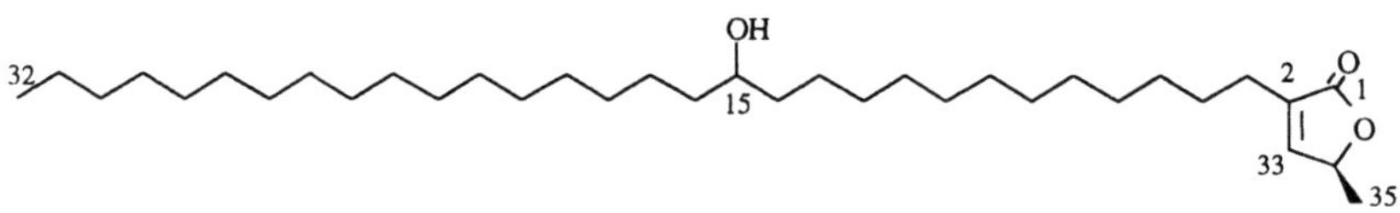

$C_{35}H_{66}O_3$; M = 534
mp = 92–94 °C (heptane) (*145*)
$[\alpha]_D = +2°$ (c = 1.0, $CHCl_3$) (*145*), $[\alpha]_D$ synthetic 15-(*SR*)-reticulatamol = + 3° (c = 1.0, $CHCl_3$) (*145*)
IR (*145*); UV (*145*)
^{1}H nmr (200 MHz) (*145*); ^{13}C nmr (50 MHz) (*145*)
FAB-Li ms, CIms (*145*)
acetate: (*145*)
synthesis 15*R*, *S* reticulatamol (*145*)
absolute stereochemistry: 34*S*

Reticulatamone

Annona reticulata (seeds) (*143*); synthesis (*143*, *145*)

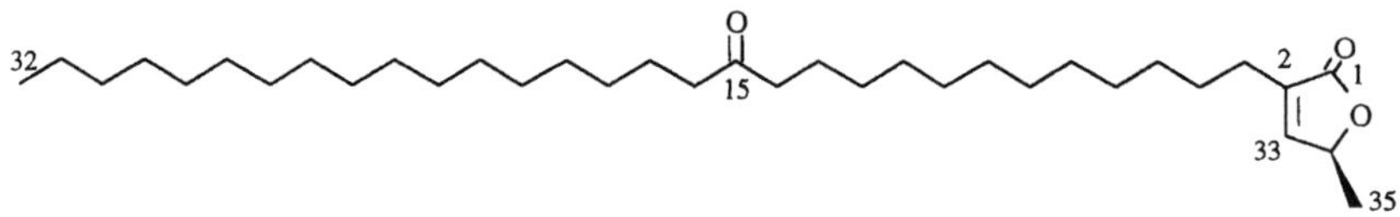

$C_{35}H_{64}O_3$; M = 532
mp = 83–85 °C (*143*); synthetic: mp = 83–84 °C (*143*)

$[\alpha]_D = +12°$ (c = 1, $CHCl_3$) (*143*), synthetic: $[\alpha]_D = +12°$ (c = 0.40, $CHCl_3$)
IR (*143*); UV (*143*)
^1H nmr (200 MHz) (*143*); ^{13}C nmr (50 MHz) (*143*)
CIms (*143*); EIms; (*143*); LSIms + Li, (*143*)
absolute stereochemistry: 34*S*

Rioclarin

Rollinia membranacea (seeds) (*128*)

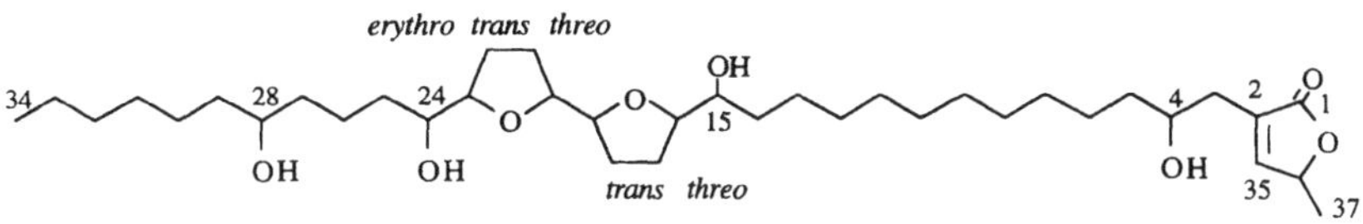

$C_{37}H_{66}O_8$; M = 638
$[\alpha]_D = +20°$ (c = 0.4, MeOH) (*128*)
IR (*128*)
^1H nmr (200 MHz) (*128*); ^{13}C nmr (50 MHz) (*128*)
CIms (*128*)
tetraacetate: IR (*128*), ^1H nmr (200 MHz) (*128*), CIms (*128*)

Rolliniastatin-1 = 4-hydroxy-25-Desoxyneorollinicin (false structure) (*146*), rev. struct. (*12*)

Annona purpurea (leaves) (*85*); *Annona reticulata* (*147*); *Rollinia membranacea* (seeds) (*128*); *Rollinia mucosa* (seeds) (*148*); *Rollinia papillionella* (roots) (*146*); *Rollinia sericea* (seeds) (*149*); synthesis (*150*)

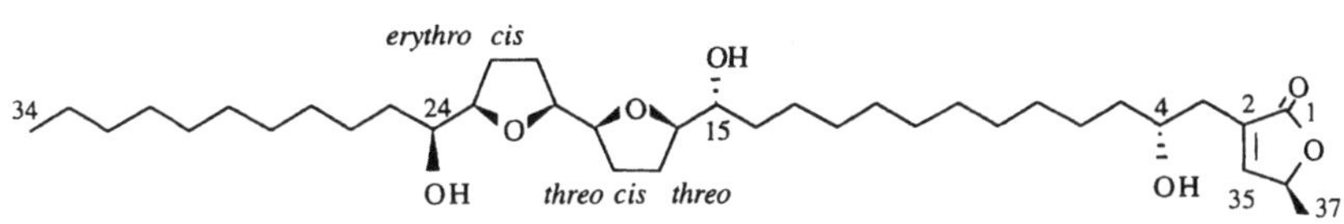

$C_{37}H_{66}O_7$; M = 622
mp = 81–83 °C (acetone) (*148*), mp = 78–80 °C (acetone) (*150*), mp = 25 °C (*146*), mp = 25 °C (*149*)
$[\alpha]_D = +19°$ (c = 1.03, $CHCl_3$) (*147*), $[\alpha]_D = +25°$ (c = 0.51, CH_2Cl_2) (*150*), $[\alpha]_D = +20.6$ (c = 0.31, MeOH) (*85*), $[\alpha]_{589} = +25.2°$, $[\alpha]_{578} = +26.2°$, $[\alpha]_{546} = +30.1°$, $[\alpha]_{436} = +48°.5$, $[\alpha]_{365} = +76.7°$ (c = 1.03, CH_2Cl_2) (*148*); CD (*83*)
IR (*146, 148*); UV (*85, 148*)

^1H nmr (300 MHz) (*65*, *146*, *148*), (500 MHz) (*150*); ^{13}C nmr (50 MHz) (*85*, *148*), (75 MHz) (*146*, *150*)
SO-SIms (*148*), SP-HRSIms (*148*), HREIms (*148*), FABms (*146*), HRFAB-Lims (*146*), CIms (*146*), EIms (*146*)
triacetate: ^1H nmr (300 MHz) (*146*), (200 MHz) (*128*), CIms (*128*)
RX: 15-O-*p*-bromophenylurethane (*148*)
relative stereochemistry C(4)-C(36) *RS* (*45*)
(*R*)- and (*S*)-tri-MTPA esters: ^1H nmr (500 MHz); ^{19}F nmr → 4*R*, 15*R*, 24*S* (*46*)
total synthesis (*150*)
absolute stereochemistry: 4*R*, 15*R*, 16*R*, 19*S*, 20*S*, 23*R*, 24*S*, 36*S*

Rolliniastatin-2 or **Bullatacin** = Squamocin-G (*64*), = 14-hydroxy-25-Desoxyrollinicin (false struct.) (*151*), rev. struct. (*5*, *12*), = Annonin-VI (false struct.) (*14*), rev. struct. (*12*)

Annona atemoya (seeds), (*78*, *129*); *Annona bullata* (bark) (*83*); *Annona cherimolia* (seeds) (*16*, *130*); *Annona glauca* (roots) (*24*), (seeds) (*120*); *Annona purpurea* (leaves) (*85*); *Annona reticulata* (bark) (*142*), (leaves) (*27*), (stem bark) (*131*); *Annona senegalensis* (seeds) (*68*); *Annona squamosa* (bark) (*84*), (seeds) (*14*, *64*); *Asimina parviflora* (twigs) (*43*); *Asimina triloba* (bark) (*72*), (seeds) (*73*); *Rollinia mucosa* (seeds) (*152*), synthesis (*74*, *153*)

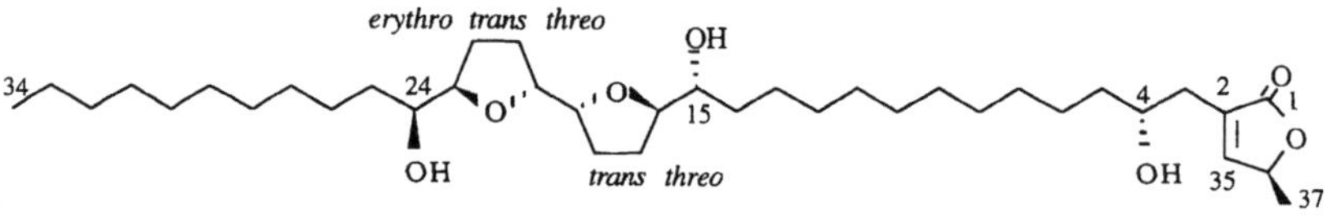

The name of rolliniastatin-2 should be retained since the correct stereochemistry (*130*), was corrected *threo-trans-threo-trans-erythro* (*12*, *130*), and reported before the stereochemistry of bullatacin (*6*, *46*). However the name bullatacin can also be used.

$C_{37}H_{66}O_7$; M = 622
mp = 73–76 °C (*152*), mp = 69–70 °C (*83*), mp = 69–70 °C (*84*), mp = 72–74° (*27*), mp = 77–78 °C (*64*), mp = 68.5–69 °C (*74*), mp = 77–78 °C (*14*), mp = 78–83 °C (*153*)
$[\alpha]_D = +5.3°$ (c = 0.23, $CHCl_3$) (*152*), $[\alpha]_{589} = +17°$ (*84*), $[\alpha]_D = +8°$ (c = 0.2, $CHCl_3$) (*27*), $[\alpha]_D = +12.8°$ (c = 0.49, MeOH) (*85*), $[\alpha]_D = +15°$ (0.27, $CHCl_3$) (*68*), $[\alpha]_D = +28.5°$ (c = 0.50, MeOH) (*64*), $[\alpha]_D = +12.8°$ (c = 0.26, $CHCl_3$) (*74*), $[\alpha]_{589} = +13°$, $[\alpha]_{578} = +14.7°$, $[\alpha]_{546} = +19.04°$, $[\alpha]_{436} = +36.63°$, $[\alpha]_{365} = +66.99°$ (c = 0.04, $CHCl_3$) (*83*), $[\alpha]_D = +14.2°$

(c = 0.23, $CHCl_3$) (*14*), $[\alpha]_D = +8.22°$ (c = 0.05, $CHCl_3$) (*153*)
CD (*64, 83*) (negative Cotton effect → 36*S*) (*64*)
IR (*24, 64, 83, 152*); UV (*83, 152*)
^1H nmr ($CDCl_3$ and C_6D_6, 470 MHz) (*83*), (500 MHz) (*14, 64, 85*), (400 MHz) (*152*), (200 MHz) (*24, 131*); ^{13}C nmr (50 MHz) (*83*), (125 MHz) (*14, 64, 85*), (100 MHz) (*152*), (50 MHz) (*24, 131*)
HREIms (*152*), FABms (*14*), FAB-Li ms (*131*), EIms (*64, 83, 131*), CIms, (*24, 83, 85*)
triacetate: ^1H nmr (470 MHz) (*83*), (500 MHz) (*14, 64*); EIms (*83*)
2,35-dihydrobullatacin: EIms (*83*), CIms (*83*); ^1H nmr (C_6D_6, 470 MHz) (*83*)
bullatacin (rolliniastatin-2) (KOH/*t*-BuOH) → bullatacinone (*83*)
relative stereochemistry of tetrahydrofuran pattern *threo-trans-threo-trans-erythro* (*130*)
absolute stereochemistry 36-*S* deduced from negative Cotton effect (*64*)
(*R*)- and (*S*)-bis-MTPA esters: ^1H nmr (500 MHz) (*46*); ^{19}F nmr (*46*) → 15*R*, 24*S* (*46*)
tri-(*R*)-MTPA ester: ^1H nmr (500 MHz) (*64*) → 15*R*, 16*R*, 19*R*, 20*R*, 23*R*, 24*S* (*64*)
total synthesis (+)-Rolliniastatin-2 (*74, 153*), (−)-Rolliniastatin-2 (*217*)
absolute stereochemistry: 4*R*, 15*R*, 16*R*, 19*R*, 20*R*, 23*R*, 24*S*, 36*S*

Rollinone = Isorolliniastatin-1

Rollinia papillonella (roots) (*2*), revised structure (*154*); *Rollinia membranacea* (seeds) (*111*)

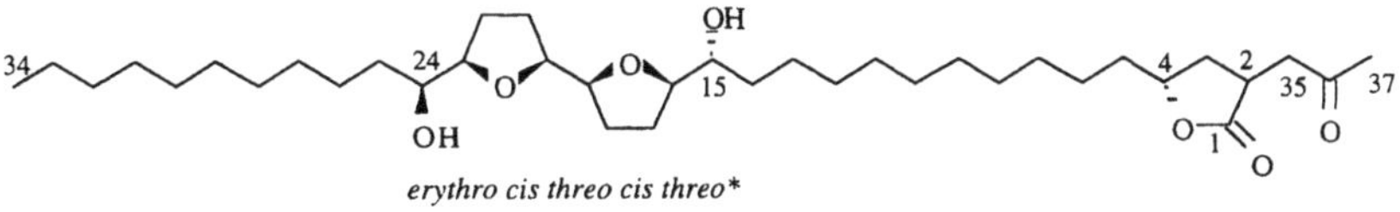

*erythro cis threo cis threo**

mixture of 2,4-*cis*- and 2,4-*trans*-isomers

$C_{37}H_{66}O_7$; M = 622
mp = 54–56 °C (*2*)
$[\alpha]_D = +25°$ (c = 0.1371, $CHCl_3$) (*2*)
IR (*2*); UV (*2*)
^1H nmr ($CDCl_3$ and C_6D_6), (90 MHz) (*2*), (300 MHz) (*154*); ^{13}C nmr (25 MHz) (*2*), (75 MHz) (*154*)

* The relative and absolute stereochemistries can be deduced from the translactonization of rolliniastatin-1 to rollinone: 4*R*, 15*R*, 16*R*, 19*S*, 20*S*, 23*R*, 24*S*

EIms, (*2*), CIms (*2*)
diacetate: ^{1}H nmr (300 MHz) (*154*)
hemisynthesis: translactonization from 4-hydroxy-25-desoxy-neorollinicin (rolliniastatin-1) (KOH/*t*-BuOH) (*154*)

Senegalene

Annona senegalensis (seeds) (*68*)

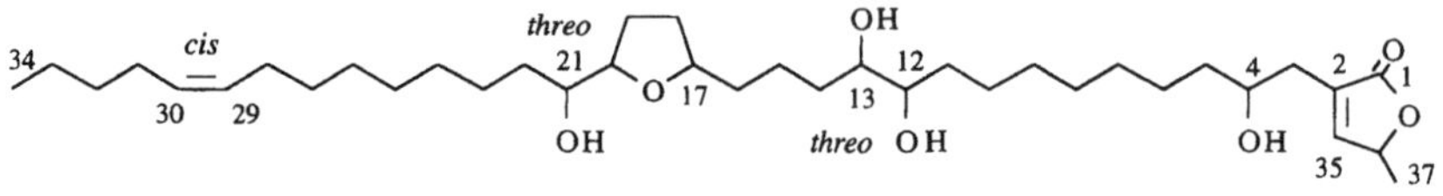

$C_{37}H_{66}O_7$; M = 622
$[\alpha]_D = +16°$ (c = 0.21, $CHCl_3$) (*68*)
IR (*68*); UV (*68*)
^{1}H nmr (200 MHz) (*68*); ^{13}C nmr (50 MHz) (*68*)
HRCIms (*68*), CIms (*68*), EIms (*68*), FAB-Li (*68*)
tetraacetate: ^{1}H nmr (200 MHz) (*68*); FABms (*68*)
2,35,29,30-tetrahydrosenegalene: CIms (*68*); FABms (*68*)
2,35,29,30-tetrahydro-tetraacetyl senegalene: CIms (*68*); FABms (*68*)
12,13-acetonide: CIms (*68*)
29,30-epoxy-tetraacetyl-: CIms (*68*)

Solamin = 4-Desoxyhowiicin-B (*41*)

Annona glabra (bark) (*36*); *Annona muricata* (seeds) (*41, 65*), (stem bark) (*109*); *Annona reticulata* (leaves) (*27*); synthesis (*144, 155, 156*)

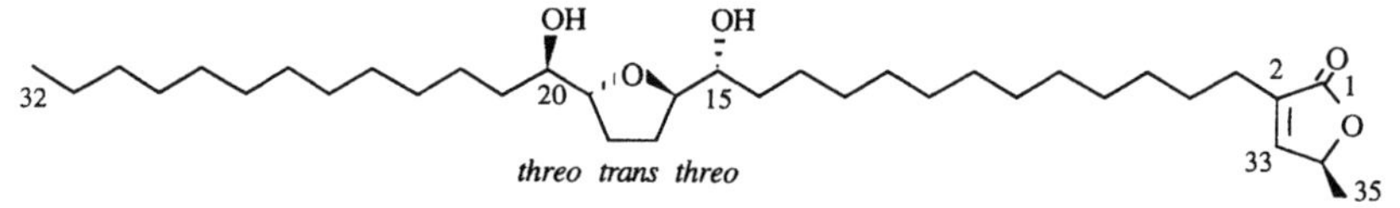

$C_{35}H_{64}O_5$; M = 564
mp = 64–68 °C (*65*), mp = 76–77 °C (*155*), mp = 65–68 °C (*27*), mp = 66–69 °C (*144*), mp = 78–79 °C (*156*), mp = 57–59 °C (*36*), mp = 62–64 °C (*41*)
$[\alpha]_D = +21.2°$ (c = 0.16, MeOH) (*65*), $[\alpha]_D = +22°$ (c = 0.2, MeOH) (*155*), $[\alpha]_D = +15.6°$ (c = 0.15, $CHCl_3$) (*27*), $[\alpha]_D = +22°$ (c = 0.1, MeOH) (*144*), $[\alpha]_D = +22.2°$ (c = 0.30, MeOH) (*156*), $[\alpha]_D = +19°$ (c = 0.30, MeOH) (*36*), $[\alpha]_D = +25.08°$ (c = 1.08, $CHCl_3$) (*41*)
IR (*36, 41, 65*); UV (*36, 65*)
^{1}H nmr (200 MHz) (*65, 109*), (400 MHz) (*34, 41*); ^{13}C nmr (50 MHz) (*65, 109*), (100 MHz) (*41*)

CIms (*65*), EIms (*36, 41, 109*), FAB-Li ms (*109*), FABms (*36*)
diacetate: IR (*65*); ^{1}H nmr (200 MHz) (*41, 65*); HRCIms (*65*), CIms (*65*), EIms (*41*)
2,33-dihydrosolamin: ^{1}H nmr (200 MHz) (*65*); CIms (*65*)
TMSi: EIms (*109*)
synthesis of solamin (*threo-trans-threo*) (*144, 155, 156*)
synthesis of 15,16-diepi-solamin (*threo-cis-threo*) (*144*): mp = 66–69 °C (*144*), $[\alpha]_D = +22°$ (c = 0.1, MeOH) (*144*)
absolute stereochemistry: 4*R*, 15*R*, 16*R*, 19*R*, 20*R*, 34*S*

Squamocin = Annonin-I, = Rollinicin (inexact struct.) (*2, 3*) (revised struct.) (*5, 12*)

Annona atemoya (seeds) (*78, 129*); *Annona bullata* (bark) (*90, 93*); *Annona cherimolia* (seeds) (*16, 130*), (roots) (*17*); *Annona glabra* (seeds) (*67, 92*); *Annona reticulata* (seeds) (*147*); *Annona senegalensis* (seeds) (*68*); *Annona squamosa* (seeds) (*10, 64, 138, 157, 158*); *Asimina triloba* (stem bark) (*82*); *Rollinia membranacea* (seeds) (*128*); *Rollinia papillionella* (roots) (*2, 3*); *Rollina sericea* (seeds) (*149*); *Uvaria hookeri* (root bark) (*137*); *Uvaria narum* (root bark) (*137, 139*)

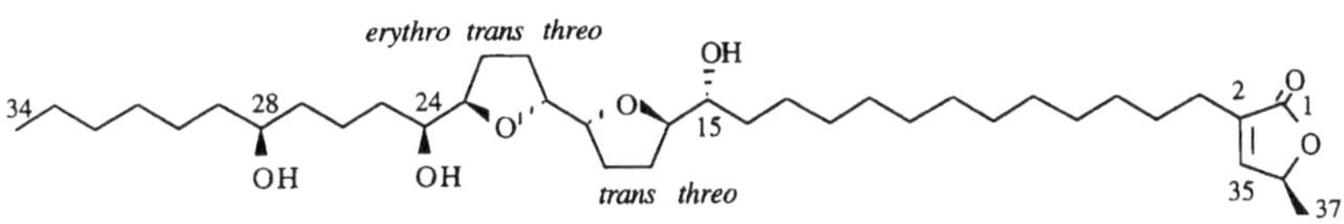

$C_{37}H_{66}O_7$; M = 622
mp = < 30° (*157*), mp = < 30° (*139*), mp = 43 °C (*67*), mp = < 30° (*128*), mp = 30–32 °C (*3*), mp = < 30° (*149*)
$[\alpha]_D = +0.15°$ (c = 1.7, MeOH) (*157*), $[\alpha]_D = +21.25°$ (c = 0.4, $CHCl_3$) (*90*), $[\alpha]_D = +16.2°$ (c = 1.01, $CHCl_3$) (*147*), $[\alpha]_D = +14°$ (c = 0.13, $CHCl_3$) (*68*), $[\alpha]_D = +6.8°$, (c = 0.2411, $CHCl_3$) (*3*)
CD (*64, 90, 173*): negative Cotton effect → 36*S* (*64, 173*)
IR (*3, 64, 90, 139, 157*); UV (*3, 157*)
^{1}H nmr (300 MHz) (*139*), (360 MHz) (*3, 157*), (500 MHz) (*15*), (C_6D_6 and $CDCl_3$, 500 MHz) (*90*); ^{13}C nmr (*157*), (22.5 MHz) (*3*), (50 MHz) (*139*), (125 MHz) (*15, 90*), (new assignment) (*64*)
HRCIms (*3*), EIms (*2, 139, 157*), FABms (*90, 139, 157*), HRFABms (*90*), CIms (*2*), ms (*159*): structural studies of polyhydroxy-bis-tetrahydrofuran using a combination of chemical derivatization and precursor-ion scanning mass spectrometry (*159*).

Jones oxidation → 5-oxoundecanoic acid: [1]H nmr (360 MHz) (*157*), methyl ester EIms (*157*) + 3-(12-carboxy-dodecanyl)-5-methylfuran-2-(5H)-one: [1]H nmr (360 MHz) (*157*), methyl ester EIms (*157*)
chemical degradation → → → methyl lactate (R)-MTPA ester identified to (2*S*)-antipode → 36*S* configuration (*64*)
triacetate: IR (*3*), ^{1}H nmr (200 MHz) (*128*), (300 MHz) (*139*), (500 MHz) (*90*); ^{13}C nmr (50 MHz) (*139*); CIms (*3, 90, 128*)
tri-TMSi derivative: CIms (*90, 3*), EIms (*90, 139*)
tri-MTPA ester: [1]H nmr (500 MHz) (*15*)
relative stereochemistry fixed as *threo-trans-threo-trans-erythro* (*130*)
24,28-formaldehyde acetal derivative: HRFABms, TMSi derivative (EIms), ^{1}H nmr (500 MHz), ^{13}C nmr (125 MHz) → 24,28-*cis* (*18*)
(*R*)- and (*S*)-MTPA esters of 28,24-formaldehyde acetal derivative: [1]H nmr (500 MHz) → 15*S* (*18*)
(*R*)- and (*S*)-MTPA esters; ^{1}H nmr permitted determination of the absolute stereochemistry at C-28 thanks to the chemical shift of the terminal methyl (*18*)
advanced Mosher method → 28*S* (*19*)
RX of derivate of squamocin (resulting from the opening of the dihydrolactone and salt formation of the carboxyl: K salt: → relative stereochemistry and two possible absolute configurations: 2*R*, 15*R*, 16*R*, 19*R*, 20*R*, 23*R*, 24*S*, 28*S*, 36*S* or 2*S*, 15*S*, 16*S*, 19*S*, 20*S*, 23*S*, 24*R*, 28*R*, 36*R* (*158*)
The first alternative has been shown to be correct: 15*R*, 16*R*, 19*R*, 20*R*, 23*R*, 24*S*, 28*S*, 36*S* (*19*)
oxidative cleavage derivative: squamostanal-A (*160*)

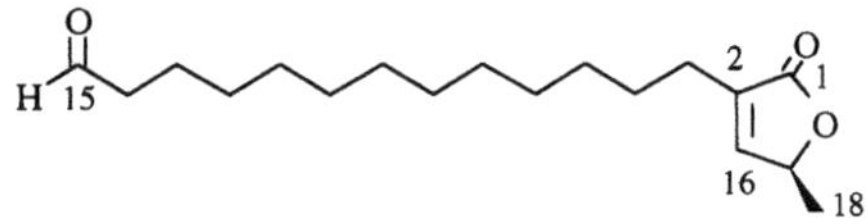

$C_{18}H_{30}O_3$; M = 294
CD: max à 237 nm (Δε 0.39, MeOH) → absolute stereochemistry 17*S* (*160*)
[1]H nmr (270 MHz) (*160*); ^{13}C nmr (67.5 MHz) (*160*), HRFABms (*160*), EIms (*160*)

Squamocin-B

Annona squamosa (seeds) (*64*)

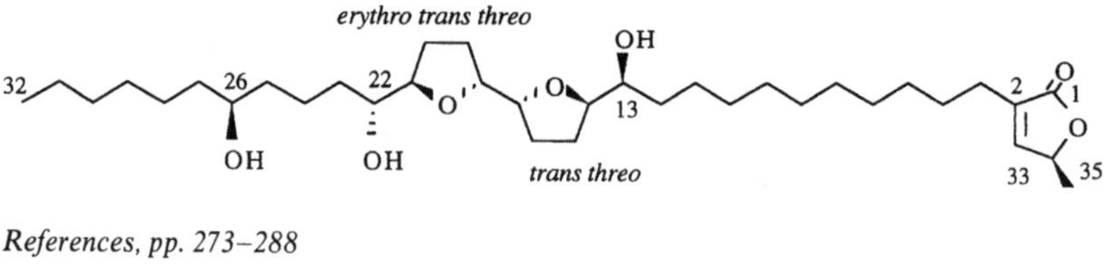

$C_{35}H_{62}O_7$; M = 594
$[\alpha]_D = +27.6°$ (c = 0.2, MeOH) (*64*)
IR (*64*)
^{1}H nmr (500 MHz) (*64*); ^{13}C nmr (125 MHz) (*64*)
HRFABms (nitrobenzylalcohol) (*64*), FABms (*64*)
tri-(*R*)-MTPA ester: ^{1}H nmr (500 MHz) (*64*), absolute stereochemistry deduced from analogy with squamocin (*64*)
advanced Mosher method → 26*S* (*19*)
absolute stereochemistry: 13*R*, 14*R*, 17*R*, 18*R*, 21*R*, 22*S*, 26*S*, 34*S*

Squamocin-D (*64*) or **Asiminacin** (*76*)

Squamocin-E (*64*) or **Parviflorin** (*43*) or **Atemoyacin-A** (*77*)

Squamocin-F

Annona squamosa (seeds) (*64*)

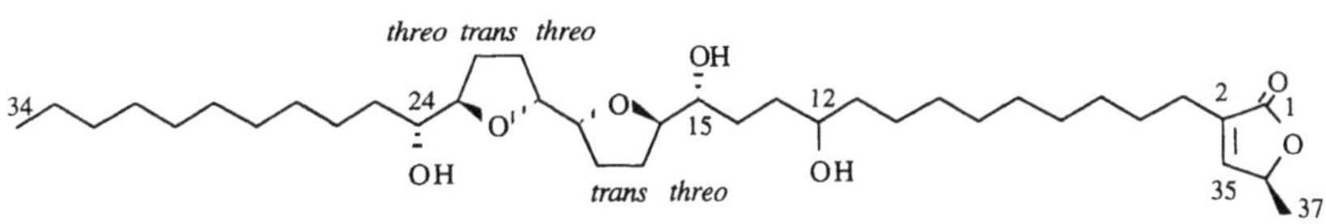

$C_{37}H_{66}O_7$; M = 622
$[\alpha]_D = +21°$ (c = 0.58, MeOH) (*64*)
IR (*64*)
^{1}H nmr (500 MHz) (*64*); ^{13}C nmr (125 MHz) (*64*)
HRFABms (nitrobenzylalcohol) (*64*), EIms (*64*)
triacetate: ^{1}H nmr (500 MHz) (*64*)
N,N-dimethylethylenediamine derivative: precursor Ion Scanning ms (*64*)
tri-(*R*)-MTPA ester: ^{1}H nmr (500 MHz); absolute stereochemistry presumed by analogy with the other squamocins (*64*)

Squamocin-I

Annona squamosa (seeds) (*64*)

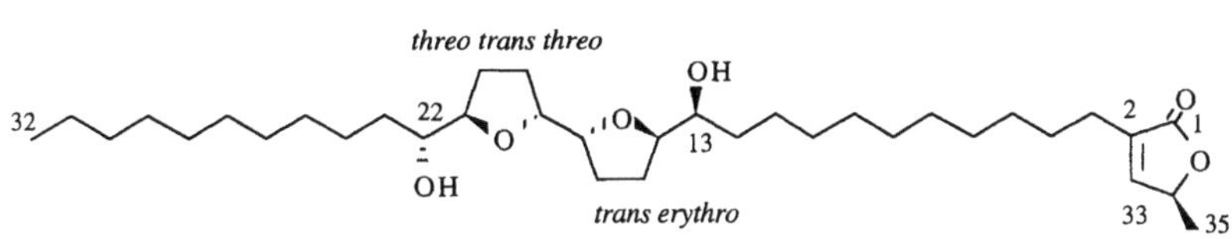

$C_{35}H_{62}O_6$; M = 578
mp = 68.5–71 °C (*64*)
$[\alpha]_D$ = + 22.2° (c = 0.50, MeOH) (*64*)
CD (*64*): stereochemistry 34*S* deduced from the negative Cotton effect (*64*)
IR (*64*); UV (*64*)
^{1}H nmr (500 MHz) (*64*); ^{13}C nmr (125 MHz) (*64*)*
HRFABms (nitrobenzylalcohol) (*64*), FABms (*64*), EIms (*64*)
diacetate: ^{1}H nmr (500 MHz) (*64*)
di-(*R*)-MTPA ester: ^{1}H nmr (500 MHz) → absolute stereochemistry 13*S*, 14*R*, 17*R*, 18*R*, 21*R*, 22*R* although it could not be established which of the two isomers, neoannonin (squamocin-J) and squamocin-I, has the 13*R*, 14*R*, 21*R*, 22*S* or 13*S*, 14*R*, 21*R*, 22*R* absolute stereochemistry (*64*)

Squamocin-K (*64*) or **Atemoyin** (*78*)

Squamocin-N

Annona squamosa (seeds) (*64*)

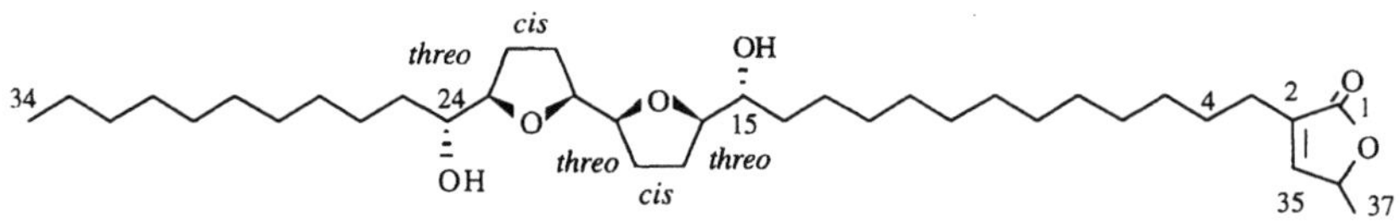

$C_{37}H_{66}O_6$; M = 606
$[\alpha]_D$ = + 40.6° (c = 0.43, MeOH) (*64*)
CD (*64*): stereochemistry 36*S* deduced from the negative Cotton effect (*64*)
IR (*64*)
^{1}H nmr (500 MHz) (*64*); ^{13}C nmr (125 MHz) (*64*)
HRFABms (*64*), EIms (*64*)
diacetate: ^{1}H nmr (500 MHz) (*64*)
di-(*R*)- and (*S*)-MTPA esters: ^{1}H nmr (500 MHz) → 15*R*, 24*R* (*64*)
absolute stereochemistry: 15*R*, 16*R*, 19*S*, 20*R*, 23*R*, 24*R*, 36*S*

The arguments for assigning the relative stereochemistry which is based on the δ values in the ^{1}H nmr spectra of the acetates are not very reliable since the ^{1}H nmr spectra of acetylsquamocin-N and acetylatemoyin (squamocin-K) are very close

---

* The ^{13}C nmr spectra of squamocin-I and of neoannonin (squamocin-J) are very close

Squamocinone = Squamocin-28-one (*139*)

Uvaria hookeri (root bark) (*137*); *Uvaria narum* (root bark) (*137, 139*)

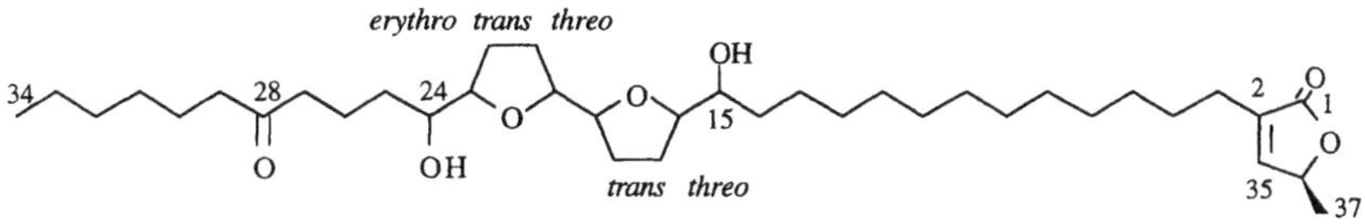

$C_{37}H_{64}O_7$; M = 620
IR (*139*)
^{1}H nmr (300 MHz) (*137, 139*); ^{13}C nmr (50 MHz) (*139*)
CIms (*137, 139*), FABms (*137, 139*), EIms (*137, 139*)
diacetate: ^{1}H nmr (500 MHz) (*139*); EIms (*139*)
di-TMSi derivative: EIms (*139*)

Squamone

Annona reticulata (leaves) (*27*); *Annona squamosa* (bark) (*84*)

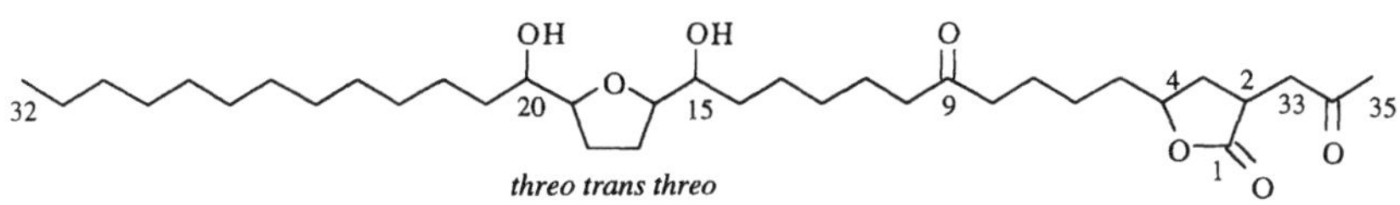

mixture of 2,4-*cis*- and 2,4-*trans*- isomers

$C_{35}H_{62}O_7$; M = 594
mp = 89 °C (*84*), mp = 87–89 °C (*27*)
$[\alpha]_D = +7.0°$ (c = 0.12, $CHCl_3$) (*27*)
IR (*84*)
^{1}H nmr (C_6D_6, 500 MHz) (*84*); ^{13}C nmr (50 MHz) (*84*)
HRCIms (*84*), EIms (*84*)
diacetate: ^{1}H nmr (200 MHz) (*84*); CIms (*84*), EIms (*84*)
di-TMSi and di-$TMSi_{d9}$ derivatives: CIms (*84*), EIms (*84*)
tetrahydrosquamone: ^{1}H nmr (500 MHz) (*84*); HRCIms (*84*); TMSi derivative: EIms (*84*)
hemisynthesis of squamone from annoreticuin-9-one (KOH/*t*-BuOH) (*27*)

Squamosinin A

Annona squamosa (seeds) (*169*)

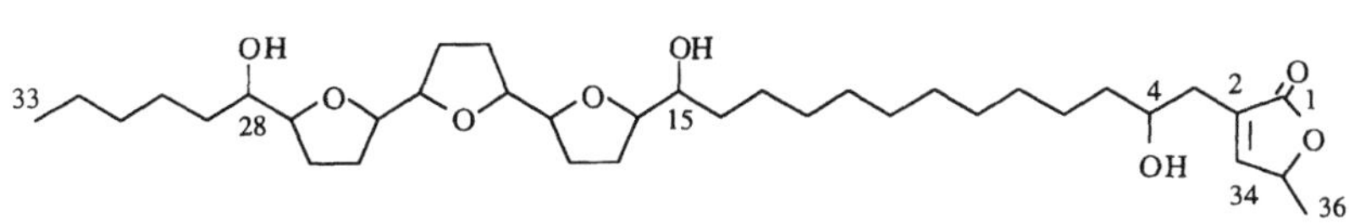

$C_{36}H_{62}O_8$; M = 622
$[\alpha]_D = +24.7°$ (c = 1.4, $CHCl_3$) (*169*)
IR (*169*)
^{1}H nmr (300 MHz) (*169*); ^{13}C nmr (75 MHz) (*169*)
EIms (*169*)
triacetate: ^{1}H nmr (300 MHz) (*169*); EIms (*169*)

This structure is doubtful since biogenetically, an acetogenin with a 36-carbon chain is improbable.

Squamostanal-A

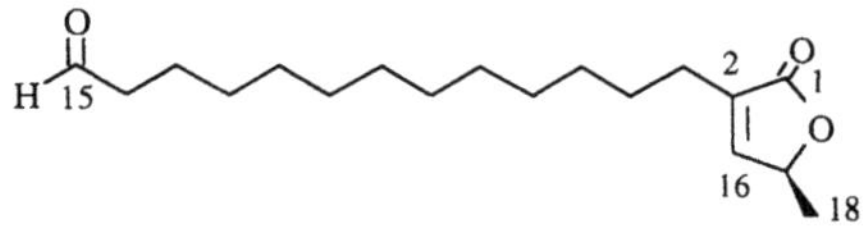

oxidative cleavage derivative from squamocin: see squamocin

Squamostatin-A or **Almunequin** (*17*), = Annonin XVI (inexact structure) (*10*), revised structure, (*6, 11, 12*), = Squamostatin-B "bis"* (*13, 14*)

Squamostatin-D

Annona squamosa (seeds) (*20*)

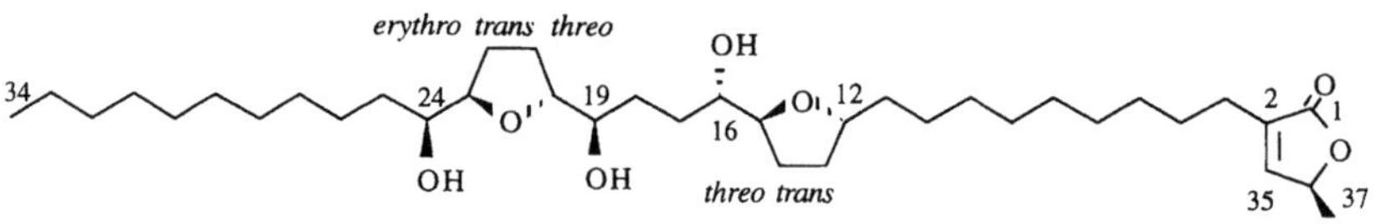

$C_{37}H_{66}O_7$; M = 622
mp = 112–113.5 °C (AcOEt) (*20*)
$[\alpha]_D = +7.9°$ (c = 0.51, MeOH) (*20*)
IR (*20*)
^{1}H nmr (500 MHz) (*20*); ^{13}C nmr (125 MHz) (*20*)
EIms (*20*), FAB ms (*20*), HRFAB ms (*20*)
tri-(*R*)-MTPA esters: ^{1}H nmr (500 MHz) (*20*)

The relative and absolute stereochemistry is based on the correspondence of the nmr data with those of squamostatin-B (= bullatalicin = cherimolin-1) (*20*); configuration 36*S* is assigned by analogy with other acetogenins of *Annona squamosa* (*20*)

Squamostatin-E = 4-Desoxycherimolin-2 (*92*)

Annona glabra (seeds) (*92*); *Annona squamosa* (seeds) (*20*)

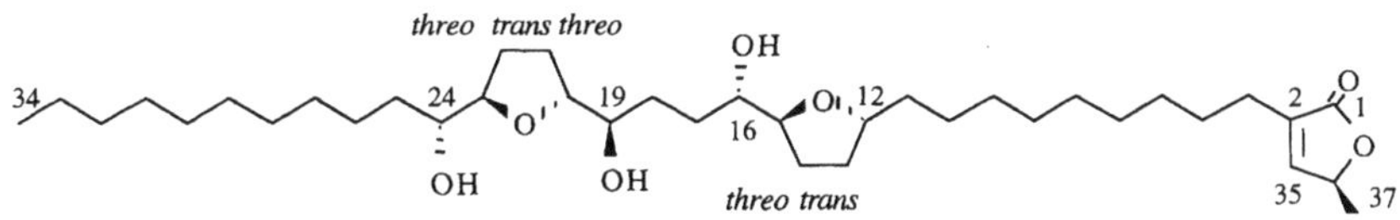

$C_{37}H_{66}O_7$; M = 622
mp = 105–106 °C (MeOH-H_2O) (*20*)
$[\alpha]_D$ = + 14.7° (c = 0.51, MeOH) (*20*)
IR (*20*)
^1H nmr (500 MHz) (*20*); ^{13}C nmr (125 MHz) (*20*)
EIms (*20*), FAB ms (*20*)
tri-(*R*)-MTPA ester: ^1H nmr (500 MHz) indicated the identity of the absolute stereochemistry of the tetrahydrofuranic pattern with squamostatin-C (= bullatanocin, = cherimolin-2) (*20*)
the 36*S* stereochemistry was assigned by analogy with the majority of *Annona squamosa* acetogenins (*20*)
absolute stereochemistry: 12*S*, 15*S*, 16*S*, 19*R*, 20*R*, 23*R*, 24*R*, 36*S*

Squamosten-A

Annona squamosa (seeds) (*161*)

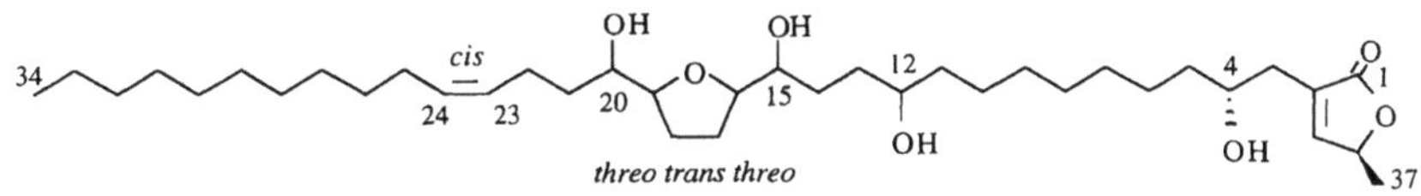

$C_{37}H_{66}O_7$; M = 622
mp = 64–67 °C (*161*)
$[\alpha]_D$ = + 9° (c = 0.1, MeOH) (*161*)
CD: stereochemistry 36*S* (*161*)
IR (*161*); UV (*161*)
^1H nmr (500 MHz) (*161*); ^{13}C nmr (125 MHz) (*161*)
EIms (*161*), FABms (*161*)
tetra-(*R*)-MTPA ester: ^1H nmr (500 MHz) → stereochemistry C-4*R* (*161*)
placement of double bond: treatment with $NaIO_4$ and $RuCl_3$ → undecanoic acid (*161*)

Sylvaticin = Uleicin-C (*162*), rev. struct. (*12*)

Annona purpurea (leaves) (*85*); *Rollinia membranacea* (seeds) (*21, 111*); *Rollinia mucosa* (leaves) (*87*); *Rollinia sylvatica* (fruit) (*163*); *Rollinia ulei* (leaves) (*162*)

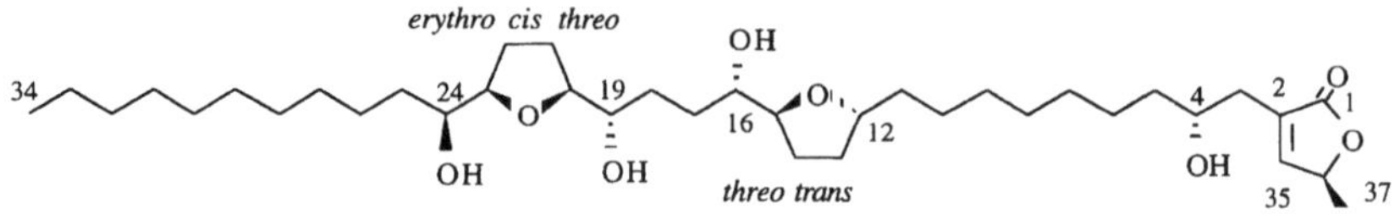

$C_{37}H_{66}O_8$; M = 638
mp = 48–50 °C (*163*), mp = 63–64 °C (*87*)
$[\alpha]_D$ = + 5.9° (c = 0.524, $CHCl_3$) (*163*), $[\alpha]_D$ = + 4.4° (c = 0.27, MeOH) (*85*), $[\alpha]_D$ = + 5.0° (c = ?, CH_2Cl_2) (*87*)
IR (*87, 162, 163*); UV (*85, 87*)
^1H nmr (300 MHz) (*163*), (200 MHz) (*85*), (500 MHz) (*87*), (350 MHz) (*162*); ^{13}C nmr (50 MHz) (*85, 163*), (100 MHz) (*162*), (125 MHz) (*87*)
HRFABms (*87, 163*), EIms (*85, 87*), CIms (*85*)
tetraacetate: ^1H nmr (300 MHz) (*163*); ^{13}C nmr (50 MHz) (*163*)
TMSi: EIms (*87, 163*), TMSi-$_{d9}$ derivative: EIms (*163*)
formal acetal: HRFABms, (*87*), EIms, (*87*)
TMSi derivative of formal acetal: EIms, (*87*)
isosylvaticin (mixture of 2,4-*cis*- and 2,4-*trans*-isosylvaticin) obtained from sylvaticin (diethylamine/CH_2Cl_2) (*21, 111*): $[\alpha]_D$ = + 23° (c = 0.30, MeOH) (*111*); IR (*111*); UV (*111*); ^1H nmr (200 MHz) (*111*); ^{13}C nmr (50 MHz) (*111*); FAB-Li ms (*111*)
The 20,23-relative stereochemistry was given as *trans* in (*85*) and *cis* in (*87*) but SHI *et al.* apparently were not aware of the publication of COPLEANU *et al.* (*85*)
(*R*)- and (*S*)-di-MTPA esters of formal acetal: ^1H nmr (500 MHz) → 4*R*, 24*S*, 36*S* (*87*)
absolute stereochemistry: 4*R*, 12*R*, 15*S*, 16*S*, 19*S*, 20*S*, 23*R*, 24*S*, 36*S*

12,15-*cis*-Sylvaticin

Rollinia mucosa (leaves) (*87*)

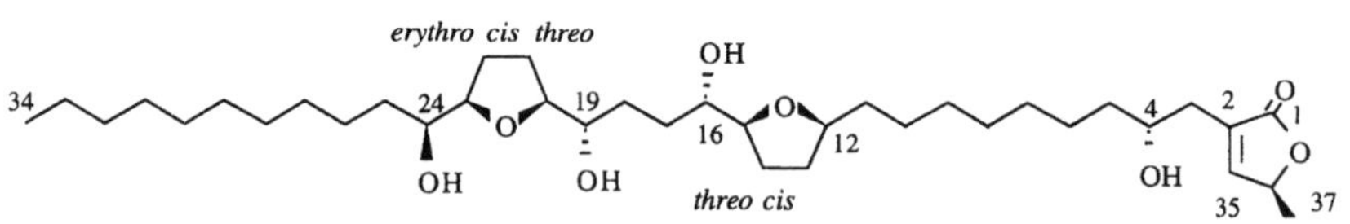

$C_{37}H_{66}O_8$; M = 638
mp = 63–64 °C (*87*)
$[\alpha]_D$ = + 5.2° (c = ?, CH_2Cl_2) (*87*)
IR (*87*); UV (*87*)
^1H nmr (500 MHz) (*87*); ^{13}C nmr (125 MHz) (*87*)
HRFABms (*87*); EIms (*87*)
TMSi: EIms (*87*)
(*R*)- and (*S*)-tetra-MTPA esters: ^1H nmr (500 MHz) (*87*)
formal acetal: HRFABms, (*87*), EIms, (*87*)
TMSi derivative of formal acetal: EIms, (*87*)
(*R*)- and (*S*)-di-MTPA esters of formal acetal: ^1H nmr (500 MHz) → 4*R*, 24*S*, 36*S* (*87*)
absolute stereochemistry: 4*R*, 12*S*, 15*S*, 16*S*, 19*S*, 20*S*, 23*R*, 24*S*, 36*S*

Trilobacin

Asimina triloba (bark) (*72*), (seeds) (*73*), (stem bark) (*164*)

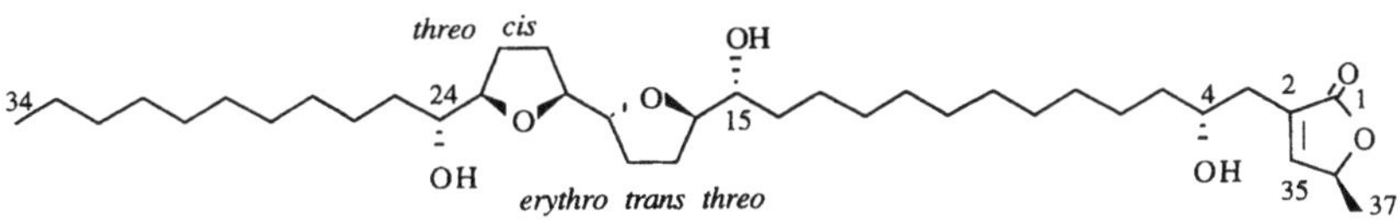

$C_{37}H_{66}O_7$; M = 622
$[\alpha]_D$ = + 10° (c = 0.5, $CHCl_3$) (*72*)
CD (*164*): → 36*S* configuration (*164*)
IR (*72*); UV (*72*)
^1H nmr (500 MHz) (*72*), (reassignment) (*164*); ^{13}C nmr (125 MHz) (*72*)
CIms (*72*), HRCIms (*72*), EIms (*72*),
triacetate (*72*): IR (*72*); UV (*72*); ^1H nmr (500 MHz) (*72, 164*); CIms (*72, 164*)
trilobacin-4,24-diacetate: ^1H nmr (C_6D_6, 500 MHz) (*164*)
trilobacin-4,24-diacetate-15-TMSi derivative: EIms (*164*), HREIms (*164*)
relative stereochemistry revised: 20,23-*cis* (*164*)
(*R*)- and (*S*)-tri-MTPA esters: ^1H nmr (500 MHz) → 15*R*, 24*R* (*164*)
absolute stereochemistry: 4*R*, 15*R*, 16*R*, 19*R*, 20*S*, 23*R*, 24*R*, 36*S*

Trilobin

Asimina triloba (seeds) (*73*), (stem bark) (*164*)

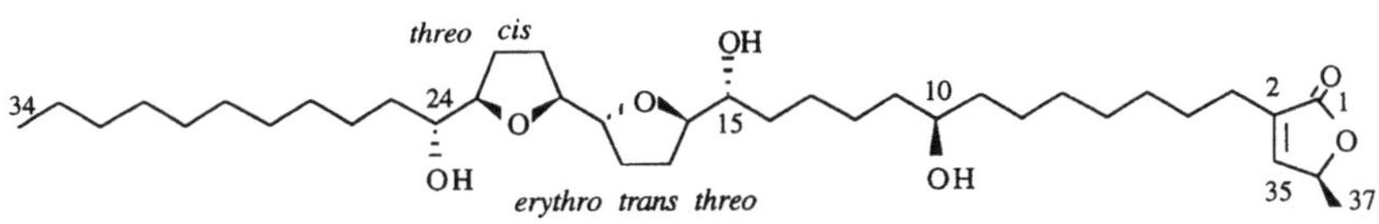

$C_{37}H_{66}O_7$; M = 622
$[\alpha]_D = +33.3°$ (c = 0.15, MeOH) (*164*)
CD (*164*)
IR (*164*); UV (*164*)
^1H nmr (500 MHz) (*164*); ^{13}C nmr (125 MHz) (*164*)
FABms (*164*)
triacetate (*164*): ^1H nmr (500 MHz) (*164*), HREIms (*164*)
tri-TMSi derivative: EIms (*164*)
(*R*)- and (*S*)-tri-MTPA esters: ^1H nmr (500 MHz) → 15*R*, 24*R* (*164*); 10*R* configuration proposed by comparing the $\Delta\delta_H$ values of two methyl groups of the (R)-MTPA ester with the model compounds, 6-undecanol and 8-pentadecanol (*164*), negative Cotton effect → 36*S* (*164*)
absolute stereochemistry: 10*R*, 15*R*, 16*R*, 19*R*, 20*S*, 23*R*, 24*R*, 36*S*

Tripoxyrollin = Trieporeticanin[†] (*110*)

Annona reticulata (seeds) (*110*); *Rollinia membranacea* (seeds) (*165*)

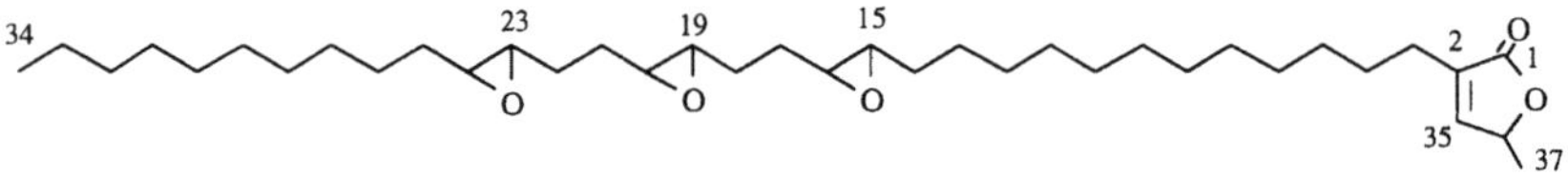

$C_{37}H_{64}O_5$; M = 588;
$[\alpha]_D = +10.3°$ (c = 0.68, $CHCl_3$) (*165*), $[\alpha]_D = +13°$ (c = 1, $CHCl_3$) (*110*)
IR (*110, 165*); UV (*110, 165*)
^1H nmr (200 MHz) (*110, 165*); ^{13}C nmr (50 MHz) (*110, 165*)
HRFABms (*165*), FAB-Li ms (*110, 165*), CID of $[M + Li]^+$ (*165*), CIms (*110*)
hemisynthesis of isodesacetyluvaricin from tripoxyrollin ($HClO_4$/acetone) (*165*)

Uleicin-C = Sylvaticin

Uleirollin

Rollinia ulei (leaves) (*166*)

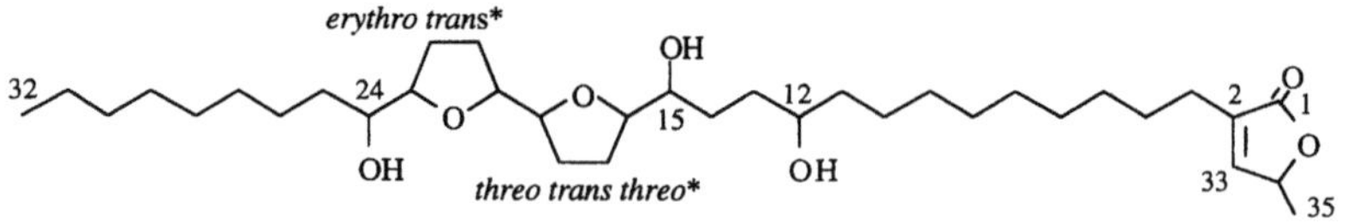

† The differences in the physical and spectral data are not sufficiently significant to differentiate trieporeticanin and tripoxyrollin
* We have proposed the relative stereochemistry shown in the formula

$C_{37}H_{66}O_7$; M = 622
IR (*166*); UV (*166*)
^{1}H nmr (250 MHz) (*166*); ^{13}C nmr (60 MHz) (*166*)
EIms (*166*), CIms (*166*), tandem ms (*166*)

Uvariamicin-I

Uvaria narum (root bark) (*137, 141*); *Uvaria hookeri* (root bark) (*137*); *Annona bullata* (bark) (*93*)

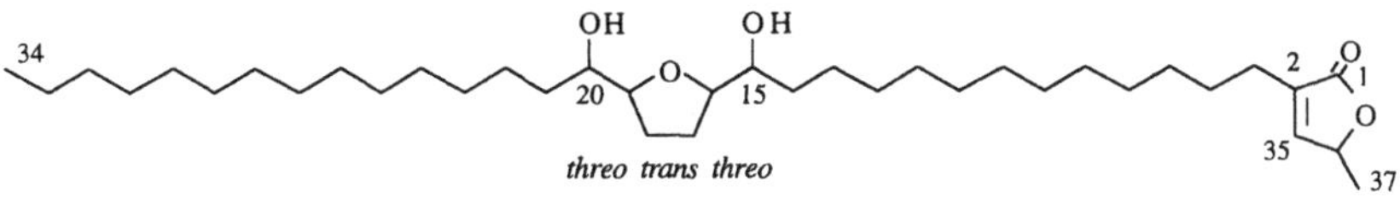

mixture uvariamicin I, II (reticulatacin) and III (*137, 141*)
mixture uvariamicin I, II (reticulatacin), III and IV (*93*)
data in (*93*) correspond to the mixture uvariamicin-I, -II (reticulatacin), -III, -IV
data in (*141*) correspond to the mixture uvariamicin-I, -II (reticulatacin), -III

$C_{37}H_{68}O_5$; M = 592
$[\alpha]_D = +13.75°$ (c = 0.4, $CHCl_3$) (*93*)
CD (*93*)
IR, (*93*); UV (*93*)
^{1}H nmr (*141*), (500 MHz) (*93*), (300 MHz) (*137*); ^{13}C nmr (*141*), (125 MHz) (*93*), (50 MHz) (*137*)
CIms (*93, 141*), EIms (*93, 137, 141*), HREIms (*141*), FABms (*93, 141*), HRFABms (*93*)
diacetate: ^{1}H nmr (500 MHz) (*93*); EIms (*93, 141*)
di-TMSi derivative: EIms (*93, 141*)

Uvariamicin-III

Annona atemoya (seeds) (*63*); *Annona bullata* (bark) (*93*); *Annona reticulata* (seeds) (*143*); *Uvaria hookeri* (root bark) (*137*); *Uvaria narum* (root bark) (*137, 141*)

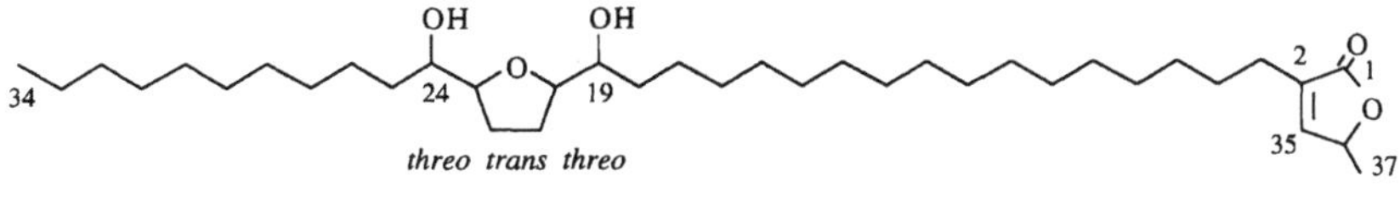

mixture uvariamicin I, II and III (*141*)
mixture uvariamicin I, II, III and IV (*93*)
pure (*63*, *143*)
data in (*93*) correspond to the mixture uvariamicin-I, -II, -III and IV
data in (*141*) correspond to the mixture uvariamicin-I, -II, -III

$C_{37}H_{68}O_5$; M = 592
$[\alpha]_D = +13.75°$ (c = 0.4, $CHCl_3$) (*93*)
CD (*93*)
IR (*93*); UV (*93*)
^{1}H nmr (*141*), (500 MHz) (*93*), (300 MHz) (*137*); ^{13}C nmr (*141*), (125 MHz) (*93*), (50 MHz) (*137*)
CIms (*93*, *141*), EIms (*93*, *137*, *141*), HREIms (*141*), FABms (*93*, *141*), HRFABms (*93*)
diacetate: ^{1}H nmr (500 MHz) (*93*); EIms (*93*, *141*)
di-TMSi derivative: EIms (*93*, *141*)

data in (*143*) correspond to pure uvariamicin-III

$[\alpha]_D = +19°$ (c = 1, $CHCl_3$) (*143*)
IR (*143*); UV (*143*)
^{1}H nmr (200 MHz) (*143*); ^{13}C nmr (50 MHz) (*143*)
CIms (*143*); EIms; (*143*); LSIms, L-SIms (CID *B*/*E* + Li) (*143*)
By analogy to reticulatacin the absolute stereochemistry can be assumed to be 19*R*, 20*R*, 23*R*, 24*R*, and 36*S* (*143*)

Uvariamicin-IV

Annona bullata (bark) (*93*)

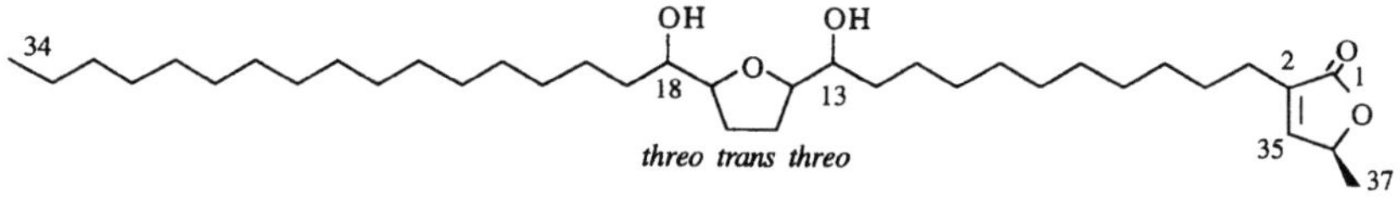

data in (*93*) correspond to the mixture uvariamicin-I, -II, -III, -IV

$C_{37}H_{68}O_5$; M = 592
$[\alpha]_D = +13.75°$ (c = 0.4, $CHCl_3$) (*93*)
CD (*93*)
IR, (*93*); UV (*93*)
^{1}H nmr (500 MHz) (*93*); ^{13}C nmr (125 MHz) (*93*)
CIms (*93*), EIms (*93*), FABms (*93*), HRFABms (*93*)
diacetate: ^{1}H nmr (500 MHz) (*93*); EIms (*93*)
di-TMSi derivative: EIms (*93*)

Uvaricin

Uvaria acuminata (roots) (*1*, *105*)

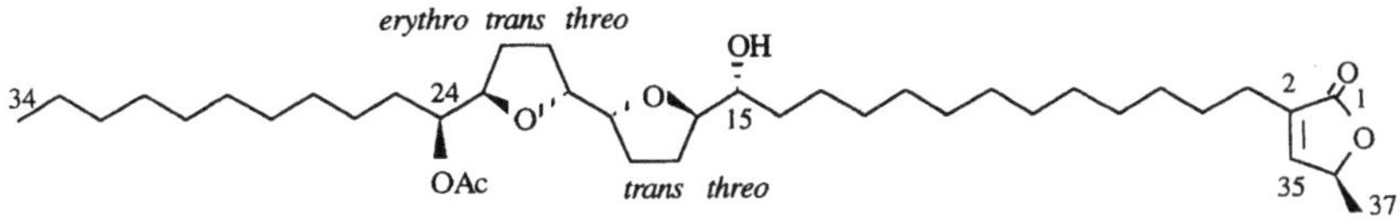

$C_{39}H_{68}O_7$; M = 648
$[\alpha]_D = +11.3°$ (MeOH) (*1*)
IR (*1*); UV (*1*)
^1H nmr ($CDCl_3$ and C_6D_6) (*1*); ^{13}C nmr ($CDCl_3$ and C_6D_6) (*1*)
EIms (*1*), CIms (*1*),
TMSi-derivative: ms (*1*)
uvaricinone: UV, IR, ^1H nmr, ms: (*1*)
ozonolysis: → (*S*)-lactic acid (*105*) → stereochemistry C-36*S* (*105*)
(*R*)- and (*S*)-bis-MTPA esters: ^1H nmr (500 MHz) (*46*); ^{19}F nmr (*46*): absolute stereochemistry: (*46*)
synthesis of hexepi-uvaricin (*216*): mp = 23–33 °C (*216*), $[\alpha]_D = +9.5°$ (c = 1.07, MeOH) (*216*); IR, ^1H nmr, and ^{13}C nmr identical with those of uvaricin (*216*)
absolute stereochemistry: 15*R*, 16*R*, 19*R*, 20*R*, 23*R*, 24*S*, 36*S*

Venezenin

Xylopia aromatica (barks) (*30*)

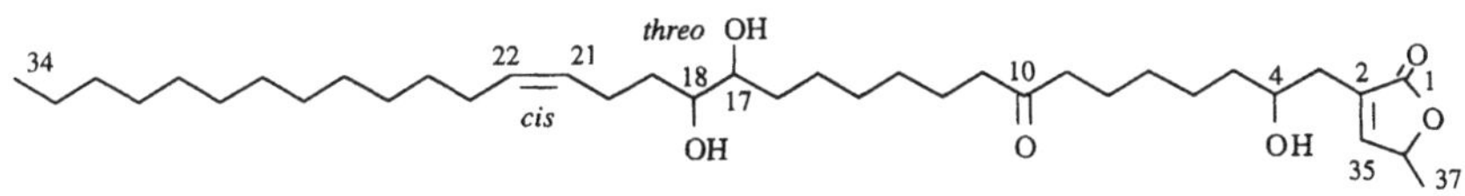

$C_{37}H_{66}O_6$; M = 606
mp = 72–73 °C (*30*)
$[\alpha]_D = +16.8°$ (c = 0.06, MeOH) (*30*)
IR (*30*); UV (*30*)
^1H nmr (500 MHz) (*30*); ^{13}C nmr (125 MHz) (*30*)
CIms (*30*), HRFABms (*30*)
acetonide: ^1H nmr (500 MHz) (*30*)
tri-TMSi: EIms (*30*)
dihydrovenezenin: ^1H nmr (500 MHz) (*30*); tetra-TMSi derivative: EIms (*30*)

oxidation and cyclization (*m*-CPBA; perchloric acid) → annomontacin-10-one (*threo-trans-threo*) and 18,21-*cis*-annomontanacin-10-one (*threo-cis-threo*) (*30*)

Xylomatenin or **Annogalene** (*21*, *22*)

Annona senegalensis (seeds) (*21*, *22*); *Xylopia aromatica* (barks) (*23*)

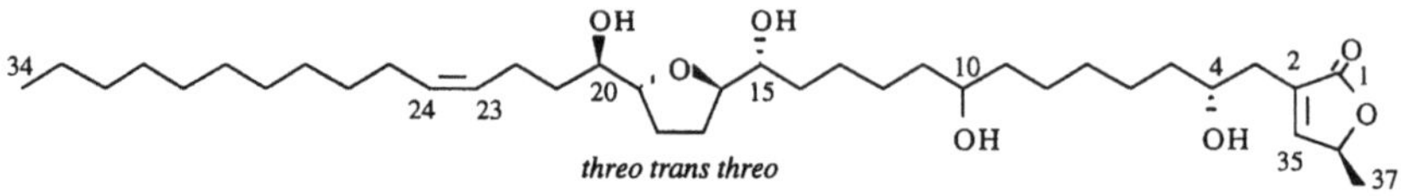

$C_{37}H_{66}O_7$; M = 622
mp = 52–53 °C (*23*)
$[\alpha]_D = +19°$ (c = ?, MeOH) (*23*), $[\alpha]_D = +14°$ (c = 0.31, $CHCl_3$) (*22*)
IR (*22*, *23*); UV (*23*)
^{1}H nmr (500 MHz) (*23*), (400 MHz) (*22*); ^{13}C nmr (125 MHz) (*23*), (100 MHz) (*22*)
FABsm (glycerol) (*23*), HRFABsm (glycerol) (*23*), EIms (*22*, *23*), FAB-Li ms (*22*)
tetraacetate: ^{1}H nmr (500 MHz) (*23*)
tetra-TMSi derivative: EIms (*23*)
(*R*)- and (*S*)-MTPA esters: ^{1}H nmr (500 MHz) (*23*) → 4*R*, 15*R*, 20*R* (*23*)
absolute stereochemistry: 4*R*, 15*R*, 16*R*, 19*R*, 20*R*, 36*S*

Xylomaticin

Xylopia aromatica (*bark*) (*167*); *Asimina longifolia* (leaves and twigs) (*42*)

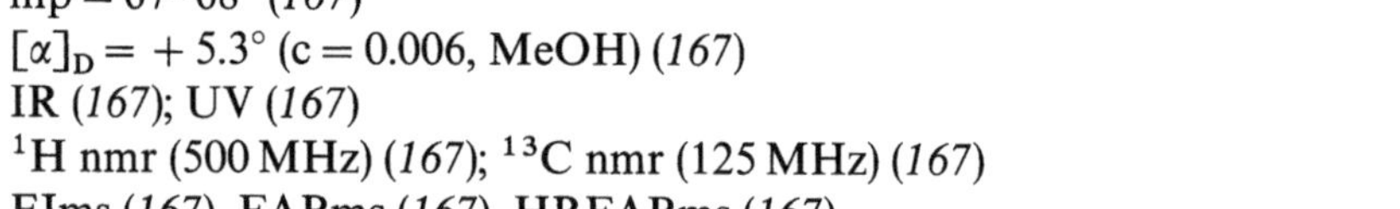

$C_{37}H_{68}O_7$; M = 624
mp = 67–68° (*167*)
$[\alpha]_D = +5.3°$ (c = 0.006, MeOH) (*167*)
IR (*167*); UV (*167*)
^{1}H nmr (500 MHz) (*167*); ^{13}C nmr (125 MHz) (*167*)
EIms (*167*), FABms (*167*), HRFABms (*167*)
tetra-TMSi derivative: EIms (*167*)

Xylopiacin

Xylopia aromatica (bark) (*167*)

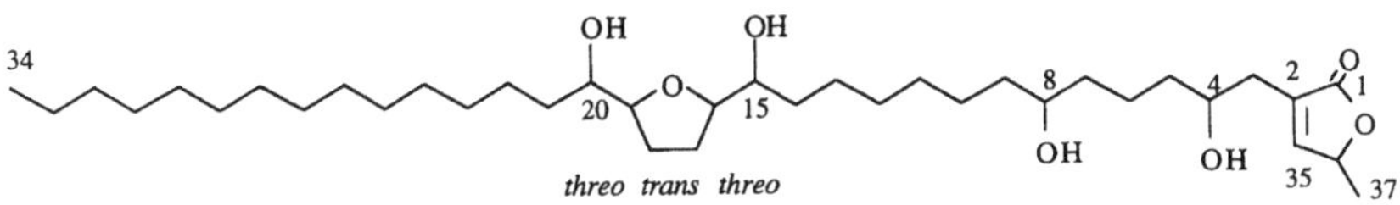

$C_{37}H_{68}O_7$; M = 624
mp = 90–91 °C (*167*)
$[\alpha]_D = +24°$ (c = 0.006, MeOH) (*167*)
IR (*167*); UV (*167*)
^1H nmr (500 MHz) (*167*); ^{13}C nmr (125 MHz) (*167*)
EIms (*167*), FABms (*167*), HRFABms (*167*)
tetra-TMSi derivative: HRFABms (*167*), EIms (*167*)

Xylopianin

Xylopia aromatica (bark) (*167*)

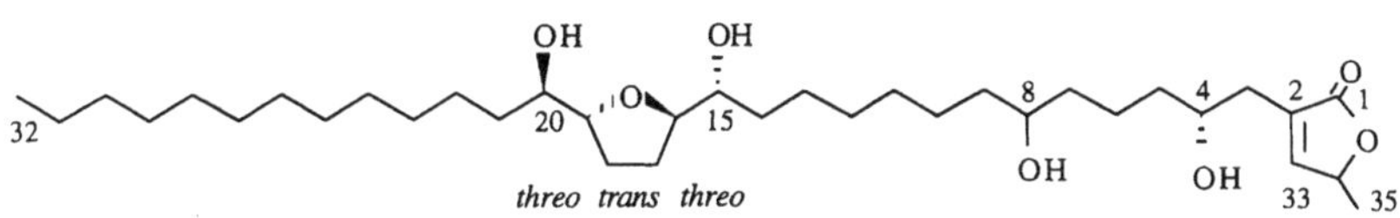

$C_{35}H_{64}O_7$; M = 596
mp = 78–79 °C (*167*)
$[\alpha]_D = +23.3°$ (c = 0.008, MeOH) (*167*)
IR (*167*); UV (*167*)
^1H nmr (500 MHz) (*167*); ^{13}C nmr (125 MHz) (*167*)
EIms (*167*), FABms (*167*), HRFABms (*167*)
tetra-TMSi derivative: EIms (*167*), FABms (*167*), HRFABms (*167*)
The absolute stereochemistry shown in the formula is based on the close similarity to annonacin (*167*)

Xylopien

Xylopia aromatica (barks) (*23*)

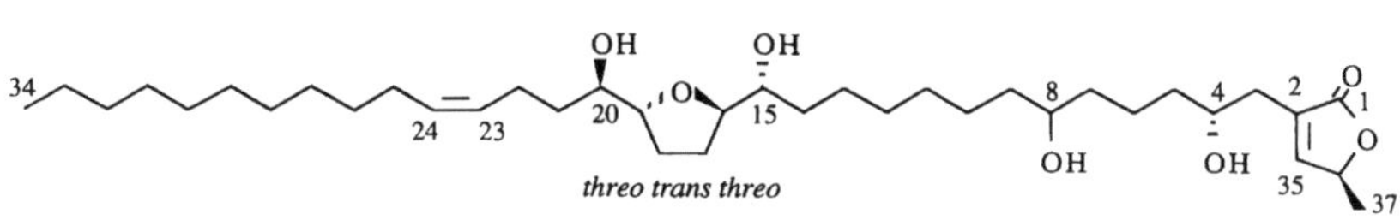

$C_{37}H_{66}O_7$; M = 622
mp = 48–49 °C (*23*)
$[\alpha]_D = +15°$ (c = 0.001, MeOH) (*23*)
IR (*23*); UV (*23*)
^{1}H nmr (500 MHz) (*23*); ^{13}C nmr (125 MHz) (*23*)
FABms (glycerol) (*23*); HRFABms (glycerol) (*23*), EIms (*23*)
tetraacetate: ^{1}H nmr (500 MHz) (*23*)
tetra-TMSi derivative: EIms (*23*); HRFABms (*23*)
(*R*)- and (*S*)-MTPA esters: ^{1}H nmr (500 MHz) (*23*) → 4*R*, 15*R*, 20*R* (*23*)
absolute stereochemistry: 4*R*, 15*R*, 16*R*, 19*R*, 20*R*, 36*S*

II.c. Distribution of Acetogenins in the Annonaceae

Plant source	Name of compound	References
Annona atemoya	Annotemoyin-1	(*63*)
	Annotemoyin-2	(*63*)
	Atemoyin	(*77, 78*)
	Desacetyluvaricin	(*78*)
	Isodesacetyluvaricin	(*78*)
	Molvizarin	(*129*)
	Neoannonin	(*78, 129*)
	Reticulatain-1	(*63*)
	Rolliniastatin-2	(*78, 129*)
	Squamocin	(*78, 129*)
	Uvariamicin-III	(*63*)
Annona bullata	Bullacin	(*79*)
	Bulladecinone	(*80*)
	Bullatacin	(*83*)
	Bullatacinone	(*83*)
	Bullatalicin	(*86*)
	C-12,15-*cis*-Bullatalicin	(*88*)
	Bullatalicinone	(*90*)
	C-12,15-*cis*-Bullatalicinone	(*88*)
	Bullatanocin	(*89*)
	Bullatanocinone	(*88, 89*)
	Bullatencin	(*93*)
	C-12,15-*cis*-Bullatanocin	(*88*)
	C-12,15-*cis*-Bullatanocinone	(*88*)
	4-Deoxyasimicin	(*93*)
	Desacetyluvaricin	(*89*)
	30-Hydroxybullatacin	(*81*)
	31-Hydroxybullatacin	(*81*)
	32-Hydroxybullatacin	(*81*)
	10-Hydroxybullatacinone	(*121*)
	12-Hydroxybullatacinone	(*121*)

Plant source	Name of compound	References
	28-Hydroxybullatacinone	(*81*)
	29-Hydroxybullatacinone	(*121*)
	30-Hydroxybullatacinone	(*122*)
	31-Hydroxybullatacinone	(*122*)
	32-Hydroxybullatacinone	(*122*)
	Parviflorin	(*79*)
	Squamocin	(*90, 93*)
	Uvariamicin I, II, III, IV (mixture)	(*93*)
Annona cherimolia	Almunequin	(*16, 17*)
	Annonacin-A	(*22*)
	Annosenegalin	(*22*)
	Asimicin	(*66, 130*)
	Bullatacinone	(*17*)
	Cherimolin-1	(*4, 16, 66*)
	Cherimolin-2	(*16*)
	Dihydrocherimolin* = Otivarin	(*4, 16, 66, 127*)
	Gigantetrocin-A	(*22*)
	Isocherimolin-1	(*17*)
	Isomolvizarin-1	(*17*)
	Isomolvizarin-2	(*17*)
	Itrabin	(*16, 127*)
	Jetein	(*16, 127*)
	Laherradurin	(*16, 66, 127*)
	Molvizarin	(*16, 130*)
	Motrilin	(*16, 130*)
	Otivarin	(*16, 127*)
	Rolliniastatin-2	(*16, 130*)
	Squamocin	(*16, 17, 130*)
Annona coriacea	Coriacin	(*95*)
	Coriadienin	(*96*)
	4-Deoxycoriacin	(*95*)
	4-Deoxygigantecin	(*95*)
	Gigantecin	(*114*)
	Gigantetronenin	(*96*)
Annona crassiflora	Crassiflorin	(*91*)
Annona densicoma	Annonacin	(*35*)
	Annonacin-10-one	(*7*)
	Densicomacin 13,14-*erythro*	(*101*)
	Densicomacin 13,14-*threo*	(*101*)
	Goniothalamicin	(*101*)
	8-Hydroxy-annonacin	(*101*)
	Isoannonacin	(*7*)
	Isoannonacin-10-one	(*7*)

* Wrong structure

Plant source	Name of compound	References
Annona glabra	Annonacin	(*36*)
	Asimicin	(*67*)
	Cherimolin-2	(*92*)
	Corossolone	(*36*)
	Desacetyluvaricin	(*67, 92*)
	Solamin	(*36*)
	Squamocin	(*67, 92*)
	Squamostatin-E	(*92*)
Annona glauca	Annoglaucin	(*24*)
	Glaucanisin	(*120*)
	Rolliniastatin-2	(*24, 120*)
Annona montana	Annomonicin	(*26*)
	Annomontacin	(*28*)
	Annonacin	(*26, 28*)
	Annonacinone	(*28*)
	Diepoxymontin	(*112*)
	Goniothalamicin	(*26*)
	Montanacin	(*26*)
Annona muricata	Annohexocin	(*25*)
	Annomuricin-A	(*31*)
	Annomuricin-B	(*31*)
	Annomuricin-C	(*32*)
	Annomutacin	(*33*)
	Annonacin	(*31, 38, 39, 40*)
	Annonacin-10-one	(*31, 40*)
	Annonacin-A	(*59*)
	10*R*-Annonacin-A-one	(*33*)
	Annonacinone	(*38, 40*)
	Corepoxylone	(*94*)
	Corossolin	(*38*)
	Corossolone	(*38*)
	Desacetyluvaricin	(*174*)
	4-Desoxyhowiicin B	(*41*)
	Diepomuricanin = Diepomuricanin-A	(*107, 108, 109*)
	Epomuricenin A	(*108*)
	Epomuricenin B	(*108*)
	Epoxymurin-B	(*109*)
	Gigantetrocin A	(*31, 40, 118*)
	Gigantetrocin B	(*119*)
	Gigantetronenin	(*32*)
	Goniothalamicin	(*31, 40*)
	Howiicin-A	(*41*)
	Howiicin-B	(*41*)
	Howiicin-F	(*174*)
	Howiicin-G	(*174*)

Plant source	Name of compound	References
	Isoannonacin	(*40, 59*)
	Isoanonacin-10-one	(*40*)
	Isoneoannonacinone	(*126*)
	Muricatacin	(*39*)
	Muricatalin	(*134*)
	Muricatetrocin-A	(*31, 119*)
	Muricatetrocin-B	(*31, 119*)
	Muricatin-A	(*118*)
	Muricatin-B	(*118*)
	Muricatin-C	(*118*)
	Muricatocin-A	(*59*)
	Muricatocin-B	(*59*)
	Muricatocin-C	(*32*)
	Murihexocin-A	(*135*)
	Murihexocin-B	(*135*)
	Murisolin	(*125*)
	Neoannonacinone	(*126*)
	Neoisoannonacinone	(*126*)
	Solamin	(*65*)
Annona purpurea	Bullatacin	(*85*)
	Cherimolin	(*85*)
	Purpureacin-1	(*85*)
	Purpureacin-2	(*85*)
	Rolliniastatin-1	(*85*)
	Sylvaticin	(*85*)
Annona reticulata	Annomonicin	(*27*)
	Annoreticuin	(*62*)
	Annoreticuin-9-one	(*27*)
	Bullatacin	(*142*)
	Diepomuricanin	(*110*)
	Dieporeticanin-1	(*110*)
	Dieporeticanin-2	(*110*)
	Dieporeticenin	(*110*)
	14-Hydroxy-25-desoxyrollinicin*	(*151, 168*)
	Isoannoreticuin	(*62*)
	Molvizarin	(*131*)
	Reticulacinone	(*131*)
	Reticulatacin	(*142, 143*)
	Reticulatain-1	(*143*)
	Reticulation-2	(*143*)
	Reticulatamol	(*145*)
	Reticulatamone	(*143*)
	Rolliniastatin-1	(*147*)
	Rolliniastatin-2	(*27, 131*)

* Wrong structure

Plant source	Name of compound	References
	Solamin	(*27*)
	Squamocin	(*147*)
	Squamone	(*27*)
	Trieporeticanin	(*110*)
	Uvariamicin III	(*143*)
Annona senegalensis	Annogalene	(*21, 22*)
	Annonacin	(*22*)
	Annonacin-A	(*22*)
	Annosenegalin	(*21, 22*)
	Asimicin	(*68, 69*)
	Gigantetrocin-A	(*22*)
	Molvizarin	(*68, 69*)
	Rolliniastatin-2	(*68*)
	Senegalene	(*68*)
	Squamocin	(*68*)
Annona squamosa	Annonacin = Annonacin I	(*10, 37*)
	Annonacin-A = Annonacin II	(*10, 37*)
	Annonastatin	(*10, 37*)
	Annonin III* = Motrilin	(*64, 132*)
	Annonin IV* = Bullatanocin	(*10, 11*)
	Annonin VI* = Rolliniastatin-2	(*10, 14, 158*)
	Annonin VIII* = Bullatalicin	(*10, 11*)
	Annonin XIV*	(*10*)
	Annonin XVI* = Squamostatin A	(*10, 11*)
	Annonsilin A	(*61*)
	Asimicin	(*10, 37*)
	Bullatacin	(*84*)
	Bullatacinone	(*84*)
	Desacetyluvaricin	(*14*)
	Neoannonin	(*138*)
	Neo-Annonin B	(*104*)
	Neo-desacetyluvaricin	(*104*)
	Neo-reticulatacin	(*104*)
	Squamocin	(*10, 64, 138, 157, 158*)
	Squamocin-B	(*64*)
	Squamocin-C	(*64*)
	Squamocin-D	(*64*)
	Squamocin-E	(*64*)
	Squamocin-F	(*64*)
	Squamocin-G	(*64*)
	Squamocin-H	(*64*)
	Squamocin-I	(*64*)
	Squamocin-J	(*64*)
	Squamocin-K	(*64*)

* Wrong structure

References, pp. 273–288

Plant source	Name of compound	References
	Squamocin-L	*(64)*
	Squamocin-M	*(64)*
	Squamocin-N	*(64)*
	Squamone	*(84)*
	Squamosinin	*(169)*
	Squamostanal-A	*(160)*
	Squamostatin-A	*(15, 64)*
	Squamostatin-B*	*(13, 14)*
	Squamostatin-B*	*(20)*
	Squamostatin-C	*(20)*
	Squamostatin-D	*(20)*
	Squamostatin-E	*(20)*
	Squamosten-A	*(161)*
Asimina longifolia	Annonacin	*(42)*
	Gigantetrocin-A	*(42)*
	Gigantetrocin-B	*(42)*
	Gigantetrocinone	*(42)*
	Goniothalamicin	*(42)*
	Goniothalamicinone	*(42)*
	Isoannonacin	*(42)*
	Longicin	*(42)*
	Muricatetrocin-A	*(42)*
	Muricatetrocin-B	*(42)*
	Xylomaticin	*(42)*
Asimina parviflora	Annonacin	*(43)*
	Asimicin	*(70)*
	Bullatacin	*(43)*
	Goniothalamicin	*(43)*
	Molvizarin	*(43)*
	Parvifloracin	*(43)*
	Parviflorin	*(43)*
Asimina triloba	Annonacinone	*(60)*
	Annonacin-A-one	*(60)*
	Asimicin	*(71, 72, 73)*
	Asimin	*(76)*
	Asiminacin	*(76)*
	Asiminecin	*(76)*
	Asiminenin-A	*(73)*
	Asiminenin-B	*(73)*
	Asitribin	*(73)*
	Bullanin	*(82)*
	Bullatacin	*(72, 73)*

* Two different acetogenins have been described under the name squamostatin-B; the first one is probably squamostatin-A and the other one was identified as bullatalicin.

Plant source	Name of compound	References
	Bullatacinone	(*72*)
	Bullatin	(*82*)
	Gigantetrocinone	(*60*)
	Isoannonacin	(*60*)
	Motrilin	(*82*)
	Murisolin	(*136*)
	16,19-*cis*-Murisolin	(*136*)
	Murisolin-A	(*136*)
	Parviflorin	(*73*)
	Squamocin	(*82*)
	Trilobacin	(*72, 73, 164*)
	Trilobin	(*73, 164*)
Goniothalamus giganteus	Annomontacin	(*29*)
	Annonacin	(*44*)
	Asimilobin	(*75*)
	4-Deoxygigantecin	(*103*)
	Giganenin	(*103*)
	Giganin	(*29, 113*)
	Gigantecin	(*115*)
	Gigantetrocin	(*116*)
	Gigantetronenin	(*29*)
	Gigantriocin	(*116*)
	Gigantrionenin	(*29*)
	Goniocin	(*99*)
	Goniodenin	(*75*)
	Gonionenin	(*100*)
	Goniothalamicin	(*44*)
Goniothalamus howii	Goniothalamicin	(*34*)
	Howiicin-A	(*34*)
	Howiicin-B	(*34*)
	Howiicin-C	(*34*)
	Howiicin-D	(*117*)
	Howiicin-E	(*117*)
	Howiicin-F	(*117*)
	Howiicin-G	(*117*)
Polyalthia plagioneura	Plagionicin-A	(*140*)
Rollinia membranacea	Diepomuricanin-A	(*111*)
	Diepomuricanin-B	(*111*)
	Dieporeticanin-1	(*111*)
	Dieporeticanin-2	(*111*)
	Diepoxyrollin	(*111*)
	Membranacin	(*128*)
	Rioclarin	(*128*)
	Rolliniastatin-1	(*128*)
	Rollinone	(*111*)

Plant source	Name of compound	References
	Squamocin	(*128*)
	Sylvaticin	(*21, 111*)
	Tripoxyrollin	(*165*)
Rollinia mucosa	Bullatalicin	(*87*)
	Mucocin	(*133*)
	Muricatetrocin-B	(*87*)
	Rolliniastatin-1	(*148*)
	Rolliniastatin-2	(*152*)
	Sylvaticin	(*87*)
	12,15-*cis*-Sylvaticin	(*87*)
Rollinia papilionella	4-Hydroxy-25-desoxyneorollinicin* = Rolliniastatin-1	(*146*)
	Isorollinicin*	(*2, 3*)
	Rollinicin	(*2, 3*)
	Rollinone	(*2, 154*)
Rollinia sericea	Rolliniastatin-1	(*149*)
	Squamocin	(*149*)
Rollinia sylvatica	Sylvaticin	(*163*)
Rollinia ulei	Epoxyrollin A* = Dieporeticanin-1	(*106*)
	Epoxyrollin B* = Diepomuricanin-A	(*106*)
	Uleicin A*	(*106, 162*)
	Uleicin B*	(*106, 162*)
	Uleicin C* = Sylvaticin	(*106, 162*)
	Uleicin D*	(*106, 162*)
	Uleicin E*	(*106, 162*)
	Uleirollin	(*106*)
Uvaria acuminata	Desacetyluvaricin	(*105*)
	Uvaricin	(*1, 105*)
Uvaria grandiflora	Isodesacetyluvaricin	(*123*)
Uvaria hookeri	Isodesacetyluvaricin	(*137*)
	Narumicin I, II (mixture)	(*137*)
	Squamocin	(*137*)
	Squamocinone	(*137*)
	Uvariamicin-I, II, III (mixture)	(*137*)
Uvaria narum	Isodesacetyluvaricin	(*124*)
	Narumicin I, II (mixture)	(*124, 137*)
	Panalicin	(*139*)
	Squamocin	(*137, 139*)
	Squamocinone	(*137, 139*)
	Uvariamicin I, II, III (mixture)	(*137, 141*)

* Wrong structure

Plant source	Name of compound	References
Xylopia aromatica	Annomontacin	(*167*)
	Annonacin	(*167*)
	Asimicin	(*30*)
	Gigantetrocin	(*167*)
	Gigantetronenin	(*167*)
	Venezenin	(*30*)
	Xylomatenin	(*23*)
	Xylomaticin	(*167*)
	Xylopiacin	(*167*)
	Xylopianin	(*167*)
	Xylopien	(*23*)

III. Extraction and Isolation

The isolation of acetogenins from plant materials (*e.g.* seeds, leaves, roots, barks) is carried out by successive solvent extraction (with increasing polarity of the solvents) or by liquid/liquid partition from a first alcoholic extract. The separation of acetogenins is then performed by solvent partition and chromatographies on silica gel (on columns and/or on thin layer plates), guided by TLC and bioassay.

Analytical TLC plates are generally sprayed by Kedde's reagent (characteristic of an unsaturated γ-lactone), but sulphomolybdic reagent may be used as well.

Activity-guided fractionation using lethality to brine shrimp larvae appears to be very useful owing to its rapidity, its low cost and its good correlation with antitumour activity (*170*).

Care must be taken during the extraction, and also during the isolation process, to avoid the occurrence of artefacts (*8*). Indeed, the acetogenins belonging to sub-type 1b may easily be translactonized. Translactonization is carried out by subjecting a C-4-hydroxylated acetogenin to potassium hydroxide in butanol (*7*). This reaction is also feasible by using a weak base such as diethylamine and also, but more slowly, by using alkaloids, either as bases or as weak acid salts, this last reaction being faster if the solution is heated. It has also been found that translactonization occurs in neutral or acidic media by heating 4-hydroxyacetogenins in methanol solution (*9*). Therefore long or strong heating of the alcoholic solutions and use of alkaline media during the extraction and isolation process should be avoided (*8*).

The isolation of acetogenins can be performed by column or circular partition chromatography. Isolation of pure products is not easy because acetogenins exist as complex mixtures of positional or stereochemical isomers of similar polarities and most of them do not crystallize. The purity of the isolated acetogenins must be controlled by high performance liquid chromatography (HPLC) which is the best, if not the only, method for assuring the purity of the natural products. Indeed, a single spot produced on thin layer chromatography often corresponds to a mixture of several compounds as detected by HPLC.

HPLC serves to check the purity of isolated compounds and can also be used for analytical and preparative purposes. In spite of the absence of a large chromophore, UV detection at 210–220 nm is satisfactory for acetogenins of sub-type-1, but UV detection is not convenient for acetogenins of sub-type-2 and sub-type-3 because of the lack of a chromophore in these substances. For these reasons refractometric detection and evaporative light scattering detection (ELSD) (*171*) are very useful, the last method being more interesting as it allows the use of a gradient of solvents (*172*).

Normal or reverse stationary phases are used. The chromatographic behaviour of acetogenins in reverse phase HPLC generally depends on their substitution pattern. The higher the number of oxygenated groups, *i.e.* hydroxyls, the shorter the retention time. However, other factors, such as the position and the type of substituent, can change the elution order (*172*).

As was pointed out above, HPLC serves to separate not only acetogenins of different polarities but also permits separation of positional and stereochemical isomers such as reticulatain-1 and reticulatain-2 (*143*), rolliniastatin-1 and rolliniastatin-2 (*172*) or 2,4-*cis* and 2,4-*trans* acetogenins (*89*). It also permits separation of *pseudo*-enantiomers which differ from each other in the absolute configurations of the stereogenic centers of the THF backbone but with identical configuration in the lactone ring, *e.g.* corossolone and its 15,16,19,20-epimer (*94*).

IV. Structure Elucidation

Two reviews appeared in 1990 (*5*) and 1993 (*175*) dealing with methods used for structure elucidation of acetogenins. Structure elucidation of acetogenins of Annonaceae can be performed in a five-step sequence, namely

1 – molecular weight determination.
2 – characterization of the five-membered ring lactone (γ-lactone).
3 – identification of the functional groups on the alkyl chain, *e.g.* THF rings, epoxides, hydroxyls, unsaturations...

4 – location of the functional groups.
5 – determination of the stereochemical relationships of the stereogenic centres.

IV.1. Molecular Weight Determination

Mass spectrometry by chemical ionization (CIms), or FABms (fast atom bombardment) or preferably FAB-Lims in the presence of lithium (*107, 176*) is used for determination of the molecular weight. High resolution mass spectrometry by chemical ionization or FAB is also used in order to obtain the molecular formula of the compound under study.

IV.2. Characterization of the Terminal γ-Lactone

All acetogenins possess a characteristic butyrolactone moiety. As described in Section II, four sub-types 1a, 1b, 2 and 3 exist (Fig. IV.1.).

All four sub-types 1a, 1b, 2, 3 can be easily differentiated by these simple observations:

1) A positive reaction with the Kedde reagent indicates the presence of acetogenins of sub-type 1 (*5*).

2) In the infrared spectrum the α,β-unsaturated γ-lactones (sub-type 1) exhibit an intensive absorption due to the conjugated carbonyl at 1745 cm^{-1}, whereas saturated γ-lactones (sub-types 2 and 3) have a peak at 1770 cm^{-1}.

3) ^{1}H and ^{13}C nmr spectra are helpful for confirming the different γ-lactone sub-types.

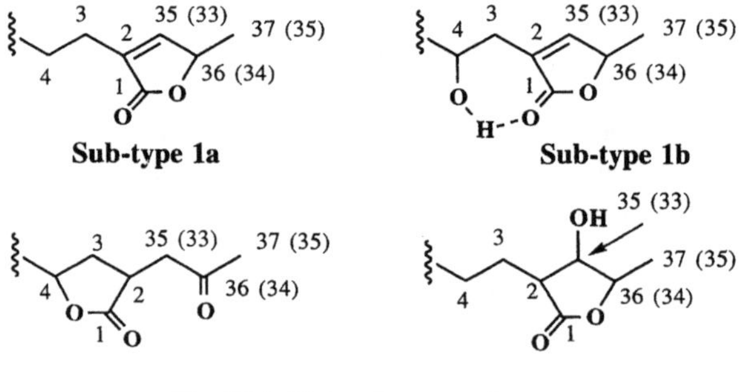

Fig. IV.1. Butyrolactone sub-types

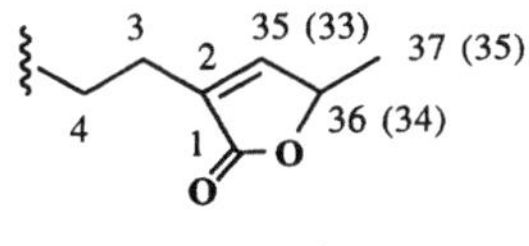

Sub-type 1a

Fig. IV.2.

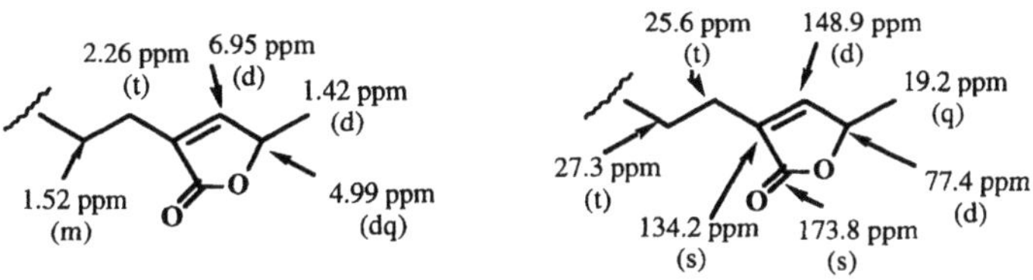

Fig. IV.3. Partial nmr (^{1}H nmr and ^{13}C) data of corossolone, acetogenin of sub-type 1a (*38*)

IV.2.1. Sub-type 1a

Acetogenins of sub-type 1a (Fig. IV.2.) possess an α,β-unsaturated γ-methyl-γ-lactone ring bearing at the α position of the carbonyl group of the lactone a long alkyl chain of 32 or 30 carbon atoms. In the ^{1}H nmr spectrum the resonance of a single vinylic proton, H-35 (33), is seen at δ 6.95 ppm as a doublet ($^3J_{35-36}$ = 1.7 Hz, characteristic for the ^{3}J coupling constant of a vinylic proton and a hydrogen atom in a five-membered ring).

The H-36 (34) proton, under the terminal methyl and the oxygen atom of the lactone, appears at δ 4.99 ppm as a doublet of quadruplets with $^3J_{H36-35}$ = 1.7 Hz and $^3J_{H36-37}$ = 7 Hz. The terminal methyl group C-37 (35), under the oxygen atom of the lactone, then appears at δ 1.42 ppm as a doublet with a vicinal constant $^3J_{H37-36}$ = 7 Hz. The H-3 protons are equivalent and appear as a triplet at 2.26.

For instance the partial ^{1}H and ^{13}C nmr data of the lactone moiety of corossolone (*38*) are as shown in Fig. IV.3.

IV.2.2. Sub-type 1b

Acetogenins of sub-type 1b (Fig. IV.4.) possess, along with the α,β-unsaturated γ-methyl-γ-lactone, a hydroxyl group at C-4. This is easily detected by ^{1}H nmr spectrometry because of the presence of a characteristic ABX system between the H-3 and H-4 protons. Both protons, H-3a and H-3b appear separately as two doublets of doublet $^2J_{3a-3b}$ = 15 Hz,

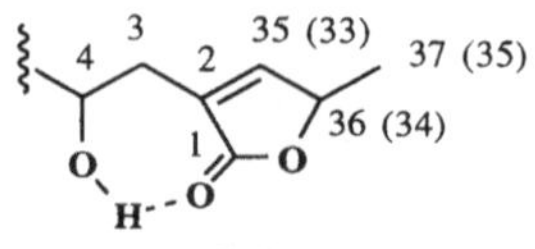

Sub-type 1b

Fig. IV.4.

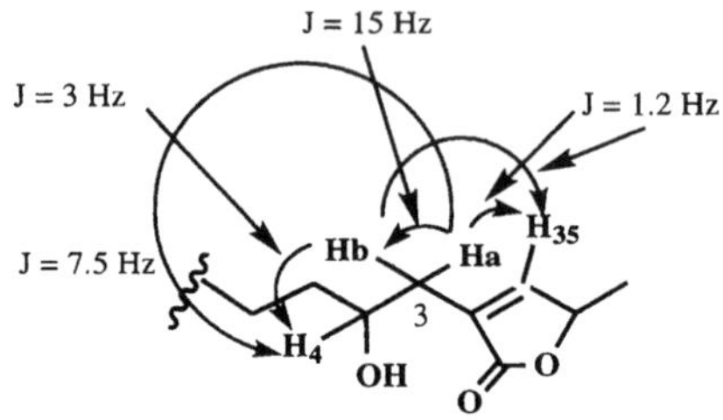

Fig. IV.5. Coupling constants in the sub-type 1b

and $^4J_{3a-35} = {^4J_{3b-35}} \approx 1.2$ Hz (Fig. IV.5.), one at δ 2.40 ppm, the other at δ 2.50 ppm. The non-equivalence of these H-3 protons is probably due to the formation of a pseudo-ring through hydrogen bond formation as shown in Fig. IV.4.

Furthermore, the H-4 proton under the hydroxyl appears at δ 3.83 ppm as a multiplet ($^3J_{4-3a} = 7.5$ Hz and $^3J_{4-3b} = 3$ Hz) (Fig. IV.5.) and the vinylic proton at C-35 (33) appears now at δ 7.17 ppm, shifted downfield compared with sub-type 1a.

The presence of the hydroxyl group at C-4 is confirmed by ^{13}C nmr spectrometry. The carbon atom bearing the OH group appears at δ69.8 ppm and the C-3 is now shifted downfield by 8 ppm because of the substituent at the α position. For instance, in gigantetrocin (sub-type 1b) (*116*) (Fig. IV.6.), the 3-CH_2 methylene appears at δ 33.3 ppm, compared with the 3-CH_2 methylene of corossolone (*38*) (sub-type 1a) at δ 25.6 ppm.

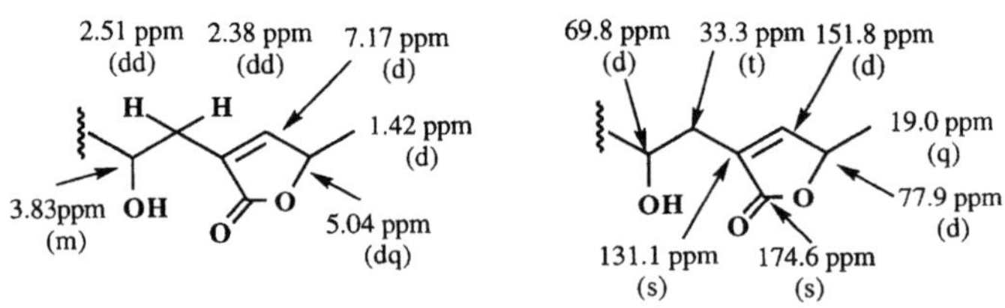

Fig. IV.6. Partial ^1H nmr and ^{13}C nmr data of gigantetrocin, acetogenin of sub-type 1b (*116*)

References, pp. 273–288

IV.2.3. Sub-type 2

Acetogenins of sub-type 2, named isoacetogenins (Fig. IV.7.), belong to a group of compounds bearing a saturated γ-lactone which is substituted at C-2 by an acetonyl function. In the IR spectrum, two characteristic peaks appear at 1770 and 1715 cm^{-1} indicating the presence of two different carbonyl groups. The terminal methyl at C-37 (35) is characterized in the ^{1}H nmr spectrum by the presence of a singlet at δ 2.20 ppm. These compounds always exist as a mixture of two diastereomers *cis* and *trans* with respect to the five-membered ring. Thus signals of the carbon atoms close to C-2 as well as of the H-2, H-3, H-4 and H-35 (33) protons appear as two peaks in the ^{13}C and the ^{1}H nmr spectra respectively (Fig. IV.8).

It is worth noting that the *cis* and *trans* compounds can be separated by means of HPLC (*89*).

In order to distinguish between the two different relative configurations, *cis* and *trans*, the two models were prepared and analyzed by nmr (*47*) (Fig. IV.9.).

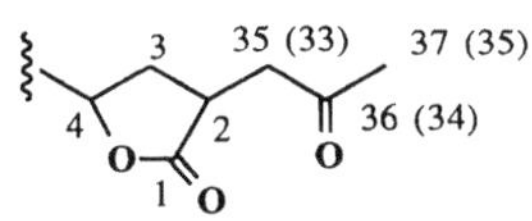

Sub-type 2

Fig. IV.7.

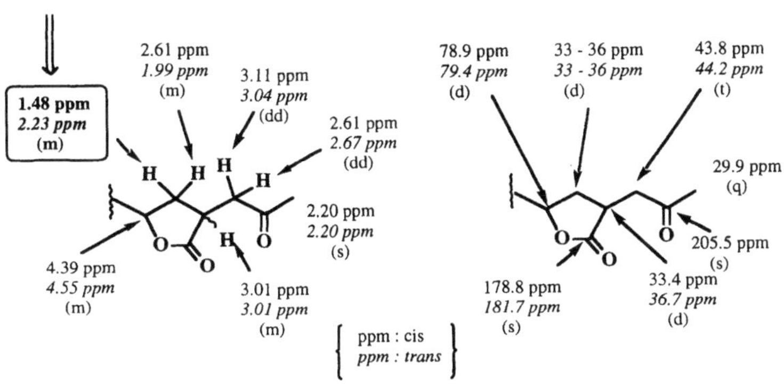

Fig. IV.8. Partial ^{1}H and ^{13}C nmr data of bullatacinone (*90*), acetogenin of the sub-type 2

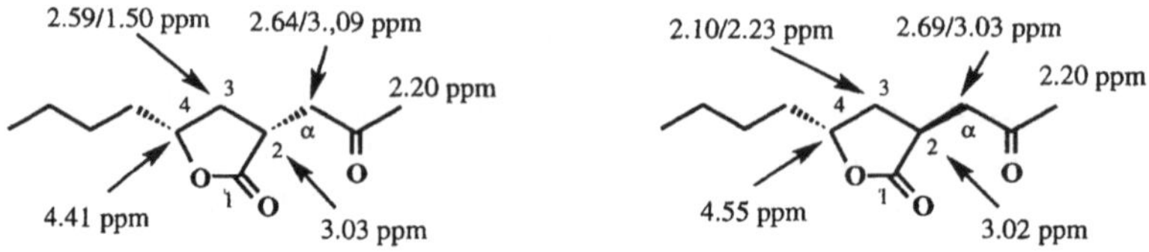

Cis *Trans*

Fig. IV.9. Synthetic *cis*- and *trans*- 2-acetonyl-4-butyl-γ-butyrolactones (*47*)

Table IV. 1. [1]*H nmr of cis- and trans- 2-Acetonyl-4-butyl-γ-butyrolactones*

	Ha-α	Hb-α	Ha-3	Hb-3
Cis	2.64	3.09	2.59	1.50
Trans	2.69	3.03	2.10	2.23

Table IV.1 shows that in the *trans* compound the Hb-3 signal is shifted downfield (δ 2.23 ppm) compared with the *cis* product (δ 1.50 ppm).

Recently DURET *et al.* (*9*) showed that compounds of sub-type 2 do not occur naturally but are obtained by translactonization of acetogenins of sub-type 1b (possessing an OH at C-4), a process that occurs during the extraction and/or purification steps (*9*). Several mechanisms have been proposed to explain the experimental results observed *e.g.* use of methanol as a solvent and/or use of a weakly alkaline medium (Fig. IV.10.).

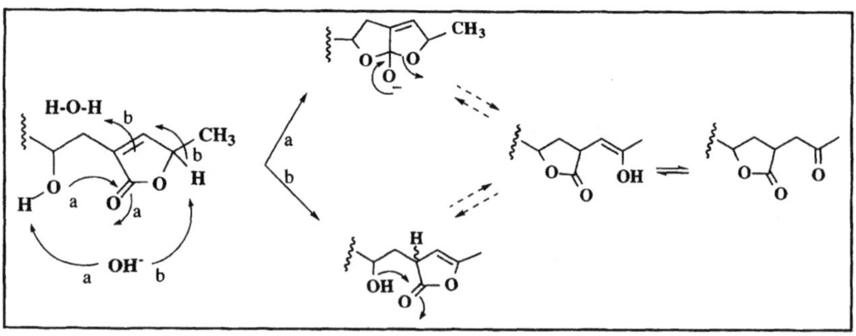

Fig. IV.10. Proposed translactonization mechanism (*9*)

IV.2.4. Sub-type 3

Acetogenins of sub-type 3 (Fig. IV.11) possess a saturated γ-methyl γ-lactone with an OH at the β position of the carbonyl (*127*). These compounds are probably the natural precursors of compounds of sub-types 1a and 1b. Only four acetogenins of this sub-type are known to date, *i.e.* itrabin, jetein, laherradurin and otivarin, all isolated from the same plant (*127*).

The structure of these acetogenins was determined without ambiguity by ^{1}H and ^{13}C nmr spectrometry. A close examination of the ^{1}H nmr spectra of these acetogenins of sub-type 3 shows, for the H-35, H-36 and H-2 protons of the lactone ring, an ABX system with $^3J_{AB} = J_{H-35-H-36} \leqslant 1$ Hz and $^3J_{AX} = J_{H-35-H-2} = 5.5$ Hz. These three protons are *cis* to each other as shown by the NOE effects which are positive between H-35 and H-2, and between H-35 and H-36 (Fig. IV.12.).

All four sub-types can be differentiated easily by nmr analysis from the following simple observations:

1) Resonances at δ 6.95 ppm and at δ 134 and 148 ppm in the ^{1}H and ^{13}C nmr spectra respectively confirm the presence of an unsaturated lactone of sub-type 1a: a slight downfield shift of these values indicates presence of sub-type 1b. Absence of these chemical shifts confirms the presence of a saturated lactone.

2) The presence of an OH at C-4 in sub-type 1b induces a shift downfield, in both the ^{1}H and ^{13}C nmr spectra, for C-1, C-35 (33), C-36

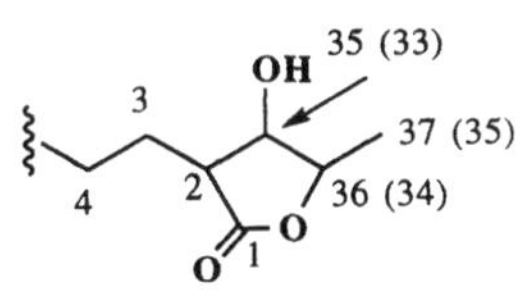

Fig. IV.11.

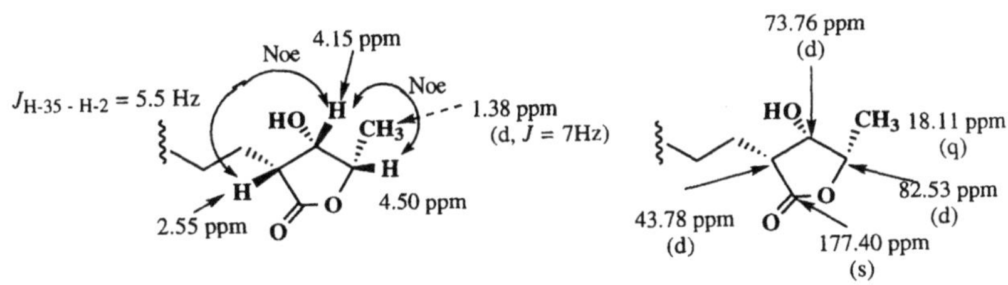

Fig. IV.12. Partial ^{1}H and ^{13}C nmr data of jetein (*127*), acetogenin of sub-type 3

Table IV. 2. *^{1}H and ^{13}C nmr Comparative Data of the Lactonic Moieties*

	Unsaturated				Saturated			
Atom	Sub-type **1a** (corossolone) (lack of OH at C-4)		Sub-type **1b** (murisolin) (OH at C-4)		Sub-type **2** (isocherimolin) (iso)		Sub-type **3** (otivarin) {OH at C-35 or (C-33)}	
	^{1}H	^{13}C	^{1}H	^{13}C	^{1}H	^{13}C	^{1}H	^{13}C
1		173.9		174.6		178/181		177.4
2		134.2		131.1		78.9/79.4		43.78
35 (33)	6.95	148.9	7.17	151.8	3.11/3.04	43.8/44.2	4.15	73.7
36 (34)	4.99	77.4	5.04	77.9		205.5	4.50	82.5
37 (35)	1.42	19.1	1.42	19.1	2.20	29.9	1.38	18.1

(34) compared with the corresponding values in sub-type 1a, and an opposite effect for C-2.

3) Nmr spectra of compounds of sub-type 2 (isoacetogenins) show a typical singlet at δ 2.20 ppm and a characteristic signal at δ 205.5 ppm in the ^1H and ^{13}C nmr spectrum respectively, corresponding to the methyl ketone pattern. Furthermore, the carbonyl signal of the lactone appears at δ 178/181 ppm, shifted downfield in comparison with the unsaturated lactones in which the carbonyl appears at δ 174 ppm (Table IV.2.).

Finally, mass spectrometric observation of the relative intensity of the principal diagnostic ions produced by fragmentation of the lactone ring of the $[M + Li]^+$ ions leads to unambiguous structural identification of the sub-types (*177*).

The EIms of the terminal α,β-unsaturated γ-lactone of sub-type 1a (with a CH_2 at C-4) contains a peak of weak intensity at m/z 111 corresponding to bond fission between C-3 and C-4. When there is an OH at C-4 (sub-type 1b), a peak at m/z 141 corresponding to the bond fission between C-4 and C-5 is observed. Another fragment appears at m/z 123 due to the loss of one molecule of water from the fragment at m/z 141 (Fig. IV.13.) (*175, 177*).

In unsaturated γ-lactones of sub-types 1a and 1b, fragments due to the loss of 28 mass units (m.u.) corresponding to loss of CO and loss of 44 m.u. corresponding to CO_2 from the ion $[M + Li]^+$ are observed. The fragment at m/z $[M + Li - 112]^+$ is formed by cleavage at the β position of the unsaturated γ-lactone, followed by transfer of one hydrogen atom from C-5 toward the lactone ring (Fig. IV.14.): this peak is more intensive (relative abundance: 100%) in the case of compounds of sub-type 1b compared with products of sub-type 1a (relative abundance 20 to 50%).

In the case of acetogenins of sub-type 2, the corresponding peak due to the loss of 112 m.u. is no longer observed, whereas a peak at m/z $[M + Li - 44]^+$ corresponding to the loss of CO_2 is now very intense. Two more peaks of weaker intensity appear at m/z $[M + Li - 58]^+$ and $[M + Li - 114]^+$ which correspond to the loss of the acetonyl substituent and to loss of the lactone ring.

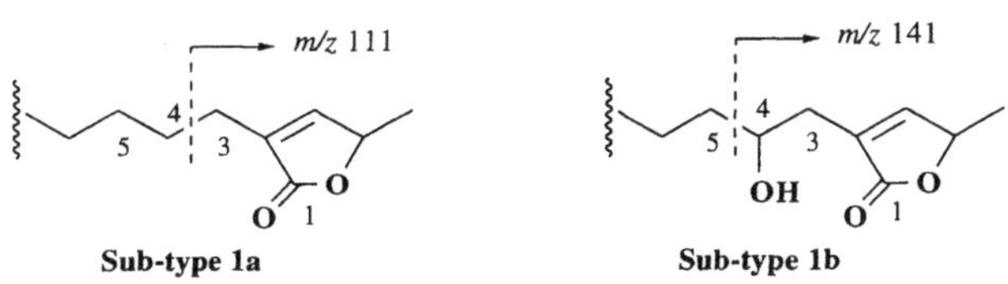

Fig. IV.13. Fragmentation of the lactonic moiety of acetogenins of sub-type 1

Table IV.3. *Intensity of Typical Peaks in FAB-ms-B/E Li⁺ of the Different Lactonic Sub-types (177), Numbers in Bold Type Correspond to the More Intensive Peaks*

Ions	Lactone sub-type **1a** CH_2 at C-4	Lactone sub-type **1b** CHOH at C-4	Lactone sub-type **2** Iso-	Lactone sub-type **3** OH at C-33 (C-35)
$[M + Li - H_2O]^+$	**75–100**	6–12	15–20	**40–100**
$[M + Li - CO]^+$	45–70	< 10	< 10	5–15
$[M + Li - CO_2]^+$	**85–100**	8–15	**100**	**90–100**
$[M + Li - 58]^+$	–	–	7–20	–
$[M + Li - 112]^+$	20–50	**100**	–	–
$[M + Li - 114]^+$	–	–	8–14	–
$[M + Li - 116]^+$	–	–	–	5–15

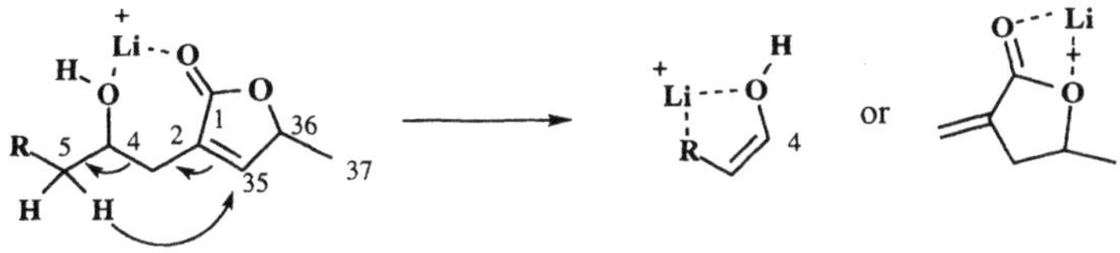

[M+Li]$^+$ [M+Li-112]$^+$ *m/z* 119

Fig. IV.14. Fragmentation leading to $[M + Li - 112]^+$ ion

Compounds of sub-type 3 (which possess a hydroxyl at C-33 or C-35) exhibit a peak at m/z $[M + Li - 116]^+$ corresponding to the cleavage at the α position of the lactone.

IV.3. Identification of the Tetrahydrofuran Rings

The five different types related to either the presence and positions or the lack of tetrahydrofuran rings (THF) on the alkyl chain are easily characterized by ^{13}C nmr analysis.

IV.3.1. Type A (mono THF)

About seventy acetogenins which possess a single THF ring (Fig. IV. 15) have so far been isolated and characterized.

Determination of the presence of the THF ring(s) on the alkyl chain is performed by nmr analysis. In the ^{13}C nmr spectrum the two carbon atoms bound to the oxygen of the THF ring appear between δ 81 and 83 ppm. Furthermore the two α,α' carbon atoms bearing the hydroxyls appear between δ 73 and δ 76 ppm, except in the case of pseudo-symmetry when a single peak appears in each region (see later for the relative stereochemistry). Finally, the signals near δ 3.80 ppm in ^1H nmr spectra correspond to the protons bound to the oxygenated bridge of the THF ring. The CH–OH signals appear at δ 3.40 or δ 3.80 ppm depending on the stereochemistry.

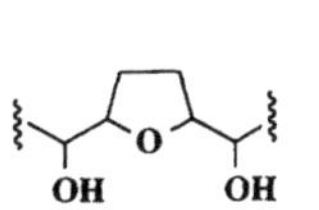

1 or 2 C between 81 and 83 ppm

Fig. IV.15.

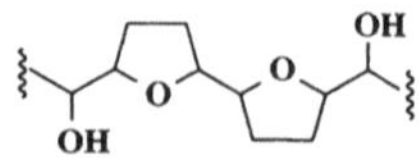

2 or 4 C between 81 and 83 ppm

Fig. IV.16.

Therefore, the first evidence of the presence of an acetogenin of type A (mono THF) in the ^{13}C and ^{1}H nmr spectra is the presence of one or two signals between δ 81–83 ppm and a signal at δ 3.8–3.9 ppm integrating for two or three protons, with additional signals, due to the protons on the hydroxylic carbons at the α and α' position.

A few acetogenins of type A possess only a single hydroxyl at the α position. In this case, while the chemical shift of the proton under the hydroxyl remains the same, integration reveals the presence of only one proton and the oxygenated carbon atom of the THF ring, which does not possess an α substituent now appears at δ 79 ppm.

IV. 3.2. Type B (Adjacent bis-THF)

To date, about forty acetogenins with adjacent bis-THF rings have been isolated and characterized (Fig. IV.16.).

The main difference between acetogenins of type B and of type A is the presence of two or four signals between δ 81 and δ 83 ppm in the ^{13}C nmr and four to six protons at δ 3.80 ppm in the ^{1}H nmr spectrum. When only two signals appear around δ 81–83 ppm, to differentiate compounds of type A from those of type B it is necessary to consider the intensity of the proton signals at δ 3.8–3.9 ppm. When these signals integrate for two or three protons the acetogenin is of type A (mono-THF) whereas a value of four to six protons indicates the presence of an acetogenin of type B (bis-THF).

IV.3.3. Type C (Non-adjacent bis-THF)

About twenty five acetogenins with two non-adjacent THF rings separated by two methylene groups have been isolated and characterized (Fig. IV.17.). The common feature of these compounds is that the first ring, with respect to the terminal γ-lactone, possesses a single OH at the α position between the two rings, while the second THF ring bears two hydroxyls at the α and α' positions. In the ^{13}C nmr spectra, three signals between δ 81 and δ 83 ppm correspond to the carbon atoms of the THF

References, pp. 273–288

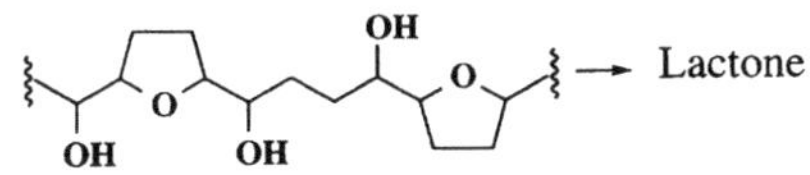

3 C between 81 and 83 and 1 C at 79 ppm

Fig. IV.17.

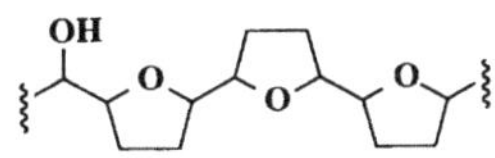

6 C between 79 and 83 ppm

Fig. IV.18.

rings close to the CH–OH group. A single signal at δ 79 ppm is typical for the carbon atom of the THF ring which does not possess an α substituent.

IV.3.4. Type D (adjacent tris THF)

A single acetogenin, goniocin, with 37 carbon atoms and three adjacent THF rings with a single hydroxyl α to the last ring, has been isolated and characterized (Fig. IV.18.) (*99*). Recently, two hemisynthetic isomers have been obtained from goniodenin, an acetogenin of type B (*75*). The ^{13}C nmr spectrum shows six signals between δ 79 and δ 83 ppm with the corresponding peaks in the ^1H nmr spectrum between δ 3.78 and δ 3.96 ppm integrating for six protons.

IV.3.5. Type E (Non-THF Acetogenins)

An increasing number of compounds without THF rings has been isolated and characterized. It is generally admitted that these compounds, classified as type E, are the natural precursors of the THF acetogenins. These substances, with 35 or 37 carbon atoms and a terminal γ-lactone of sub-type 1, may possess one, two or three epoxide rings and/or double bond(s) or other functional groups cited below. It is noteworthy that two compounds, reticulatamol and reticulatamone have been very recently isolated which possess only a hydroxyl or a ketonic function, (*143*, *145*). To this list, muricatacin (*39*) and squamostanal-A (*160*) can be added, even though these two are probably artefacts of isolation due to oxidation which, after a C–C bond fission, has led to these metabolites.

IV.4. Identification of Functional Groups

Acetogenins of types A, B, C, D and E cited above may also possess functional groups such as hydroxyl, carbonyl, vicinal diols, unsaturations and epoxides on the alkyl chain. Recently, mucocin, an acetogenin of type A bearing a hydroxy-tetrahydropyran ring, was described (*133*). The presence of such groups is evidenced by close examination of the ^{13}C and ^{1}H nmr spectra, as well as by classic spectroscopic means.

The presence of the epoxide rings is indicated by characteristic signals at δ 2.95 ppm in the ^{1}H nmr spectra and between δ 56 and δ 57 ppm in the ^{13}C nmr spectra. Other functional groups are identified by their respective typical chemical shifts in the ^{1}H and ^{13}C nmr spectra and the corresponding absorption peaks in the IR spectra as shown in the Table IV.4.

IV.5. Location of THF and Functional Groups on the Alkyl Chain

While the presence of these functional groups can easily be ascertained by nmr spectroscopy, locating them on the chain by nmr is still quite challenging and therefore is usually accomplished by different mass spectrometric techniques.

IV. 5.1. Location of the THF Pattern

Degradation methods

When large quantities of acetogenins are available, degradation studies can be used to ascertain the position of the tetrahydrofuran pattern. One method consists of Jones' oxidation of the acetogenin (*157*), followed by analysis of the EIms of the products. (Fig. IV.19.).

Thus in the case of atemoyin, oxidative degradation by $Pb(OAc)_4$ and identification by GC of undecanal led to placement of the bis-THF core (*78*).

Electron Ionization (EIms)

Conventional electron ionization mass spectrometry, especially using permethylated and perdeutero methylated derivatives at 70 eV (EIms) allows one to locate the THF rings in the chain (*159*, *178*) although in a few cases, it does not serve to elucidate completely the structure of the acetogenins. However peracetylated derivatives are useless in mass spectroscopy analysis due to their fast deacetylation. Fragments corresponding to cleavage at the α position of the CH–OR are observed (*5*, *175*) as

Table IV.4. *Characteristic Chemical Shifts of the Functional Groups*

Functional groups	^1H nmr (ppm)	^{13}C nmr (ppm)	IR (cm^{-1})
Hydroxyl: —*CH*—OH	3.5	71.4	3400–3600
—CH_2—CHOH—CH_2—	1.4	37	
Diol: —*CHOH*—*CHOH*—	3.4	77	3400–3600
—CH_2—CHOH—CHOH—CH_2—	1.30–1.50	33–35	
Ketone: —C=O—	–	211	1700–1710
—CH_2—(C=O)—CH_2—	2.4	42.7–42.8	
Double bond: —*HC*=*CH*—	5.3	128–130	–
	cis: (J = 6–12 Hz)		
	trans: (J = 12–18 Hz)		
—CH_2—HC=CH—CH_2—	*cis*: 2.20 and 2.05	*cis*: 27	–
	trans: 2.00	*trans*: 32	
Epoxides: 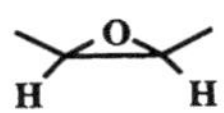	2.99	56–58	
-H_2C–O–CH_2-	1.55	26–28	

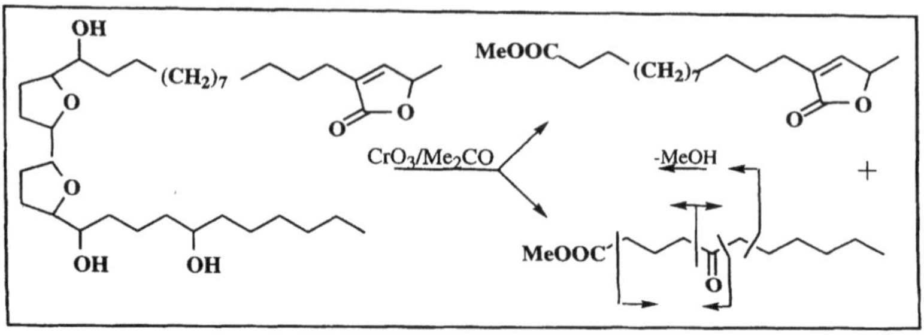

Fig. IV.19. Fragmentations observed in EIms, after oxidation

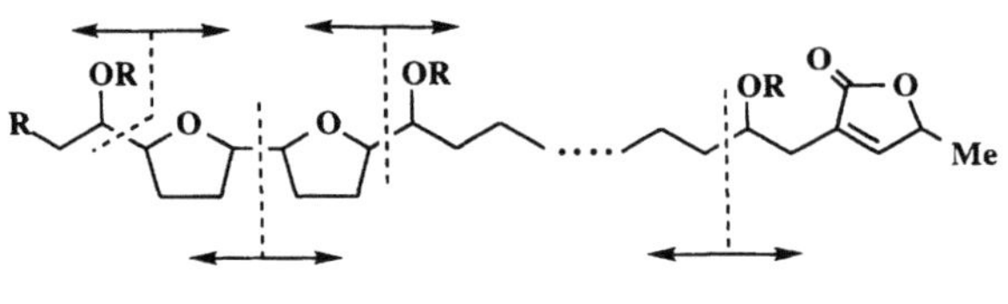

Fig. IV.20. EIms fragmentations of acetogenins and acetylated or trimethylsilyl derivatives

well as, in the case of type B acetogenins, fragments due to C–C bond fission between two THF rings (Fig. IV.20.).

Chemical Ionization (CIms)

The position of the THF ring(s) may be determined by analysis of the CIms of the natural acetogenin and its derivative obtained by hydrogenation of the unsaturated γ-lactone (*175, 175a*). Fragments corresponding to the lactone moiety increase by 2 m.u. whereas fragments containing the THF framework remain unchanged (Fig. IV.21.).

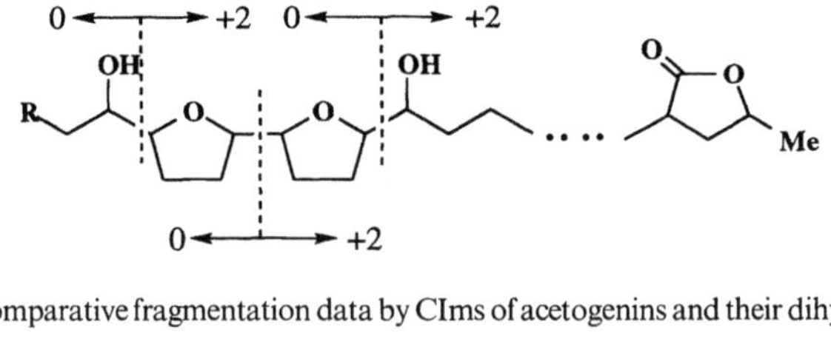

Fig. IV.21. Comparative fragmentation data by CIms of acetogenins and their dihydro-derivatives

FAB-ms-B/E Li$^+$ (FABms; FAB-Li ms; CID *B/E* Li)

More recent techniques, *e.g.* tandem mass spectrometry (ms/ms) (*106*, *162*), fast atom bombardment in the presence of lithium (FAB-Lims) and scanning (at B/E constant) of the $[M + Li]^+$ ion activated by collision (CID *B/E* Li^+), are very useful for determining the relative positions of the functional groups (*107*, *177*, *179*). Thus, analysis of the fragments obtained by linked scan at *B/E* constant of the $[M + Li]^+$ ion activated by collision (CID *B/E* Li^+) has been found to be a very efficient method for locating the THF rings (*177*). Typical fragmentations are shown in Fig. IV. 22. Ions containing the lactone part are represented by the letters A, B, and C, and ions containing the terminal alkyl chain by the letter X or Y.

Ions A and X are formed by cleavage α to a hydroxyl and retain the oxygen atom. Ions B and Y correspond to fragmentation of bonds within

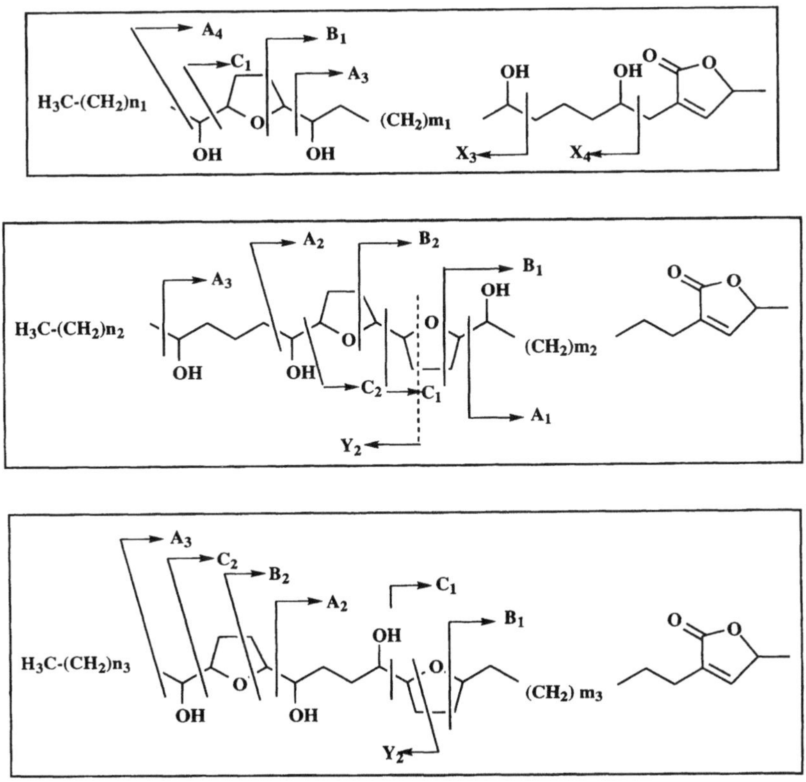

Fig. IV.22. Fragmentation scheme of acetogenins of type A, B or C by FAB-ms-*B/E* Li^+ (*177*)

the THF ring, and ions C to cleavage at the α position of the ring. The numbers accompanying these letters distinguish different ions. For instance, the B_1 ion corresponds to the fragment arising from cleavage of the first THF ring (numbered with respect to the lactone) while the B_2 ion indicates the same type of fragmentation for the second THF ring. This method allows use of the observed fragmentations as a fingerprint of an acetogenin. This technique also allows one to locate the non-adjacent THF rings (*177*). The utility of other ionization methods is very limited compared with this technique which does not require any derivatization prior to analysis (*177*).

IV.5.2. Location of the Functional Groups

IV.5.2.1. Hydroxyls

Hydroxyls flanking the THF at C-4 and on the γ-lactone ring (subtype 3), are easily located. For the other hydroxyls along the chain the problem is more delicate. Careful analysis of the EIms-induced fragmentations of acetogenins and their permethylsilylated derivatives has been used often (*5, 175*), while CID *B/E* Li^+ is also very helpful in determining the relative positions of hydroxyls (*177*). To further confirm the location of the different hydroxyls, 2D nmr techniques such as COSY ^1H, double relayed COSY, NOeSY (*6*), HMBC, HMQC, HOHAHA (*96*) *etc.* are now in constant use, while the preparation of derivatives such as acetonides allows one to establish the location of vicinal diols (*18*). Finally the ^1H nmr chemical shift of the terminal methyl (C-34) of hydroxy-bullatacinones serves to locate the hydroxyl at either C-28, C-29, C-30, C-31 or C-32 (*121*).

IV.5.2.2. Epoxides

The location of the epoxide functionality in alkyl chains has long been a major challenge (*107*). Several methods have been developed, depending on the structure of the compound in questions. In the case of epoxides formed by oxidation of a terminal double bond, EIms allows an easy determination of the location (*180*) whereas in the case of monoepoxides obtained from polyunsaturated fatty acids, the determination is best performed by charge exchange mass spectrometry at low energy (*181*). Chemical ionization (with isobutane or nitric oxide) has also been used for the determination of the location of epoxide rings (*182, 183*).

In the case of acetogenins, several methods have been used.

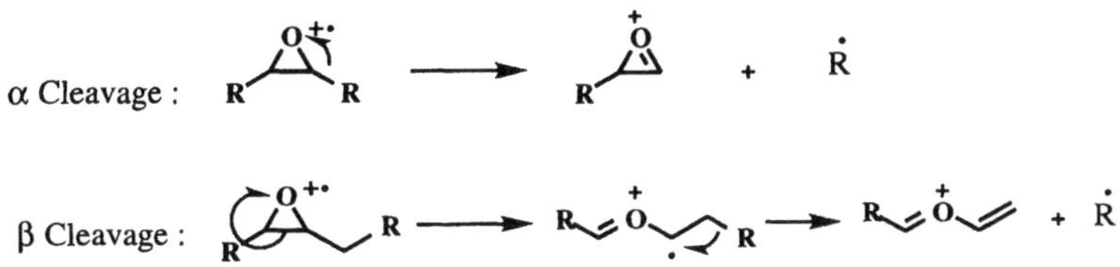

Fig. IV.23. Epoxide fragmentation described by DJERASSI *et al.* (*180*)

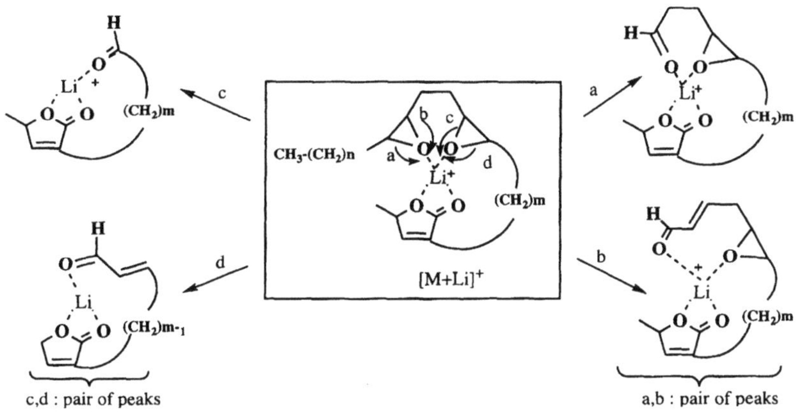

Fig. IV.24. Mechanism proposed for the formation of ions

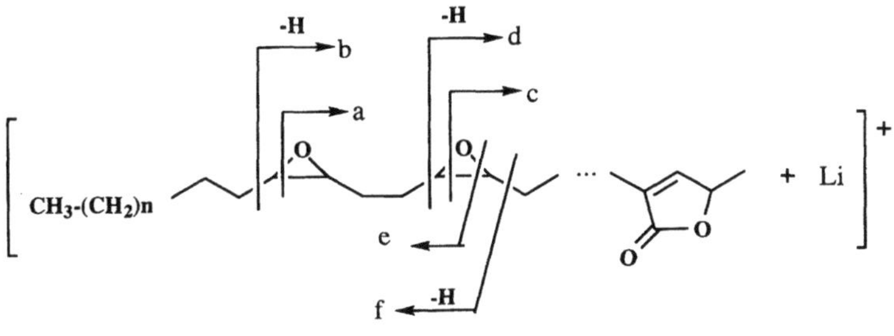

Fig. IV.25. Typical fragmentations of acetogenins of type E by FAB-ms-*B*/*E* Li^+

EIms

EIms was used in the early beginnings. The most important fragments are due to cleavages at the α and β positions of the epoxide through a mechanism proposed by DJERASSI *et al.* for epoxy-alkanes (*180*) (Fig. IV.23.).

In the case of acetogenins we have noticed that even though several epoxides may be present in the molecule a unique fragmentation occurs at the α position of the first epoxide counted from the terminal lactone.

Therefore, locating the second or even the third epoxide ring is not feasible by using this technique.

FAB-ms-B/E Li$^+$ [CID (*B/E*) Li$^+$]

We have found that the FAB-ms-*B/E* Li$^+$ method is a very powerful technique for locating epoxide rings along the alkyl chain of acetogenins. In addition to fragmentations associated with the tetrahydrofuran acetogenins (loss of CO and CO_2), the spectra contain diagnostic pairs of peaks separated by 12 m.u. The mechanism proposed for rationalizing the formation of such ions is shown below (Fig. IV.24.) (*107*).

These pairs of peaks separated by 12 m.u. are due to cleavage at the α and α' positions of the epoxide rings (cleavages **a** and **b**, and cleavages **c** and **d**) (Fig. IV.24. and IV.25.). Other peaks of weaker intensity are due to cleavages **e** and **f** of the first epoxide containing the terminal methyl (*107*).

IV.5.2.3. Other Substituents

The relative positions of carbonyl groups, are determined by classic EIms fragmentation. For double bonds, preparation of derivatives, *e.g.* epoxides, followed by fragmentation techniques described in the previous section is helpful.

IV.6. Stereochemical Relationships of Acetogenins

After determining the relative positions of the functional groups, the next step in structural elucidation of acetogenins is the determination of the relative and absolute configurations of the stereogenic centres.

Direct X-ray analysis is very difficult in the case of acetogenins because of their waxy nature. To date, a single acetogenin, gigantecin, has been analyzed by X-ray methods (*114*), and two others were studied after derivatization (*148, 158*). However, the absolute configurations could not be deduced from their X-ray analyses.

IV. 6.1 Relative Configuration

A recent review by RAMIREZ and HOYE (*184*) summarizes the state of the art for establishing the configuration of acetogenins. These authors also point out the problem of directionality of the stereochemical relationship across the THF moiety in the case of dissymmetry. Indeed, in a few cases, long-range coupling from a nearby hydroxyl or other functional group may serve to determine with certainty the order of the relative configuration, *e.g. threo-trans-erythro versus erythro-trans-threo.*

IV.6.1.1. Lactone Moiety

Models containing the terminal γ-lactone group of sub-type 1b with an OH at C-4 were synthesized (*185, 45*). These synthetic models were divided into two classes matched ***SS*** (4*S*, 36*S* or 4*R*, 36*R*) and mismatched ***RS*** (4*R*, 36*S* or 4*S*, 36*R*). These compounds were then derivatized as (*S*) and (*R*)-MTPA Mosher esters. The absolute configuration at C-4 was first deduced by the usual method, *i.e.* a determination of the difference of the chemical shifts ($\Delta\delta_H = \delta_S - \delta_R$). However the authors observed that the *magnitude* of the $\Delta\delta_H$ was quite different for the matched and mismatched compounds. For proton H-35 (or H-33), the absolute value $|\Delta\delta_H|$ was 0.23 ppm for the pair of mismatched diastereomers, whereas it was 0.32 ppm for the matched products. For proton H-36 (or H-34), the absolute value $|\Delta\delta_H|$ was 0.06 ppm for the mismatched compounds, whereas it was 0.17 ppm for the "matched" products. Therefore, comparing the absolute values $|\Delta\delta_H|$ for protons H-35 and H-36 (or H-33 and H-34) of the Mosher esters of the acetogenin in question with those of the models would allow one to deduce the "matched" or "mismatched" relationship.

The absolute values $|\Delta\delta_H|$ for H-35 (or H-33) 0.24 to 0.26 ppm and for H-36 (or H-34) 0.04 to 0.06 ppm were reported for five naturally occurring acetogenins. This indicates a mismatched relationship between C-4 and C-36 (or C-34) for all acetogenins studied.

In order to distinguish the two different relative configurations 2,4-*cis* and 2,4-*trans* in a lactone of sub-type 2, two models were synthesized and analyzed by nmr (*47*) (Fig. IV.26.).

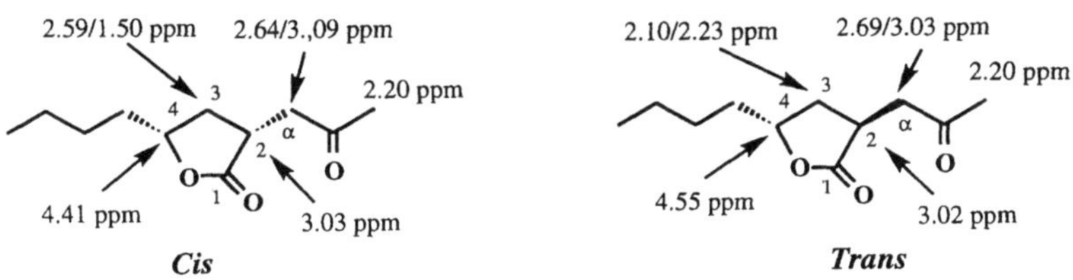

Fig. IV.26. Synthetic *cis*- and *trans*- 2-acetonyl-4-butyl-γ-butyrolactones (*47*)

Table IV.5. ^{1}H *Chemical Shifts of cis- and trans- 2-Acetonyl-4-butyl-γ-butyrolactones* (*47*)

	Ha-α	Hb-α	Ha-3	Hb-3
Cis	2.64	3.09	2.59	1.50
Trans	2.69	3.03	2.10	2.23

Note that the signal of Hb-3 is shifted downfield for the *trans* compound (δ 2.23 ppm) compared with Hb-3 of the *cis* product (δ 1.50 ppm) (Table IV.5.).

In the case of acetogenins of sub-type 3, all substituents of the lactone ring were found to be *cis* to each other by observation of a positive NoE between H-33 (35) and H-34 (36) (*127*).

IV.6.1.2. THF Pattern

Although the first acetogenin was found in 1982, it was only in 1988 that the first report concerning the relative configuration of the stereogenic centres across the THF pattern was published (*186*). This relative stereochemistry can be established by comparing the ^{13}C and ^{1}H nmr data with those of synthetic models of known relative configuration.

Hoye et al. models (*186*)

This method is based on a comparison of the chemical shifts of protons of acetylated derivatives of acetogenins with those of synthetic model **A** of known relative configuration (*186*). Compound **A** was prepared by two different pathways: a cascade reaction from a triepoxide (pathway **a**) or through pathway **b** (Fig. IV.27.). 64 diastereomers of compound **A** which possesses six asymmetric centres are theoretically possible.

Nevertheless, in view of the existence of *meso* compounds, only twenty different diastereomers are possible twelve of which were synthesized and analyzed by ^{1}H nmr spectrometry (Table IV.6.).

Using these twelve relative configurations, eight new relative configurations can be deduced by combining known configurations. For instance, configuration *er/c/th/c/th* may be obtained by combining the first part of case 1 (*er/c/th*) with the last part (*th/c/th*) of case 7 or 8, but more probably with case 7 because of the stereochemical relationship *c/th/c/th* (Table IV.7.).

Therefore, the chemical shifts of the protons concerned corresponding to the relative configuration *er/c/th/c/th* (case 13) may be obtained

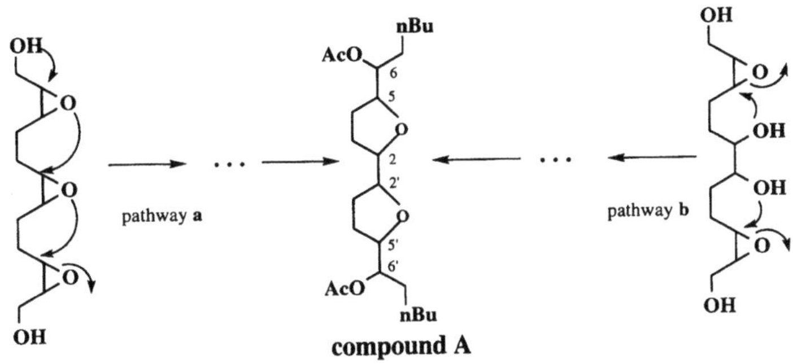

Fig. IV.27.

Table IV.6. *Chemical Shifts δ_H in ppm of 12 Diastereomers reported by* HOYE (*186*)

	H-6	**H-5**	**H-2**	**H-2′**	**H-5′**	**H-6′**
1 = *er/c/th/c/er*	4.90	3.94	3.81	3.81	3.94	4.90
2 = *er/t/th/c/er*	4.91	4.01	3.88	3.76	3.93	4.91
3 = *er/t/th/t/er*	4.91	3.98	3.88	3.88	3.98	4.91
4 = *er/c/er/c/er*	4.95	3.91	3.71	3.71	3.91	4.91
5 = *er/t/er/c/er*	4.96	3.97	3.81	3.80	3.91	4.91
6 = *er/t/er/t/er*	4.92	3.99	3.84	3.84	3.99	4.92
7 = *th/c/th/c/th*	4.94	3.93	3.86	3.86	3.93	4.94
8 = *th/t/th/c/th*	4.88	4.08	3.93	3.84	3.91	4.88
9 = *th/t/th/t/th*	4.85	3.97	3.90	3.90	3.97	4.85
10 = *th/c/er/c/th*	4.84	3.93	3.77	3.77	3.93	4.84
11 = *th/t/er/c/th*	4.85	3.97	3.82	3.82	3.93	4.85
12 = *th/t/er/t/th*	4.84	3.97	3.84	3.84	3.97	4.84

Table IV.7.

	H-6	H-5	H-2	H-2′	H-5′	H-6′
1 = *er/c/th/c/er*	4.90	3.94	3.81	3.81	3.94	4.90
7 = *th/c/th/c/th*	4.94	3.93	3.86	3.86	3.93	4.94
13 = *er/c/th/c/th*	4.90	3.94	3.86	3.86	3.93	4.94

by combining the chemical shifts of protons H-6, H-5 from case 1 and from H-2, H-2′, H-5′ and H-6′ from case 7. In the same manner, the remaining unknown configurations can be deduced and are listed below (Table IV. 8.).

To determine the relative configuration, the sum of the differences of chemical shifts ($\Sigma|\Delta\delta's|$) between identical protons of an acetylated

Table IV.8. *Deduced Relative Configurations of Eight Unknown Diastereomers by* HOYE (*186*)

13 = *er/c/th/c/th*	1 + 7
14 = *er/t/th/c/th*	2 + 8
15 = *th/t/th/c/er*	8 + 2
16 = *er/t/th/t/th*	3 + 9
17 = *er/c/er/c/th*	4 + 10
18 = *er/t/er/c/th*	5 + 11
19 = *th/t/er/c/er*	11 + 5
20 = *er/t/er/t/th*	6 + 12

acetogenin and those of the synthetic model is calculated. The value of this sum has to be the lowest one in order to determine the correct relative configuration. This method has been validated for rolliniastatin-1 (*186*), since the structure was determined by crystallographic analysis (*148*). Even though the differences in the chemical shifts of natural acetogenins are very low, the method has so far been used successfully for the determination of the relative stereochemistry of acetogenins of type B.

In order to determine the relative configurations of acetogenins of types A and C, other models of known configurations were synthesized by several authors (*20*, *158*, *187*, *188*, *189*). In the following the different models are discussed with highlights on the advantages and drawbacks of each.

Born et al. models (*158*)

One of the earliest studies is due to BORN *et al.* who synthesized two models of α-hydroxylated mono-THF compounds, with the known configuration *erythro* and *threo* (Fig. IV.28.) (*158*). The chemical shifts, both in the ^{1}H and ^{13}C nmr spectra of the synthetic products matched well with those observed for annonin I (squamocin) whose stereochemistry was further confirmed by crystallography of a derivative obtained after treatment by KOH in methanol (*158*) (Table IV.9.).

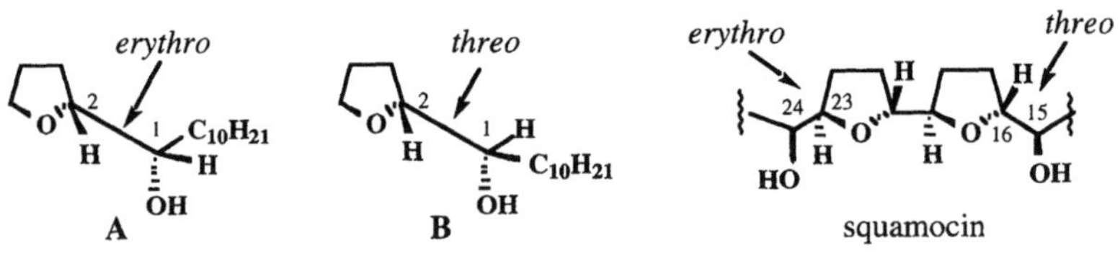

Fig. IV.28.

Table IV.9. *Chemical Shifts for THF Pattern of Synthetic Models and Squamocin* (*158*)

	Synthetic models		Squamocin	
	A: *erythro*	**B**: *threo*	*erythro*	*threo*
H-1	3.84	3.40	H-24: 3.87	H-15: 3.40
H-2	3.84	3.79	H-23: 3.87	H-16: 3.88
C-1	71.83	73.87	C-24: 71.66	C-15: 74
C-2	82.29	82.47	C-23: 82.87	C-16: 83.38

Table IV.10. *Relative Configurations versus Chemical Shifts in* ^1H *nmr by* GALE *et al.* (*187*)

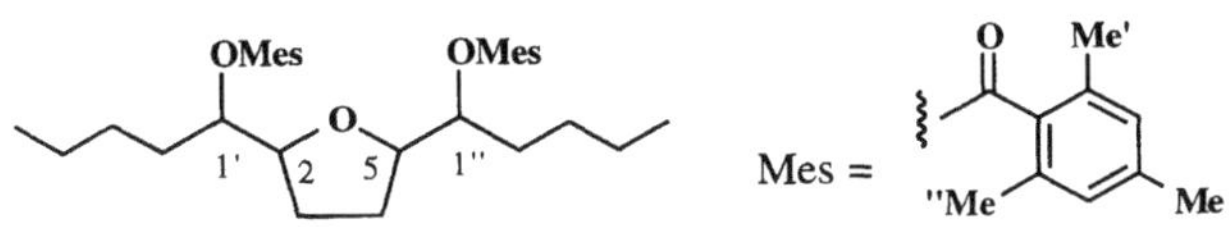

Model	**Relative configuration**	**δH (2, 5)**	**δH (1′, 1″)**	**δH (Me′, Me″)**
1	*threo-cis-threo*	3.97	5.29	2.45
2	*erythro-cis-erythro*	3.93	5.45	2.40
3	*threo-cis-erythro*	3.89; 3.97	5.37; 5.43	2.39; 2.48
4	*threo-trans-threo*	4.09	5.35	2.45
5	*threo-trans-erythro*	4.00; 4.11	5.30; 5.52	2.39; 2.51
6	*erythro-trans-erythro*	4.05	5.48	2.40

Gale et al. models (*187, 188*)

From different stereomers of 1,5-dienes six isomers of dimesitoyl-α, α′-dibutyl-2,5-tetrahydrofurandimethanol have been prepared stereospecifically. These compounds contain all possible relative configurations across the THF ring. Analysis by ^1H nmr spectrometry of all isomers allow one to correlate the typical chemical shifts with a known configuration (Table IV.10.).

In the same manner as previously, the determination of the relative configuration of an acetogenin is possible by applying the following equation and attributing the configuration to the lowest ΣΔδH value *i.e.* $\Sigma\Delta\delta H = |\delta H_{observed} - \delta H_{model}|_{H-2,5} + |\delta H_{observed} - \delta H_{model}|_{H-1',1''}$.

The ^1H nmr spectrum of the mesitoate of annonacin contains a multiplet at δ_H 4.13 ppm corresponding to the protons at positions 2 and 5 of the THF ring, and another multiplet at δ_H 5.32 ppm corresponding to

Table IV.11. *Determination of the Relative Configuration of Annonacin by* GALE *et al.* (*188*)

Annonacin C-15/C-16 and C-19/C-20	δH-2, 5 δH observed = 4.13		δH-1′, 1″ δH observed = 5.32		
Possible configuration	δH_{model}	$\|\Delta\delta H\|$	δH_{model}	$\|\Delta\delta H\|$	$\sum\Delta\delta H$
threo-cis-threo	3.97	0.16	5.29	0.03	0.19
erythro-cis-erythro	3.93	0.20	5.45	0.13	0.33
threo-trans-threo	4.09	**0.04**	5.35	0.03	**0.07**
erythro-trans-erythro	4.05	0.08	5.48	0.16	0.24

Table IV.12. *Partial* [1]*H and* [13]*C nmr Data for the Models prepared by* HARMANGE *et al.* (*189*)

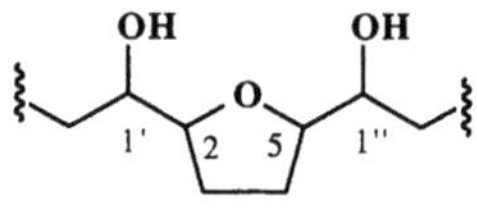

	HC-1′		HC-2		HC-5		HC-1″	
	^{1}H	^{13}C	^{1}H	^{13}C	^{1}H	^{13}C	^{1}H	^{13}C
threo-trans-threo	3.41	74.04	3.80	82.71	3.80	82.71	3.41	74.01
erythro-trans-threo	3.82	71.56	3.82	83.24	3.82	82.15	3.38	74.33

the methine protons of the ester (1′, 1″ for the model). This indicates the presence of pseudo-symmetry in the molecule; (therefore four compounds are possible: **1**, **2**, **4**, and **6**. The lowest ΣΔδH value of the annonacin derivative corresponds to compound **4** having the configuration *threo-trans-threo*. The configuration of annonacin is therefore *threo-trans-threo* (Table IV.11.).

For the asymmetric mono-THF system, the first indication is given by the presence of two different multiplets corresponding to the protons at C-2 and C-5 and two other multiplets corresponding to protons 1′ and 1″ (model numbering). The configuration is then deduced by comparison with the two models **3** and **5**.

Harmange et al. models (*189*)

Two models of mono-THF acetogenins bearing two hydroxyl groups at the α and α′ positions of the THF ring, with known configurations

Table IV. 13. ^{13}C *nmr Data of Synthetic Models prepared by* FUJIMOTO (*20*)

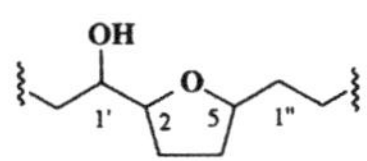

(a): *threo-trans*

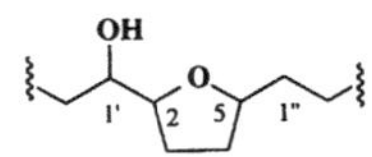

(b): *threo-cis*

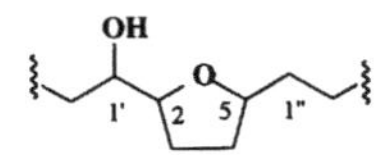

(c) :*erythro-trans*

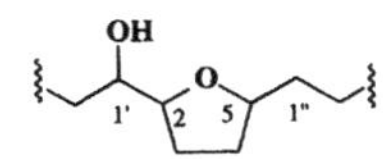

(d): *erythro-cis*

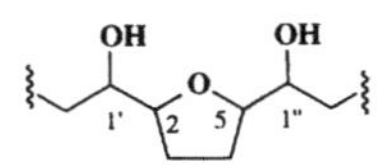

(e): *threo-trans-threo*

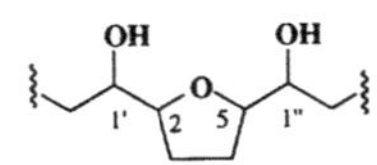

(f): *threo-cis-threo*

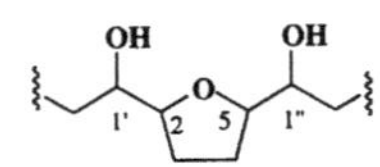

(g): *erythro-trans-threo*

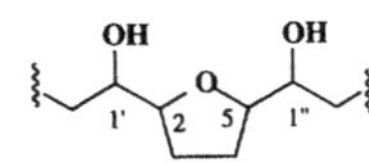

(h):*erythro-cis-threo*

Atom	**(a)** *th-t*	**(b)** *th-c*	**(c)** *ery-t*	**(d)** *ery-c*	**(e)** *th-t-th*	**(f)** *th-c-th*	**(g)** *ery-t-th*	**(h)** *ery-c-th*
1′	74.2	74.5	72.0	71.6	74.0	74.3	71.6	72.1
2	81.9	82.2	81.5	82.1	82.7	82.8	82.2	82.8
3	28.4	27.8	25.0	23.9	28.8	28.1	25.2	24.1
4	32.4	31.4	32.3	31.4	28.8	28.1	28.6	28.4
5	79.3	79.9	80.2	79.6	82.7	82.8	83.3	82.3
1″	35.7	36.1	36.1	35.8	74.0	74.3	74.3	74.2

threo-trans-threo and *erythro-trans-threo*, were stereospecifically synthesized from L-glutamic acid (*189*). The ^{1}H and ^{13}C spectra of the two exhibited significant differences of chemical shifts both for the carbon atoms bearing the hydroxyls and for the geminal protons (Table IV.12.).

Fujimoto et al. models (*20*)

Recently, FUJIMOTO *et al.* (*20*) prepared several mono-THF models of known configuration, *i.e. threo-trans* (a), *threo-cis* (b), *erythro-trans* (c) and *erythro-cis* (d) of 5-heptyl-2-(1-hydroxyheptyl)-tetrahydrofuran, as well as the isomers *threo-trans-threo* (e), *threo-cis-threo* (f), *erythro-trans-threo* (g) and *erythro-cis-threo* (h), of 2,5-di-(1-hydroxyheptyl)-tetrahydro furan. The corresponding ^{13}C nmr data are shown in the following Table IV.13.

The situation in acetogenins with two non-adjacent THF rings can be approximated by considering them as made up of two systems, an α hydroxylated mono-THF and an α, α' dihydroxylated mono-THF.

To date, only two different relative configurations for such acetogenins with non-adjacent THF rings have been encountered: *-trans-threo/threo-trans-threo* and *trans-threo/threo-trans-erythro* (*11*, *12*).

In conclusion, the various models described above are very useful for establishing the relative configuration of acetogenins, but they present drawbacks as well as advantages which are complementary to each other (Table IV.14.).

The HOYE and GALE models contain all possible configurations across the THF pattern, but they have to be derivatized (*e.g.* acetylated or permesithylated). Furthermore, the comparison is made by ^{1}H nmr spectrometers and because differences in chemical shifts are quite small, identical conditions (solvent concentration, temperature etc) have to be used.

BORN and HARMANGE have shown that direct observation of the ^{1}H and ^{13}C nmr spectra allows one to distinguish the different configurations of the acetogenins without derivatization, however, they did not study compounds with the *cis* configuration.

FUJIMOTO prepared several models of ring *e.g.* α-hydroxylated THF and α,α'-dihydroxylated THF of known configuration which, by simply comparing the ^{13}C nmr data with those of natural acetogenins, allowed the determination of relative configuration.

In the following scheme, all these methods are summarized.

Therefore, determination of the relative configurations of acetogenins now becomes straightforward, and can be deduced from the following observations:

Table IV. 14. *Advantages and Drawbacks of Methods*

Authors	Advantages	Drawbacks
HOYE	20 models	acetylated derivatives comparison only by ^1H nmr
BORN	no derivatization comparisons by ^1H and ^{13}C	only two configurations: *threo* and *erythro*
GALE	six models	mesithylated derivatives comparison only by ^1H (C_6D_6)
HARMANGE	no derivatization comparisons by ^1H and ^{13}C	only two configurations: *threo-trans-threo* *erythro-trans-threo*
FUJIMOTO	no derivatization all possible configurations	comparison only by ^{13}C

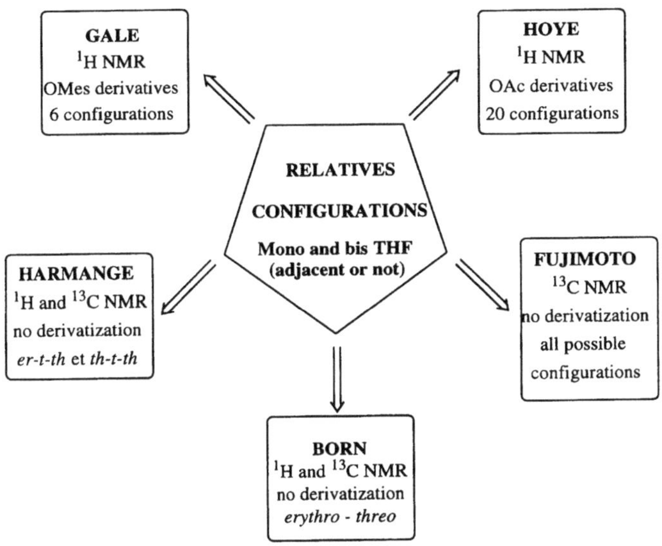

Scheme IV.1.

1) When pseudosymmetry exists across the THF pattern (Fig. IV.29.), only two signals appear between δ 71 and δ 83 ppm in the ^{13}C nmr spectrum for acetogenins of type A; these correspond to the four carbon atoms linked to the oxygen atoms. Three resonances are observed for the six carbon atoms attached to oxygen atoms in the case of acetogenins of type B,

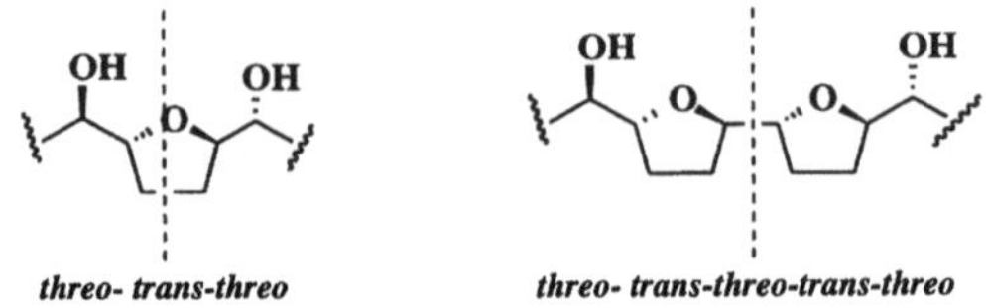

Fig. IV.29.

2) The *threo* and *erythro* stereochemical relationships are indicated by the signals at δ 74 and δ 71 ppm respectively, in the ^{13}C nmr, spectrum and by the signals at δ 3.40 and δ 3.81 ppm, respectively in the ^{1}H nmr spectrum (*20, 158, 189*).

3) Furthermore, signals of the methylene carbons of the THF rings are shifted downfield (3 ppm) for compounds with the *threo* configuration (at δ 28 ppm), compared with compounds with the *eythro* configuration (at δ 24–25 ppm).

4) Differentiation between *cis* and *trans* configuration of the THF rings is rather difficult if one limits oneself to the chemical shifts of the carbon atoms bearing the oxygen atoms (at δ 81 and δ 82 ppm). However, FUJIMOTO *et al.* noticed that the methylene of the THF rings of the synthetic models exhibit a quite remarkable difference in chemical shifts. Indeed, in the *cis* isomers, the methylene protons H_{3a} and H_{4a} appear at δ 1. 82, whereas in the *trans* compounds, δ H_{3a} and H_{4a} are now at δ 1.65 (*100*) (Fig. IV.30.).

IV. 6.1.3. 1,2-, 1,4- and 1,5-Diols

The relative configuration of vicinal diols in acetogenins can be determined by analysing the proton chemical shifts of the acetonide derivatives. In *threo* derivatives the two acetonyl methyls appear as a singlet at about δ 1.35 (6H) whereas in *erythro* compounds, two singlets

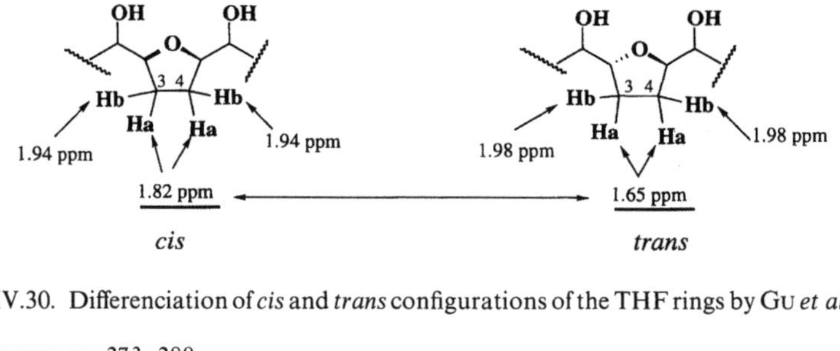

Fig. IV.30. Differenciation of *cis* and *trans* configurations of the THF rings by GU *et al.* (*100*)

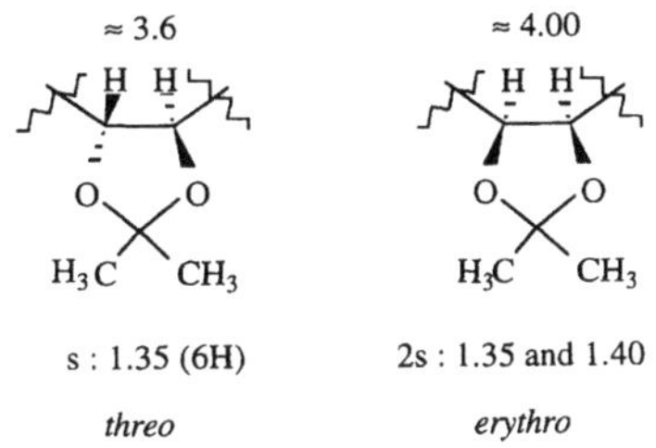

Fig. IV.31.

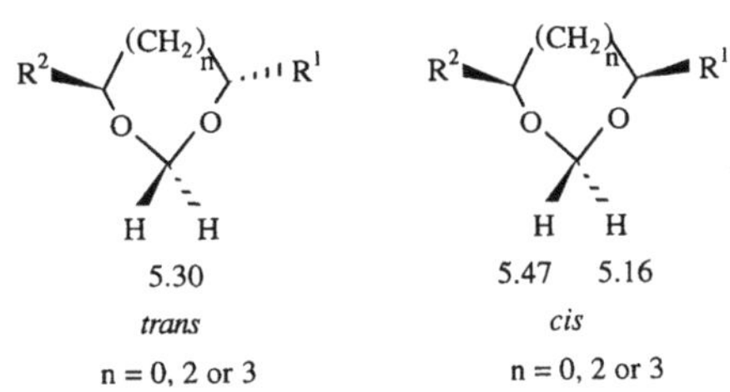

Fig. IV.32.

appear around 1.40 and 1.35. Furthermore, the methine protons of the dioxolane ring appear around δ 3.60 for the *threo* diols and around 4.00 for the *erythro* analog. (Fig. IV.31.).

Acetogenins possessing 1,2-, 1,4- or 1,5-diols have been derivatized by acetal formation with formaldehyde in the presence of chlorotrimethylsilane (Me_3SiCl) and dimethylsulphoxide (Me_2SO) in excess (*18*). This procedure affects only the diol system without racemization of the stereogenic centres. Observation of the signal corresponding to the acetal methylene in ^1H nmr serves to deduce the relative stereochemical relationship of the ring so formed. In the case of a *cis* relationship, the two acetal protons appear at δ 5.47 and δ 5.16 ppm, whereas for compounds, of *trans* configuration, a singlet now appears at δ 5.30 ppm (*18*) (Fig. IV.32.).

IV. 6.2. Absolute Configuration

The last step in structure elucidation of acetogenins is to determine the absolute configuration of the asymmetric centers. Only relative configurations have so far been obtained by X-ray analyses (*114, 148, 158*) since no study was performed on a derivatized acetogenin with a known chiral centre or heavy atom.

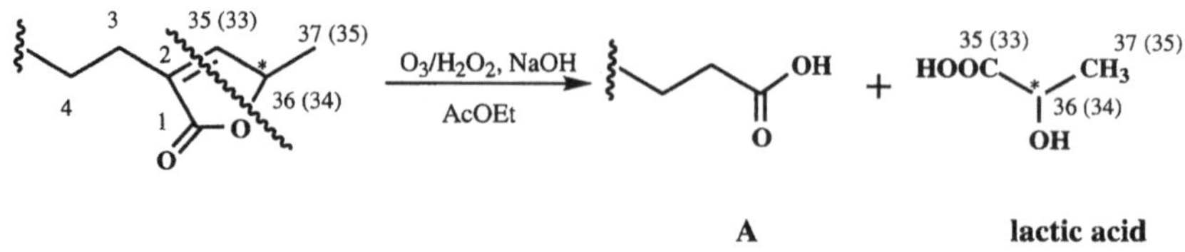

Fig. IV.33.

Degradation method

The first method used for determining the absolute configuration at C-36 (or C-34) of a γ-lactone was to oxidize uvaricin and analyze the products so formed. Ozonolysis of the acetogenin led to acid A and (*S*)-lactic acid with retention of configuration (Fig. IV. 33.); the absolute configuration of the lactic acid was established by chiral gas chromatography (*105*).

In a similar study, the methyl lactate produced was transformed as its (*R*)-MTPA ester and compared by TLC with authentic 2*S*- and 2*R*-methyl lactates (*R*)-MTPA esters. The *S* absolute configuration at C-36 was then established for squamocin (*64*).

Circular dichroism

The absolute configuration of the stereogenic centre of the γ-lactone of acetogenins of sub-type 1 has also been studied using circular dichroism (CD). Comparison of the sign of the Cotton effect with that of a synthetic model with known absolute configuration (*S*) (with negative Cotton effect) led the authors to propose the *S* configuration for squamocin (annonin-1) (*173*). However to resolve any ambiguity, it would have been useful to carry out the comparison with a synthetic model with *R* configuration as well. The existence of a number of controversial reports has so far not permitted generalization of this method for determining the absolute configuration of the lactone.

For acetogenins of sub-type 2, a comparison of the CD spectra with those of natural products (rubrenolide and rubrynolide of known absolute configurations) allowed the authors to propose the *S* configuration at C-4 for acetogenins exhibiting a negative Cotton effect (*83*).

Nmr studies of Mosher esters

The nmr method based on the analysis of the ^1H nmr spectra of secondary alcohols derivatized as their (*R*) and (*S*) 2-methoxy-2-phenyl-

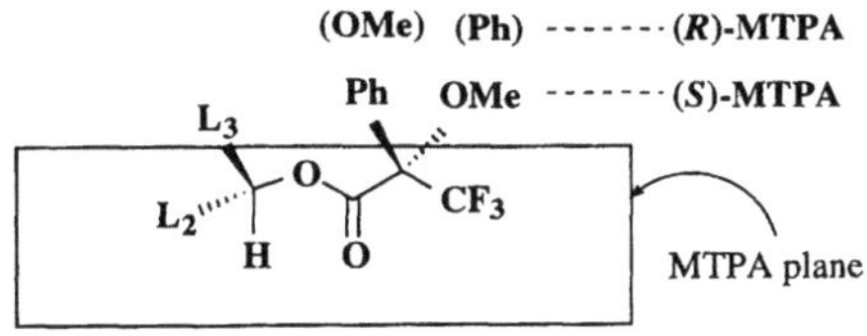

Fig. IV.34.

2-(trifluoromethyl) acetates, known as the MOSHER or YAMAGUCHI-MOSHER method is very often employed (*190, 191*). This technique is based on the analysis of the most stable conformer of the (*R*) or (*S*) MTPA esters (Fig. IV.34.).

MOSHER noted that for this probably most stable rotamer the chemical shifts of L_3 (located above the plane formed by the skeleton CF_3—C—C(O)—O—CH) appear shifted upfield in the ^1H nmr for the (*S*)-MTPA ester with respect to the (*R*)-MTPA ester because of the diamagnetic effect of the phenyl group. However at the time the nmr spectrometers limited to magnetic fields of 60–100 MHz did not allow the determination of the chemical shifts of the protons with sufficient precision. In 1983 YAMAGUCHI (*192*) used this technique with success for the determination of absolute configurations of secondary alcohols, but using higher fields, and since then this method has been proved to be quite general, even though some exceptions are known in the literature (*193*).

In the acetogenin area RIESER *et al.* used the method to determine the absolute configurations of the carbon atoms bearing a hydroxyl group, (*46*) and prepared both diastereomers from natural acetogenins of unknown configurations and the (*R*) and (*S*) Mosher acids. Analyses of the ^1H and ^{19}F nmr spectra at high field (500 MHz) then allowed assignment of (*R*) or (*S*) configuration by analysis of the chemical shifts of the α and α' protons with respect to the CHOH group (Fig. IV.35.). It should be noted that the letters defining the absolute configurations of the carboxylic acid, acid chloride, and ester are inverted because of the inversion of the Cahn-Ingold-Prelog priority order.

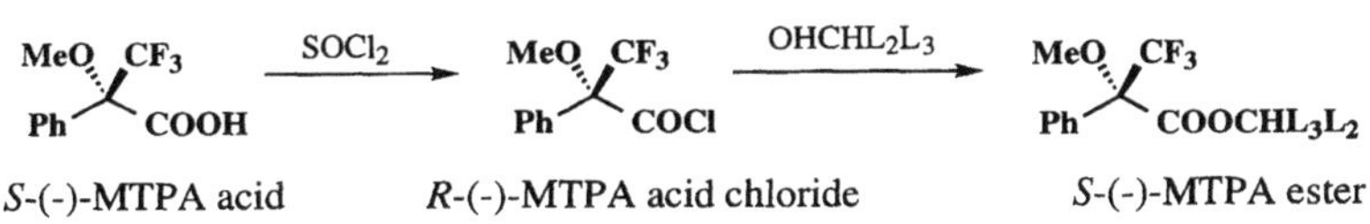

Fig. IV.35.

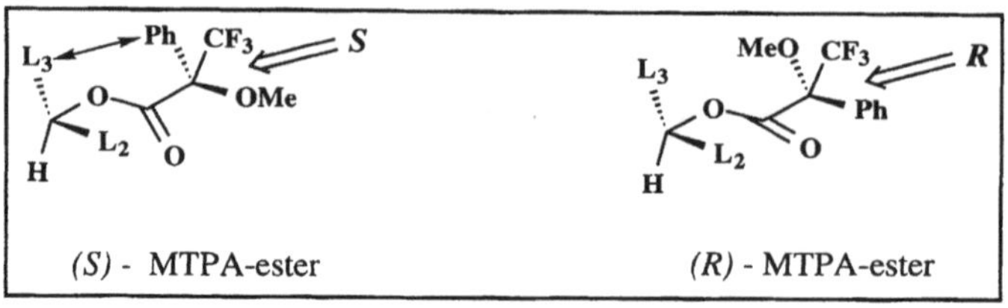

Fig. IV.36. Most stable conformations of (*S*)- and (*R*)-MTPA-esters

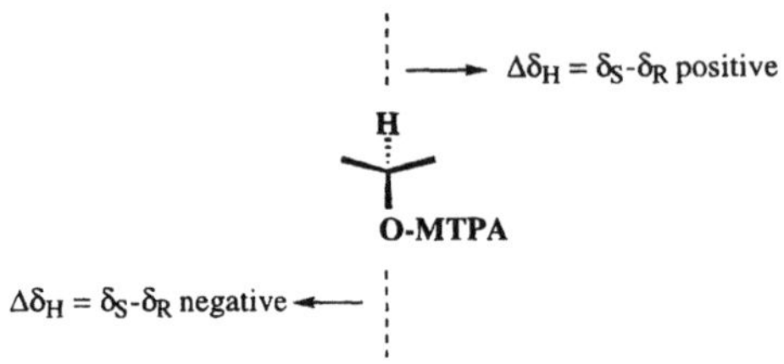

Fig. IV.37.

As Mosher predicted, analysis of the ^{1}H nmr spectrum of the (*S*)-MTPA-ester shows that L_3 protons appear at a higher field compared with L_2, because of the phenyl effect, whereas it is now the L_2 protons which appear at a higher field in the case of the (*R*)-MTPA-ester (Fig. IV.36.).

Therefore, when the value of the difference of chemical shifts is positive ($\Delta\delta_H = \delta_S - \delta_R > 0$), the protons can be placed at the right side of the plane defined above. For protons located on the left side, the difference in chemical shifts is now negative. The absolute configuration is then deduced using the Cahn-Ingold-Prelog priority order.

This method is also applicable to ^{19}F nmr spectrometry. The main problem arises from the difficulty in assigning with accuracy the chemical shifts of both protons L_2 and L_3.

A few examples of the determination of absolute configurations are given in Tables IV.15–17 with a discussion pointing out difficulties of the method.

The (*R*) absolute configuration of the CHOH at C-4 of bullatacin was successfully assigned (Table IV.17.) using the Mosher technique. But it is worth pointing out that in the case of the CHOH at C-15 both values $\Delta\delta_H = \delta_S - \delta_R$ of protons H-14 and H-16 are positive, which does not permit, only *a priori*, to apply the method to C-15 as described above. Nevertheless the authors attributed the (*R*) absolute configuration to C-15

Table IV. 15. *Reticulatacin* (46)

	H-16	H-18	H-21	H-23	Absolute Configuration	
					C-17	C-22
MTPA: *S*	–	4.01	4.01	–		
R	–	4.06	4.06	–		
$\Delta\delta_H = \delta_S - \delta_R$	–	neg.: −0.05 *left*	neg.: −0.05 left	–	*R*	*R*

(−): the chemical shifts in ^1H were not attributed

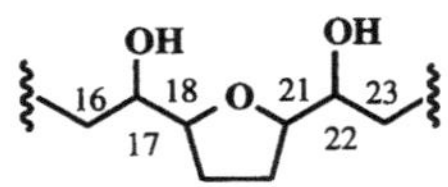

(*R*)

(*R*)

Table IV. 16. *Asimicin* (*46*)

	^{1}H nmr						Absolute Configuration		
	H-3	H-5	H-14	H-16	H-23	H-25	C-4	C-15	C-24
MTPA: *S*	2.56	1.63	1.56	3.93	3.93	1.56			
R	2.58/2.66	1.61	1.45	3.97	3.97	1.45			
$\Delta\delta_H = \delta_S - \delta_R$	neg.	pos.	pos.	neg.	neg.	pos.	*R*	*R*	*R*
	left	*right*	*right*	*left*	*left*	*right*			

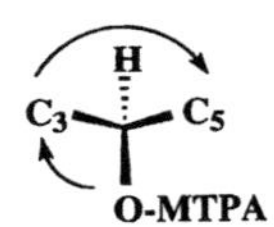

(*R*)

(*R*)

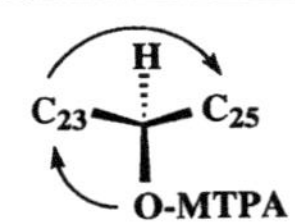

(*R*)

Table IV. 17. *Bullatacin* (*Rolliniastatin-2*) (*46*)

			^{1}H nmr				Absolute Configuration		
	H-3	H-5	H-14	H-16	H-23	H-25	C-4	C-15	C-24
MTPA: *S*	2.56	1.64	1.59	4.03	3.99	1.53			
R	2.61/2.69	1.61	1.45	3.97	3.90	1.57			
$\Delta\delta_H = \delta_S - \delta_R$	neg.	pos.	pos. 0.14	pos. 0.06	pos.	neg.	*R*	*R*	*S*
	left	*right*	*right*	*left*	*right*	*left*			

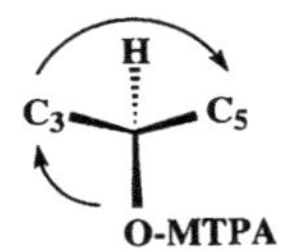

(*R*)

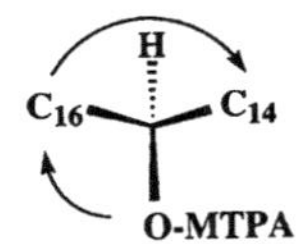

(*R*)

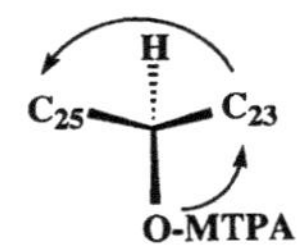

(*S*)

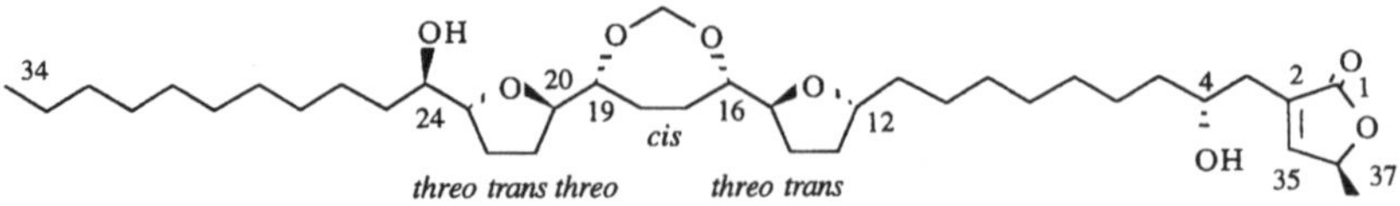

Fig. IV.38. 16,19-Formaldehyde acetal derivative of bullatanocin (*18*)

by arbitrarily giving a negative sign to the lowest $\Delta\delta$ values ($\Delta\delta_{H-16} = +0.06$ becomes ($-$); $\Delta\delta_{H-14} = +0.14$).

However, in the case of reticulatacin (Table IV.15.), only $\Delta\delta_{H-18} = -0.05$ is known. With respect to the lack of a value for $\Delta\delta_{H-16}$, the authors proposed the (*R*) absolute configuration for C-17 (as well as for C-22), by assuming that the missing value must have the opposite sign (*46*). Once again in the case of bullatacin (Table IV.17.), for determing the absolute configuration of C-15 the authors observed two positive values for $\Delta\delta_{H-16}$ and $\Delta\delta_{H-14}$ of 0.14 and 0.06 ppm. Arbitrarily a negative sign was attributed to the lowest value (*46*).

These examples show the limitations of the method. Furthermore, in the case of isolated hydroxyl groups, because it is not possible to differentiate both L_2 and L_3 groups, it remains unfeasible to determine the absolute configuration of such stereogenic centres.

If the relative configurations either around the THF backbone or *via* a formaldehyde acetal are known, determination of the absolute configuration of a single stereogenic centre allows the deduction of the remaining absolute configurations.

For instance the absolute configurations of bullatanocin have been determined *via* its 16,19-formaldehyde acetal (Fig. IV.38.) on the bis-*R*- and *S*-MTPA esters (*18*).

V. Biogenetic Hypotheses

Although no experimental work on the biosynthesis of annonaceous acetogenins has been performed hypotheses proposed as soon as the first annonaceous acetogenin uvaricin was isolated (*1*) are now supported by isolation of precursors and biomimetic hemisyntheses (*8*).

Despite the presence of numerous asymmetric centres due to the oxygenated groups, the biogenesis obviously proceeds by straightforward enzymatic processes, *e.g.* dehydrogenation and oxidation.

The lactone ring can be formed by an aldol condensation involving a three carbon precursor and a fatty acid (*194*) or from a fatty acid

2-monoglyceride. Some monoacyl-2-glycerides have been isolated from *Annona senegalensis* seeds (*21*). The acyl part consists of palmitic, oleic and linoleic acids. To our knowledge, no fatty acid made up of 32 or 34 carbons (lacceroic and ghedoic acids) has been isolated among the fatty acids of Annonaceae.

Formation of the tetrahydrofuran pattern must be subsequent to lactonization; this can be deduced from the existence of compounds such as acetogenins of type E, without any tetrahydrofuran but possessing hydroxyl(s), ketone, double bond(s) or (and) epoxy group(s). These compounds can be considered as biogenetic precursors of acetogenins of type A, B, C and D and enabled various authors to propose biogenetic pathways (*29, 75, 87, 94, 95, 96, 107, 108, 113, 145, 165*).

Tetrahydrofurans arise from compounds possessing 1,5-dienes in the case of acetogenins of type A (with two hydroxyls flanking the THF) and 1,5,9-trienes in the case of acetogenins of types B and C. The biosynthesis of the tetrahydrofurans involves oxidation of double bonds to epoxides and subsequent ring expansions of the oxiranes into THF. These might be viewed as cascade reactions which are terminated when an alkoxide ion encounters a defect, in other words when oxiranes are separated by more than two carbon atoms. The termination can also involve a hydroxyl which would explain the existence of mono-α-hydroxylated tetrahydrofurans in some cases of type A acetogenins A as well as in the acetogenins of type C. In the case of acetogenins of type A with only one hydroxyl flanking the tetrahydrofuran, the expansion of the oxirane involves a hydroxyl in the δ-position. Precedent for this pathway is the conversion of polybutadiene into the corresponding polyepoxide and then into oligo-tetrahydrofuran (*195*).

Therefore one of the early precursors might be a very long chain fatty acid terminating in a butanolide. Then the first enzymatic step is probably a dehydrogenation, which introduces either the Δ-15 or the Δ-19 double bond and then the other. This $\Delta^{n,n+4}$ pattern is practically unknown in mammalian species where fatty acid unsaturations separated by only one methylene group ($\Delta^{n,n+3}$) are found. In fact the $\Delta^{n,n+4}$pattern has been found in marine organisms, principally in marine sponges (*196*) and recently in sea anemones (*197*). These common biogenetic intermediates in very primitive living organisms such as these marine invertebrates trend to confirm the archaism of the Annonaceae family. Hutchinson (*198*) and Takhtajan (*199*) already argued that the Annonaceae is an archaic family of Angiosperms related to the Magnioliales which are considered "living fossils".

Arguments in favour of these hypotheses are provided by the presence in the same plant of products at different levels of biosynthesis. For

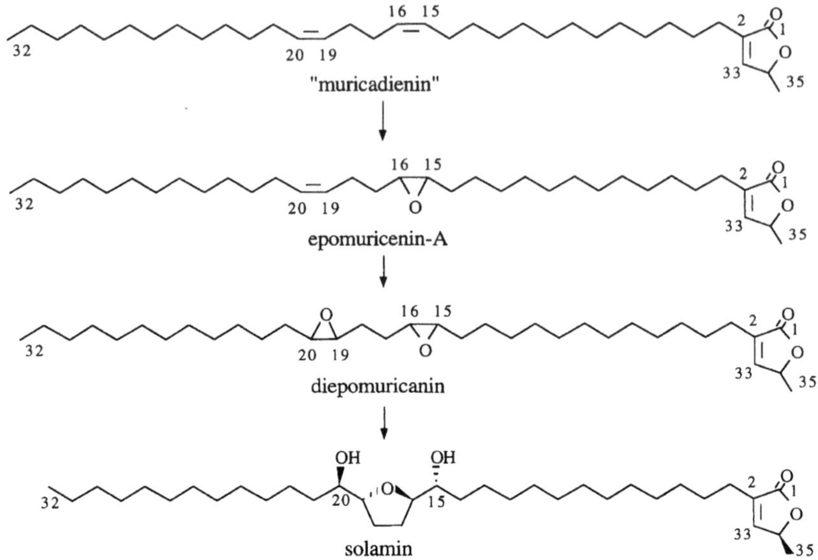

Scheme V.1.

instance (Scheme V.1), the α,α′-dihydroxylated tetrahydrofuran pattern of solamin, an acetogenin of type A, must be derived from a $\Delta^{15,19}$ diunsaturated precursor *via* an oxidation step leading to a *bis*-epoxide (isolated from *A. muricata* and named diepomuricanin (*107*)) which rearranges to a dihydroxylated THF. The fact that we successfully isolated from *A. muricata* the bis-epoxide, diepomuricanin-A (*107*, *108*) and the unsaturated epoxide, epomuricenin-A (*108*) with the double bond separated by two methylenes from the epoxide function, strongly favors our hypothesis for the biogenetic pathway.

It is noteworthy that so far the diunsaturated derivative, the hypothetical precursor, muricadienin, represents the missing link (*108*). During our search for muricadienin, we isolated reticulatamol (*145*) and reticulatamone (*143*) which are functionalized only at C-15 with a hydroxyl and a ketonic group, respectively. Reticulatamol might be formed by enzymatic hydroxylation of the corresponding saturated precursor which might further lead either to the Δ-15 double bond through enzymatic dehydration or to reticulatamone by a further oxidative process.

The same sequence can be envisaged for acetogenins of type B possessing two contiguous THF rings, but involving in this case a $\Delta^{n,n+4,n+8}$ unsaturated precursor which will lead to a *tri*-epoxide. An example is provided by dieporeticenin a δ-unsaturated diepoxide, and tripoxyrollin,

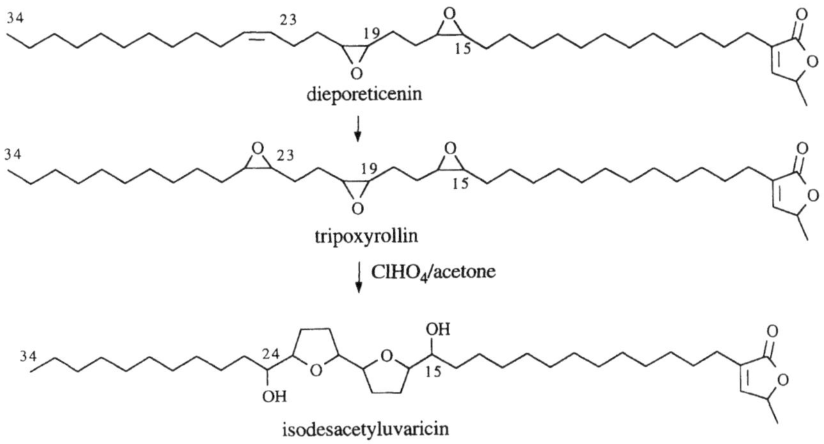

Scheme V.2.

a triepoxide, isolated from the hexane extract of the seeds of *Rollinia membranacea* (*165*) and *Annona reticulata* (*110*). Dieporeticenin on oxidation leads to tripoxyrollin which after acidic rearrangement leads to a *bis*-THF acetogenin, desacetyluvaricin (Scheme V. 2.).

For all the acetogenins found in *Goniothalamus giganteus*, an hypothetical γ-hydroxy-triene common precursor was proposed (*29*). Indeed, obvious oxidations, and acid-base rearrangements from this intermediate can explain the occurrence of the congeners with various degree of substitution, particularly diols and alkenes with a monotetrahydrofuran and adjacent or non-adjacent bistetrahydrofuran pattern.

In *Annona coriacea*, once again several structurally related products might derive from the same γ-hydroxy-triene through different processes such as dehydrogenation, oxidation and acid-base rearrangements. These transformations might occur in a random fashion, since most of the possible corresponding intermediates have been isolated from either *Goniothalamus giganteus* or *Annona coriacea*. As an illustration of this unexplained lack of order, Scheme V.3. shows the various related compounds which might be derived from the hypothetical γ-hydroxy-triene. It should be noted that from *Annona coriacea* was isolated the first acetogenin with two double bonds separated by two methylenes, *i.e.* coriadienin (*96*). The γ-hydroxy-triene could lead to coriadienin which might give rise to all types of acetogenins A, B and C. Oxidation of the Δ-13 double bond followed by acid/base rearrangement leads to coriacin (type A with a single α-hydroxyl group) which contains a Δ-17 double bond and a 21,22-diol in the appropriate places for serving as the bio-

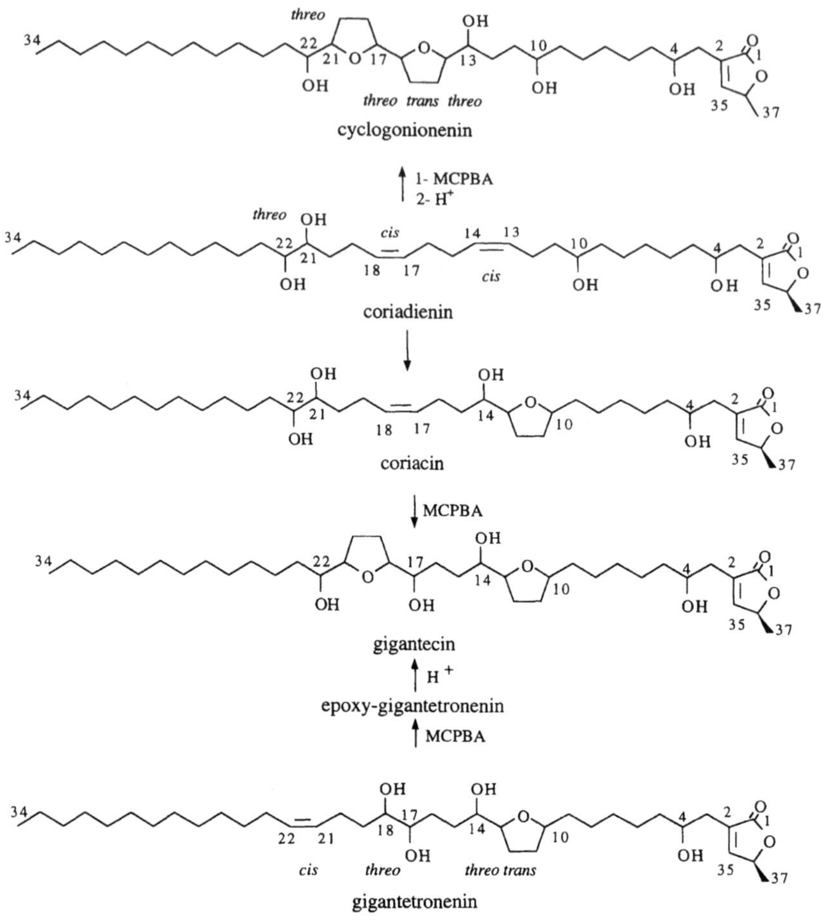

Scheme V.3.

genetic precursor of gigantecin, an acetogenin of type C after oxidation of the Δ-17 double bond and a new acid/base rearrangement (*95*). On the other hand coriadienin by oxidation of the Δ-13 and Δ-17 double bonds and acid/base rearrangement could lead to cyclogonionenin, an acetogenin of type **B**.

Finally, the simultaneous existence of all these precursors gives rise a few questions and comments:

a) All double bonds are not simultaneously oxidized to epoxides and it can be envisaged that action of deshydrogenase occurs at different stages of the biogenetic pathway; b) the transfer of oxygen leading to epoxide is stereospecifically *syn* so the relative stereochemistry of alkenes is conserved for the epoxide, but the oxidation can occur on both sides of the double bond; c) *Cis*-alkenes lead only to rings with *threo* configuration, while *trans*-alkenes produce only rings with *erythro* configuration (*195*), because the acid-base rearrangement occurs with inversion of configuration due to its S_N2 mechanism; d) finally, the cyclization processes from the epoxides can start from an *exo*-opening or from an *endo*-opening relative to the terminal γ-lactone, leading to tetrahydrofuran systems of the same relative configurations, but with opposite absolute configuration as observed in the chemical transformations of corepoxylone (*94*) and tripoxyrollin (*165*).

VI. Synthesis of Acetogenins of Annonaceae

To date total asymmetric syntheses of natural and unnatural acetogenins of only type A, B and E have been reported and reviewed (*200*, *201*). The strategies used are based on two different approaches, namely stereospecific synthesis of the THF part and the lactone moiety using either (i) as starting material a compound from the chiral pool (α-amino acids, sugars) or (ii) an asymmetric induction using homochiral catalysts (Sharpless epoxidation, Sharpless asymmetric dihydroxylation) and finally a coupling reaction between the two fragments. In addition to these total syntheses, numerous reports appeared in the literature dealing with the preparation of building blocks such as 2,5-disubstituted tetrahydrofurans (*187*, *188*, *189*, *202*, *203*, *204*) contiguous THF rings (*205*, *206*, *207*), γ-methyl-γ-lactones (*208*, *209*, *210*, *211*) and α-acetonyl-γ-lactones (*212*).

VI.1. Synthesis of Acetogenins of Type A

VI.1.1. Synthesis of ent-4-Oxo-2,33-dihydrosolamin (**1**)

The title compound was prepared from L-glutamic acid as starting material. The retrosynthetic pathway used was based on a disconnection between the C-6 and C-7 bond, which could be formed by a radical Michael type addition of an alkyl iodide to an enone (*189a*). The resulting carbonyl derivative would afford solamin after complete reduction and introduction of the unsaturation in the lactone ring, or murisolin after partial reduction. Enone **9** with the requisite configuration at C-34, and alkyl iodide **19** bearing the THF moiety with the desired relative

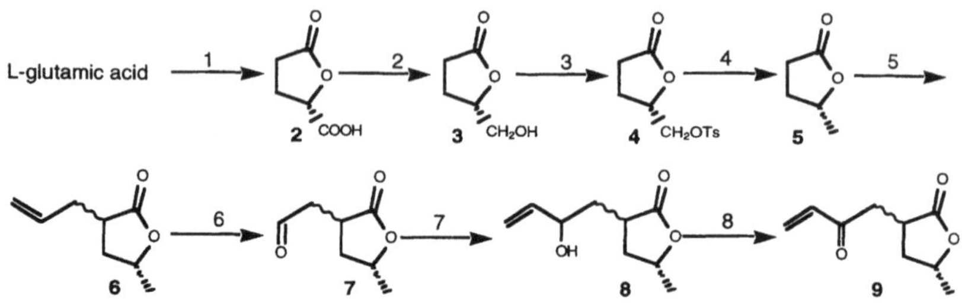

Fig. VI.1. *Reagents:* 1) $NaNO_2$, H_2SO_4, 70%; 2) $BH_3 \cdot SMe_2$, THF, 98%; 3) TsCl, pyridine, 87%; 4) NaI, *n*-Bu_3SnH, AIBN cat., THF, 80%; 5) (i) LDA, TMSCl, (ii) allyl bromide, THF, 90%; 6) OsO_4 cat., $NaIO_4$, dioxan, 70%; 7) vinylmagnesium bromide, THF, 0 °C, 51%; 8) $(COCl)_2$, DMSO, Et_3N, 96%

and absolute configurations for the four contiguous stereocentres were prepared.

Enone **9** with (*R*) or (*S*) configuration at C-34 was synthesized from pure (*R*) or (*S*)-γ-methyl-γ-lactone which can be prepared in 4 steps from L- and D-glutamic acid, respectively (Figure VI.1.). Deamination of L-glutamic acid by $NaNO_2$ in acidic medium gave rise to the carboxylic lactone **2** with complete retention of the stereogenic centre. Reduction of **2** by $BH_3 \cdot SMe_2$ and tosylation of the resulting alcohol **3** afforded tosylated **4** which was then reduced in THF under reflux in the presence of 1 eq. of sodium iodide and 1 eq. of tributyltin hydride with a catalytic amount of AIBN. The γ-valerolactone **5** was thus obtained in 80% yield for the last step (46% overall yield from glutamic acid in 4 steps and >99% ee). Alkylation of **5** by treatment with 1 eq. of LDA and allyl bromide led to a diastereomeric mixture of *cis* and *trans* alkylated products **6**. Oxidative cleavage of the double bond (OsO_4 cat. in the presence of sodium periodate in dioxan) gave the desired aldehyde **7** in 70% yield which upon addition of vinylmagnesium bromide, followed by a Swern oxidation, led to the desired enone **9** in 49% yield for the last two steps.

Alkyl iodide **19** (Fig. VI.2) was also prepared from L-glutamic acid. Acylation of dodecylmagnesium bromide at low temperature and concentration with the acid chloride **10** (obtained in 92% yield after treatment of *ent*-**2** by oxalyl chloride in dichloromethane with a catalytic amount of DMF) afforded the corresponding ketone **11** in 85% yield. Reduction of **11** with L-selectride™ gave rise to *syn* compound **12**, namely (+)-muricatacin (55), as the major product (*syn*/*anti* = 98:2). It is worth noting that use of tri-*n*-butyltin hydride with silica gel in dichloromethane at room temperature allowed formation of the *anti* compound as the major product with a 28:72 *syn*/*anti* ratio (*213*). Muricatacin **12** was then protected as a silyl ether **13** in 92% yield (*tert*-butyldimethylsilyl chloride in DMF in the presence of imidazole). Reduction of **13** by DIBAL in toluene at −78 °C afforded quantitatively the desired hemiacetal which upon addition of acetic anhydride led to a 1:1 mixture of the anomeric acetates **14** in 92% yield. This mixture when treated with trimethylsilyl cyanide in Et_2O in the presence of a catalytic amount of either trityl- or scandium perchlorate gave rise to a nearly equimolar mixture of *cis* and *trans* nitriles which were separated by flash chromatography. Treatment of the *cis* nitrile with sodium *tert*-butoxide at room temperature in *tert*-butanol for 24 h led to the *trans* product **15** in quantitative yield. DIBAL reduction of **15** at −78 °C afforded the corresponding aldehyde (*214*), whereas treatment of **15** with a functionalized Grignard reagent in the presence of trimethylsilyl chloride gave rise to the expected ketone **16** in 87% yield. Reduction of **16** with L-Selectride™ at −78 °C yielded the *syn*-*trans*-*syn* compound **17** (98:2 d.e. determined by NMR) with the (*S*) absolute configuration for all stereogenic centers. Deprotection with

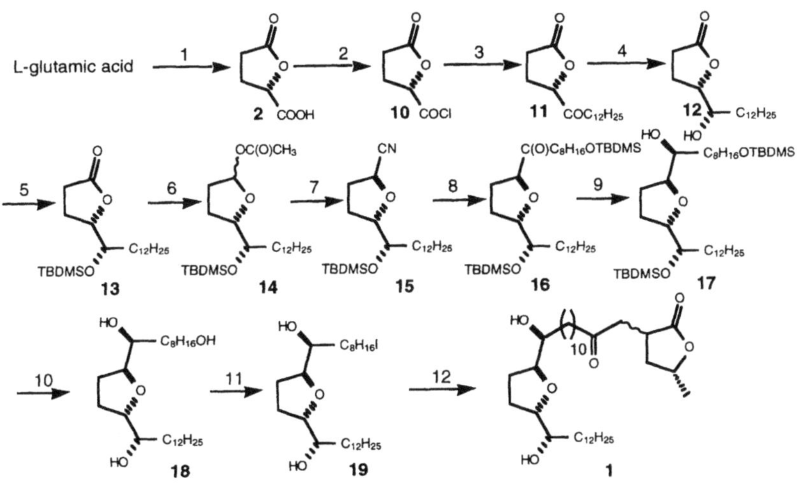

Fig. VI.2. *Reagents:* 1) $NaNO_2$, H_2SO_4, 70%; 2) $(COCl)_2$, DMF cat., CH_2Cl_2, 92%; 3) dodecylmagnesium bromide, −78 °C, THF, 85%; 4) L-Selectride™, −78 °C, THF, 88% (*syn*/*anti* = 98/2); 5) TBDMSCl, imidazole, DMF, 99%; 6) (i) DIBAL, −78 °C, toluene, 99%; (ii) $(Ac)_2O$, Et_3N, DMAP, 20 °C, 96%; 7) (i) TMSCN, $ScClO_4$ cat., Et_2O, 0 °C, 96%, $\alpha/\beta = 1{:}1$; (ii) *tert*-BuOK, *tert*-BuOH, 20 °C, 24 h, 100%, $\alpha/\beta = 100{:}1$; 8) *tert*-$BuMe_2SiO(CH_2)_8MgBr$, toluene, TMSCl, −78 °C, 75%; 9) L-Selectride™, −78 °C, THF, 71%, (*syn*/*anti* = 98:2); 10) TBAF, 20 °C, THF, 91%; 11) (i) TsCl, pyridine; (ii) NaI, acetone, 61% (for the last two steps); 12) *n*-Bu_3SnH, AIBN cat., **9**, toluene, 55%

tetrabutylammonium fluoride (TBAF) led to triol **18** which upon treatment with 1 eq. of tosyl chloride at 0 °C gave rise to the monotosylate in 61% yield. Displacement of the tosyl group by sodium iodide afforded the desired iodo compound **19** in quantitative yield. The Michael type addition of the iodo derivative **19** to enone **9** was performed under radical conditions by treatment of a stoichiometric mixture of **9** and **19** with 2 eq. of tri-*n*-butyltin hydride and a catalytic amount of AIBN in toluene under reflux. The desired coupled compound **1**, namely *ent*-4-oxo-2,33-dihydrosolamin, was then obtained in 55% yield (Fig. VI.2.). The synthesis was therefore achieved in 14 steps and 6.4% overall yield from L-glutamic acid. Two more steps, *i.e.* reduction of the carbonyl group and introduction of the unsaturation would lead to either solamin or murisolin.

VI.1.2. Synthesis of Corossolone

A hemisynthesis of corossolone was reported in 1993 (*94*), but it was only in 1995 that WU reported its total synthesis (*97*). Corossolone was described having the *threo-trans-threo* configuration across the THF ring with the absolute configuration proposed as 15*R*, 16*R*, 19*R*, 20*R*. The strategy was based on the enantiocontrolled preparation of both parts of the molecule and coupling of the two synthons by addition of a lithium acetylide to an epoxide followed by Swern oxidation of the hydroxyl at C-10.

The lactone fragment was synthesized from methyl undecenoate which upon treatment with 1 eq. of LDA and then (*S*)-O-tetrahydropyranyl lactal gave the aldol type product which was protected as its methoxymethyl ether **20** (in 55% overall yield) and then hydrolyzed by H_2SO_4 (10% in THF) to yield lactone **21** quantitatively. Epoxidation of the double bond by *m*-CPBA led to the desired epoxide **22** as an epimeric mixture of (34*S*) diastereomers at C-10, C-2 and C-33, with eventual oxidation and unsaturation expected to remove the stereogenic centres at C-2, C-10 and C-33 (Fig. VI.3.).

The THF fragment was prepared as aldehyde **30** from D-tartaric acid in 15 steps and 19% overall yield. D-tartaric acid, in three steps by known procedures, led to diol **23** which was protected as a benzyl ether **24** by the usual method in 86% yield. Swern oxidation led to the corresponding aldehyde which upon Wittig homologation with undecanyltriphenylphosphorane gave the desired alkene **25** in 63% yield for these two steps. Hydrogenation of **25** followed by treatment of the partially deprotected alkane with Li-NH_3, led to alcohol **26** which was converted to its tosylate. Deprotection of the acetonide (*p*-TsOH cat.) and subsequent treatment

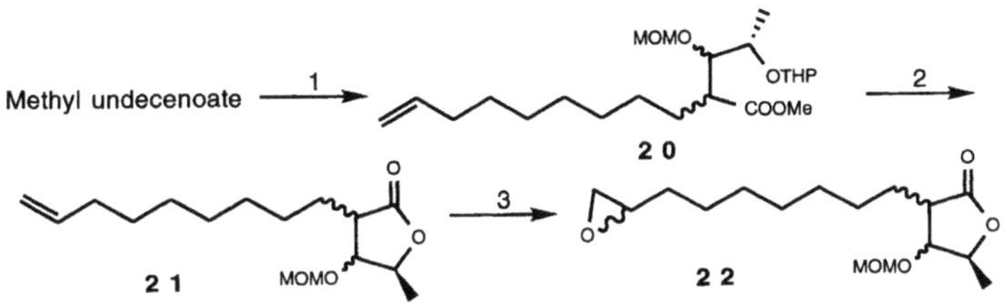

Fig. VI.3. *Reagents:* 1) (i) LDA; (ii) (*S*)-O-THPlactal, 65% (for the last two steps); (iii) MOMCl, *i*-Pr_2NEt, 85%; 2) 10% H_2SO_4, THF, 100%; 3) *m*-CPBA, 64%

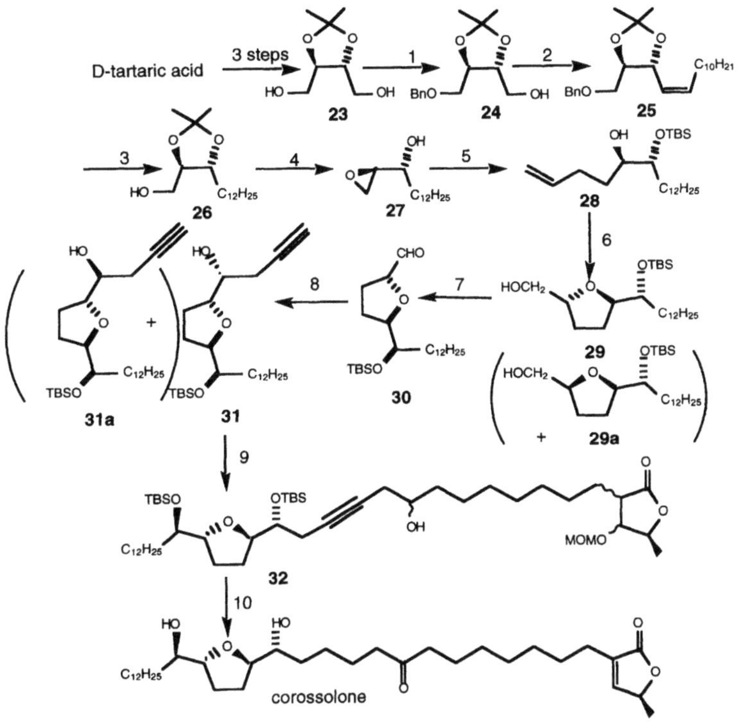

Fig. VI.4. *Reagents:* 1) NaH, BnBr, DMF, R.T., 70%; 2) (i) $(COCl)_2$, DMSO, Et_3N; (ii) $C_{10}H_{21}$ CH = PPh_3, THF, 63%; 3) (i) H_2, Pd-C, (ii) Li-NH_3, 86%; 4) (i) TsCl, Et_3N, cat. DMAP, CH_2Cl_2, 96%; (ii) TsOH, MeOH, (iii) K_2CO_3, MeOH, 75% (for the last two steps); 5) (i) TBDMSCl, $AgNO_3$, Py., THF, 93%, (ii) allylmagnesium chloride, cat.; CuBr, THF-ether, 84%; 6) (i) *m*-CPBA, CH_2Cl_2, (20 cat. CSA), 72% for **29**, 14% for **29a**; 7) $(COCl)_2$, DMSO, Et_3N; 8) 2-allenyl1,3-dioxa-2-borolane-(4*S*,5*S*)-dicarboxylic acid bis(methylethyl) ester, toluene, 68% for **31**, 4% for **31a** (*syn*/*anti* = 18.8:1); 9) (i) TBDMSCl, imid., DMF, 91%, (ii) *n*-BuLi, $BF_3 \cdot OEt_2$, **22**, −78 °C, 96%; 10) (i) H_2, Pd-C, 88%, (ii) DBU, THF, R.T., 94%, (iii) $(COCl)_2$, DMSO, Et_3N, 82%, (iv) 5% (v/v) of 40% HF in CH_3CN-THF (3:1), R.T., 83%

with K_2CO_3 afforded the epoxide **27**. Protection of the free hydroxyl of **27** was best performed by treatment with TBDMSCl in the presence of $AgNO_3$ and pyridine in THF. Addition of allylmagnesium chloride in the presence of CuBr as catalyst in THF-ether gave **28** in 84% yield. Epoxidation of **28** with *m*-CPBA provided the *trans* compound **29** in 72% yield with the undesired *cis* THF product **29a** in 14% yield (d.e. = 6:1). Alcohol **29** was then oxidized under Swern oxidation conditions to the expected aldehyde **30**. Addition of 2-allenyl-1,3-dioxa-2-borolane-(4*S*,5*S*)-dicarboylic acid bis (1′-methylethyl) ester led to the *syn* homopropargyl alcohol **31** in 68% together with the undesired *anti* product **31a** in 4% yield. Whereas addition of propargylzinc bromide to aldehyde **30′** (with the hydroxyl protected as an acetoxy) gave the *anti* homopropargyl alcohol **31′a** as the major compound in 8.3:1 *anti*/*syn* ratio (*98*). **31** was then treated with *n*-butyl lithium, followed by $BF_3 \cdot OEt_2$ and epoxide **22** to afford the coupled product **32** in 96% yield. Hydrogenation of the triple bond, dehydration of the β-hydroxyl-γ-methyl-γ-lactone by treatment with DBU in THF at room temperature, and Swern oxidation of the free hydroxyl before deprotection of the last hydroxyl groups with HF in CH_3CN–THF led to corossolone. The synthesis was achieved in 22 steps and in 7.6% overall yield from D-tartaric acid (Fig. VI.4.).

VI.1.3. Synthesis of Solamin

Keinan's synthesis (*155*)

KEINAN *et al.* used as a key step of their sequence the asymmetric dihydroxylation of an alkene, *i.e.* the very efficient Sharpless procedure for the formation of α,β-diols for preparation of the THF moiety. A subsequent cross-coupling reaction was performed by addition of an alkyne bearing the terminal lactone and a vinyl halide (bearing the THF moiety) in the presence of palladium and copper catalysts (Fig. VI.5.). Treatment of the unsaturated ester **33** (prepared in 4 steps from commercially available starting material in 65% overall yield) with AD-mix.-β in *tert*-butanol/water (1:1) with methanesulfonamide for 16 h at 0 °C afforded lactone **34** possessing 3 carbons atoms out of the 4 with the desired absolute configurations. Inversion of the fourth stereocentre after acetonide formation of the vicinal diol in two steps (tosylation, and epoxidation) afforded the lactone-THF **38**. DIBAL reduction of **38** followed by Wittig homologation with dibromomethylene triphenylphosphorane gave the bromo alkene **39**. The lactone fragment was prepared in a straightforward manner as an alkyne derivative **40** in one step and 70% yield from the mixture of (2*R*, 4*S*)- and (2*S*, 4*S*)-4-methyl-2-phenylthio-γ-butyrolactone

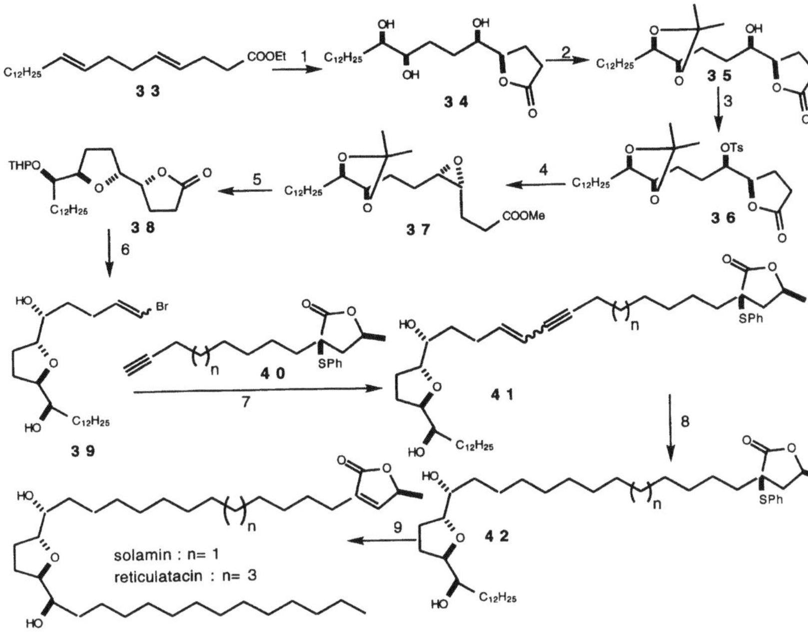

Fig. VI.5. *Reagents:* 1) AD-mix-β, 66%; 2) DMP, acetone, TsOH, 98%; 3) TsCl, Et_3N, DMAP, CH_2Cl_2, 97%; 4) K_2CO_3, MeOH, 88%; 5) $BF_3 \cdot OEt_2$,CH_2Cl_2, 75%; 6) (i) DIBAL, $-50\,°C$, THF; (ii) $BrCH_2PPh_3{}^+Br^-$, *tert*-BuOK, THF, 60% (for the last two steps); 7) **40**, $Pd(PPh_3)_4$, Et_3N, CuI, 70%; 8) H_2, $RhCl(PPh_3)_3$, 95%; 9) (i) *m*-CPBA; (ii) toluene reflux, 72% (for the last two steps)

(*215*) which upon reaction with bromoalkene **39** in the presence of palladium triphenyltetrakiscopper iodide, Et_3N in THF at 50 °C gave enyne **41** in 70% yield. Hydrogenation of **41** afforded **42** which after oxidation and thermal elimination of the phenylsulfoxide function led to the desired solamin. In conclusion, this synthesis was achieved in 14 steps and 7.7% yield from commercially available starting material. The total synthesis of reticulatacin (*142*) which differs from solamin only in the length of the alkyl chain by two extra carbon atoms has also been realized by the same authors.

Tanaka's synthesis (144)

The key steps in TANAKA's synthesis were on the one hand the very efficient asymmetric epoxidation of an allylic alcohol by the sharpless method, and on the other hand the palladium and copper catalyzed cross-coupling of an alkyne with a vinyl halide. Alkylation of propargyl alcohol **44** with dodecyl bromide **43** by lithium amide in liq. ammonia, followed by Lindlar hydrogenation gave the (*Z*) allylic alcohol **46**. Asymmetric epoxidation of the latter by the improved Sharpless procedure afforded epoxyalcohol **47** with 84% ee. Tosylation of the free hydroxyl (TsCl, pyridine) followed by iodine treatment gave iodo epoxide **48** which was then reacted with the lithium enolate of *tert*-butyl acetate to yield the alkylated product **49** before acidic hydrolysis which led to **12** (muricatacin) (*51*). Muricatacin **12** was then protected as its methoxymethyl ether **50** before reduction at low temperature with DIBAL, thus yielding hemiacetal **51** which upon reaction with pent-4-ynylidenetriphenylphosphorane gave the acyclic alkenyne **52**. Epoxidation of the double bond with *m*-CPBA, followed by acidic cyclization led to 3:2 mixture of *trans*:*cis* products (determined later in the synthesis) with the desired *trans* compound as the major product, which was separated by thin-layer chromatography. Protection of the free hydroxyl as a benzoyl ester, and deprotection of both hydroxyl groups led to the THF moiety **53** (Fig. VI.6.).

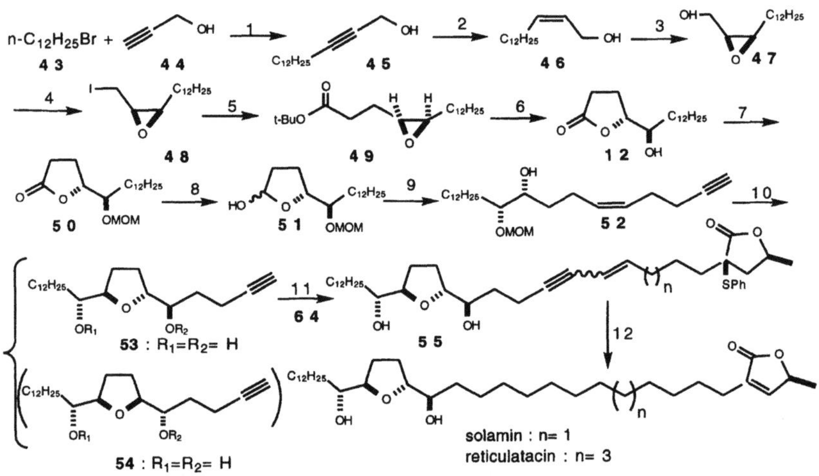

Fig. VI.6. *Reagents:* 1) $LiNH_2$, Et_2O, DMSO, 71%; 2) Lindlar hydrogenation, 91%; 3) *tert*-BuOOH, Ti(O*i*-Pr)$_4$, L-(+)-diethyl tartrate, M. S., CH_2Cl_2, 76%; 4) (i) TsCl, DMAP, Et_3N; (ii) NaI, acetone, 97% (for the last two steps); 5) *tert*-butylacetate, cyclohexylisopropylamine, *n*-BuLi, HMPA, 81%; 6) camphosulfonic acid, CH_2Cl_2, 70%; 7) MOMCl, *i*-Pr_2NEt, 94%; 8) DIBAL, −78 °C, CH_2Cl_2; 9) Pent-4-yn-1-yltriphenylphosphonium iodide, NaOEt, 0 °C, DMF; 10) (i) *m*-CPBA, CH_2Cl_2, 56% (for the last three steps); (ii) BzCl, pyridine, 0 °C; (iii) NaOH, MeOH, 79% (for the last two steps); 11) **64**, Pd(PPh$_3$)$_4$, CuI, Et_3N, 61%; 12) (i) H_2, RhCl(PPh$_3$)$_3$, 60%; (ii) *m*-CPBA, (iii) toluene, reflux, 40% (for the last two steps)

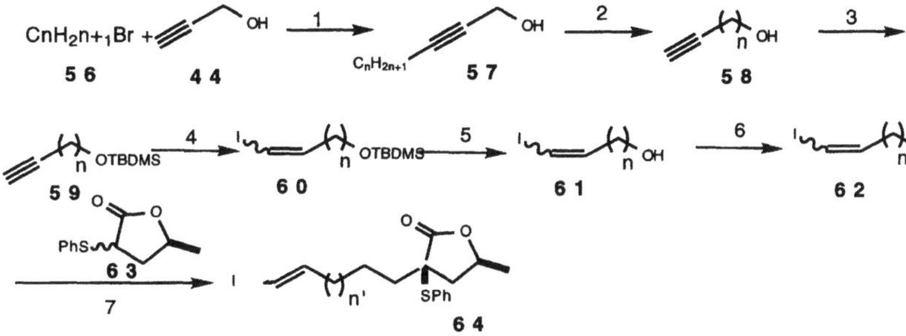

Fig. VI.7. *Reagents:* 1) *n*-BuLi, 58%; 2) KAPA, $H_2N(CH_2)_3NH_2$, 71%; 3) TBDMSCl, imidazole, DMF, 92%; 4) (i) *n*-Bu_3SnH, AIBN cat.; (ii) I_2, 70% (for the last two steps) ($E/Z = 3{:}1$); 5) TBAF, THF, 85%; 6) (i) TsCl, pyridine; (ii) NaI, acetone, 81% (for the last two steps) ($E/Z = 3{:}1$); 7) **63**, NaHMDS, 51%

Lactone fragment **64** was prepared from both (*S*) ethyl lactate (which in a few steps gave the lactone **63**) and propargyl alcohol **57**. Alkylation of lactone **63** with diiodo compound **62** (prepared in 5 steps from **57**) gave the desired furanones **64** (Fig. VI.7.). A palladium catalyzed cross-coupling reaction, was performed between the vinyl halide **64** and the alkyne **53** with $Pd(PPh_3)_4$ in the presence of CuI and Et_3N, as already described in KEINAN's synthesis, but now with the alkyne being the THF skeleton and not the lactone part. The desired coupled product **55** was then submitted to hydrogenation followed by a two step sequence involving oxidation-thermal elimination to yield the desired solamin (Fig. VI.6.). In conclusion, the synthesis was achieved in 16 steps and 1.5% overall yield, using as key steps Sharpless epoxidation and the palladium catalyzed cross-coupling reaction of an alkyne with a vinyl halide. The same authors succeeded in the total synthesis of reticulatacin by using the same strategy.

Trost's synthesis (156)

Several original approaches appear in the elegant synthesis of solamin described by TROST and coworkers such as (i) a new synthesis of a 2,5-disubstituted THF *via* a Ramberg-Backlund olefination and (ii) a ruthenium catalyzed butenolide annelation to form a direct precursor of solamin (Fig. VI. 8.). Treatment of propargyl alcohol **44** with *n*-butyl lithium in THF/HMPA followed by addition of bromododecene **65** afforded the alkylated product **66**, which upon Lindlar hydrogenation gave the (*Z*) allylic alcohol **67**. Asymmetric epoxidation of the latter (*tert*-BuOOH, Ti(O*i*-Pr)$_4$ and L-(+)-tartrate) gave the desired epoxide **68**

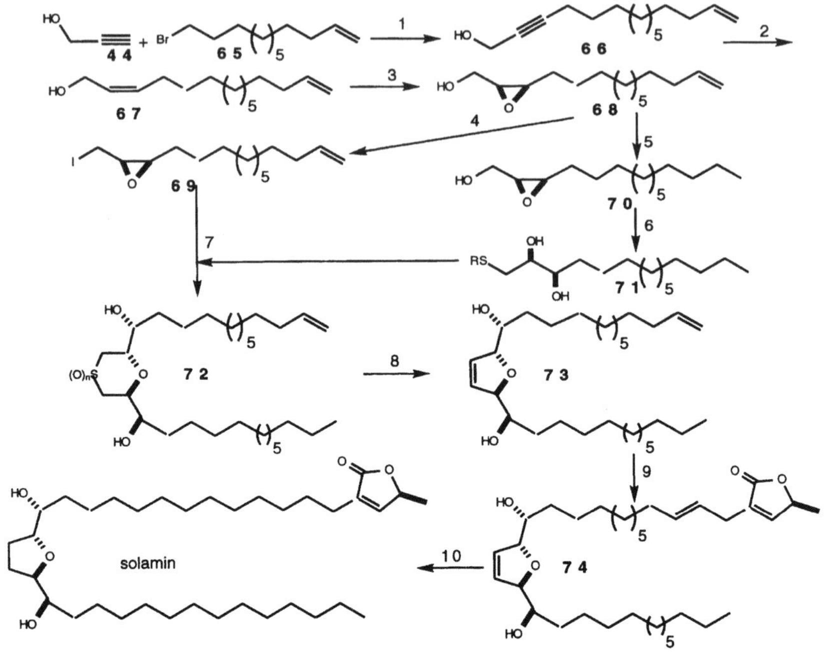

Fig. VI.8. *Reagents:* 1) *n*-BuLi, −78 °C, THF, HMPA, 76%; 2) Lindlar hydrogenation; 3) *tert*-BuOOH, $Ti(Oi\text{-}Pr)_4$, L-(+)-diethyl tartrate, M. S., −20 °C, CH_2Cl_2, 90%; 4) I_2, PPh_3, $C_3H_4N_2$, Et_3N, THF, 0 °C, 93%; 5) H_2, Pd-C, 98%; 6) (i) *tert*-BuSH, NaOH, *tert*-BuOH, H_2O, 81%; (ii) $Hg(OAc)_2$, PhOMe, CF_3COOH, 0 °C, 92%; 7) Cs_2CO_3, DMF, R.T., 92%; (ii) KOH, H_2O, *tert*-BuOH, 65%; 8) (i) TMSCl, Et_3N, CH_2Cl_2, 0 °C, R.T., 94%; (ii) *m*-CPBA, PhH-hexane, 0 °C, 95%; (iii) *tert*-BuOK, *tert*-BuOH, CCl_4, R.T., 65%; (iv) TsOH, H_2O, EtOH, R.T., 95%; 9) CpRu(COD)Cl, MeOH, EtOOCC≡CCH* (OH)CH_3, 65%; 10) H_2, $RhCl(PPh_3)_3$, PhH, EtOH, R.T., 95%

in 82% ee (after recrystallization >99% ee). At this point, this intermediate was used to prepare the two halves of the THF skeleton. First the hydroxyepoxide **68** was converted to the corresponding iodide **69** in 93% yield (I_2, PPh_3, Et_3N, THF). Second, hydrogenation of **68** led to alkane **70** which after a Payne rearrangement, on treatment with *tert*-BuSH, afforded the expected sulfide. Removal of the *tert*-butyl group with $Hg(OAc)_2$, PhOMe and CF_3COOH at 0 °C, gave thiol **71**. Coupling of **71** with **69** under basic conditions gave the 1,4-oxathiane **72** in good yield. The best procedure for the Ramberg-Backlund olefination involved treatment of the corresponding sulfone (prepared by *m*-CPBA oxidation of **72**), with the hydroxyl groups protected as their silyl ethers, with *tert*-BuOK in

References, pp. 273–288

tert-BuOH in the presence of CCl_4 at room temperature to afford the dihydrofuran **73** in 65% yield. Ruthenium catalyzed butenolide annelation of diol **73** with (4*S*)-ethyl-4-hydroxy-pentyn-2-oate, occurred chemoselectively at the less sterically demanding double bond to give the bis-dehydrosolamin **74**. Chemoselective hydrogenation of the isolated double bonds with Wilkinson's catalyst ($(Ph_3)_3PRhCl$ and H_2) gave solamin in 95% yield. This synthesis was achieved in 14 steps and 11.7% overall yield using a new and very efficient method for the synthesis of 2,5-disubstituted THF rings as well as the attractive ruthenium catalyzed butenolide annelation.

VI.2. Synthesis of Acetogenins of Type B

VI.2.1. Synthesis of Hexepi-uvaricin

In 1991 Hoye and coworkers were the first to report the total synthesis of an unnatural acetogenin, *hexepi*-uvaricin (*216*) which had the same relative stereochemical relationship across the THF core as naturally occurring uvaricin (*1*), but the opposite absolute configuration. The strategy used as a key step the asymmetric epoxidation of a homochiral bis-allylic alcohol, which allowed 5 out of 6 stereocentres to be built up at once with the desired relative configuration. Inversion of the last stereogenic centre required a few steps (Fig. VI.9.).

Diiodide **75**, derived from L-(+)-diethyl tartrate, was converted to *E*-*E*-bis-allylic alcohol **77** through the following sequence: Weiler dianion alkylation with *tert*-butyl acetoacetate followed by ketone reduction to yield the corresponding diol, dehydration *via* the bis-mesylate and finally DIBAL reduction of the ester functions. Asymmetric epoxidation of the bis-allylic alcohol **77** so obtained (*tert*-BuOOH, $Ti(Oi\text{-}Pr)_4$, D-(−)-diisopropyl tartrate) afforded bis-epoxy diol **78** possessing a C2 symmetry axis in 60% yield. Subsequent monotosylation (TsCl, Et_3N, CH_2Cl_2, 0 °C) followed by acidic treatment (Amberlyst H-15, MeOH, room temperature) led to the bis-THF skeleton **80** where one out of the two primary hydroxyl groups was tosylated. Displacement of the tosyl group with an excess of lithium dinonyl cuprate followed by acetonide protection of the vicinal diol and acetylation of the last free hydroxyl led to **81**. After removal of the acetonide, three steps were required for inversion of the configuration at C-15 (numbering as in the acetogenin series) in 78% overall yield, by protection of the primary hydroxyl group as a silyl ether (TBDPSCl, DMAP, CH_2Cl_2, Et_3N), tosylation of the secondary hydroxyl (TsCl, DMAP, pyridine), and deprotection-epoxidation with

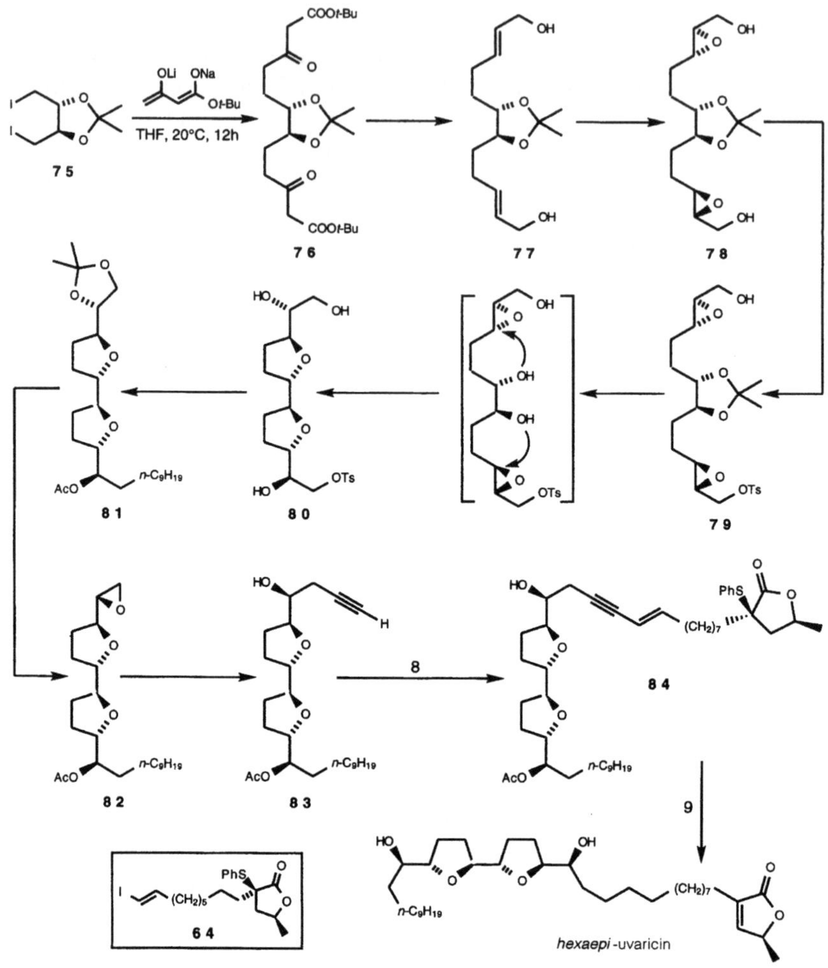

Fig. VI.9. *Reagents:* 1) (i) $NaBH_4$, MeOH, 0 °C; (ii) MsCl, Et_3N, CH_2Cl_2, then DBU, 40 °C; (iii) Dibal, Et_2O, 0 °C; 2) Sharpless epoxidation (D-(−)-diisopropyl tartrate); 3) 1 eq. TsCl, Et_3N, DMAP, CH_2Cl_2, 0 °C; 4) Amberlyst-15, MeOH, R.T.; 5) (i) $(n\text{-}C_9H_{19})_2CuLi$, THF, −10 °C, (ii) H^+, Me_2CO, R.T., (iii) Ac_2O, pyr., R.T.; 6) (i) MeOH, *p*-TsOH, R.T.; (ii) *t*-$BuPh_2SiCl$, DMAP, Et_3N, CH_2Cl_2, R.T.; (iii) TsCl, DMAP, pyr., CH_2Cl_2, R.T.; (iv) *n*-Bu_4NF, R.T.; 7) (i) LiC≡CTMS, $BF_3 \cdot OEt_2$, −78 °C, (ii) *n*-Bu_4NF, THF, R.T.; 8) **64**, $Pd(PPh_3)_4$, CuI, Et_3N, R.T.; 9) (i) $RhCl(PPh_3)_3$, H_2, PhH, (ii) OXONE™, MeOH, H_2O, 0 °C, (iii) toluene, reflux

TBAF, leading to epoxide **82**. Epoxide opening of **82** with lithium trimethylsilylacetylide/$BF_3{\cdot}OEt_2$, and removal of the TMS group by acidic treatment, then led to the terminal alkyne **83** in 58% yield for the last steps.

Preparation of the lactone fragment started with a mixture of (2*R*, 4*S*)- and (2*S*, 4*S*)-4-methyl-2-phenylsulfenyl-γ-butyrolactone (*215*) which was alkylated with (*E*)-1,9-diiodo-1-nonene. The resulting iodo compound **64** was then coupled with the alkyne **83** through an efficient palladium catalyzed reaction ($Pd(PPh_3)_4$, CuI, Et_3N, room temperature) in 86% yield. Enyne reduction of **84** with Wilkinson's catalyst, then oxidation of the sulfide to the sulfoxide and subsequent thermal elimination gave *hexepi*-uvaricin. The synthesis was achieved in 20 steps and in 0.36% yield.

VI.2.2. Synthesis of Bullatacin (= Rolliniastatin-2)

Synthesis of (−)-Bullatacin (217)

HOYE and coworkers also reported the total synthesis of unnatural (−)-bullatacin (= *ent*-bullatacin) (*217*) (Fig. VI.10.) using a strategy very similar to that employed for *hexepi*-uvaricin (*216*). The bis-THF core was prepared as already described (Fig. VI.9.). The lactone fragment was prepared from (*S*)-(−)-malic acid which was reduced with $BH_3{\cdot}SMe_2$ to afford the corresponding triol. Acetonide formation of the vicinal diol and treatment with iodine in the presence of PPh_3 and imidazole afforded iodo acetonide **85**. Alkylation of 1-lithio-1-pentyne with (*S*)-iodide **85**, followed by acetonide removal and "zipper" reaction gave the terminal alkyne **86**. The latter was then converted in 4 steps to the epoxide **87** (trimethylsilylation, hydrolysis, tosylation, epoxide formation) in 20% overall yield for the sequence. Treatment of epoxide **87** with the lithium enolate of α-phenylthioacetic followed by acidification and protection of the free hydroxyl group as a silyl ether gave the acyclic compound **88** in 52% yield for the last 3 steps. Enolization of acid **88** with 2 eq. of LDA gave the corresponding dianion which was alkylated with (*R*)-propylene oxide to give, after acidic catalyzed lactonization, the expected lactone **89**. Selective removal of the TMS group by treatment with K_2CO_3 gave the terminal alkyne which was iodinated (I_2, morpholine) to give iodoalkyne **90**. Palladium catalyzed cross-coupling reaction of the alkyne **83′** (*216*) with iodoalkyne **90** provided diyne **91** in 30–45% yield. Hydrogenation of diyne **91**, oxidation-thermal elimination and deprotection led to (−)-bullatacin. The synthesis was achieved in 20 steps and in 0.8% overall yield.

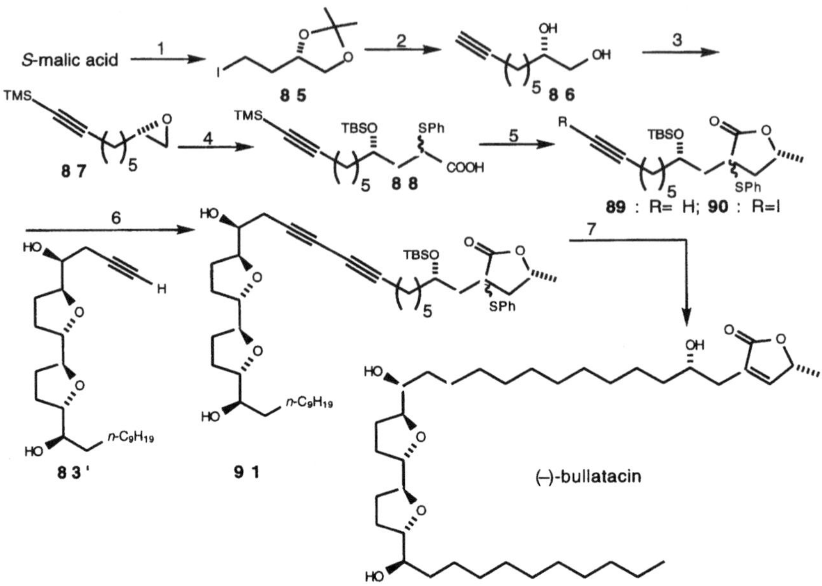

Fig. VI.10. *Reagents:* 1) (i) $BH_3 \cdot SMe_2$; (ii) acetone, TsOH; (iii) I_2, PPh_3, ImH, Et_2O/MeCN; 2) (i) 1-lithio-pentyne; (ii) CSA, MeOH; (iii) KAPA, DAP; 3) (i) EtMgBr; (ii) TMSCl; (iii) 10% HCl; (iii) TsCl, pyridine, −10 °C; (iv) NaH, THF; 4) (i) $PhSCH(Li)CO_2Li$; (ii) 10% citric acid; (iv) TBSCl, ImH; (v) MeOH; 5) (i) LDA; (ii) (*R*)-propylene oxide; (iii) 10% citric acid; (iv) CSA, PhH, reflux; (v) K_2CO_3, MeOH; (vi) I_2, morpholine; 6) **99′**, $Pd(PPh_3)_4$, CuI, *i*-Pr_2NH, THF; 7) (i) H_2, $RhCl(PPh_3)_3$; (ii) oxone™, MeOH, H_2O; (iii) PhMe, reflux; (iv) 5% HF, MeCN, THF

Synthesis of (+)*-bullatacin* (= *rolliniastatin-2*) (*74*)

For the synthesis of (+)-bullatacin, Hoye's group used a different strategy (*74*) inspired by Keinan's work (*155*). The bis-THF core was built up through asymmetric dihydroxylation of a functionalized diene **92** which was then submitted to the same sequence of transformations as ester **33** in figure VI.5., thus leading to the THF-lactone **97**. Compound **97** was then protected as a silyl ether (TBDMSCl, imidazole, DMF) before DIBAL reduction thus affording the hemiketal which was homologated by means of a standard Wittig reaction followed by another DIBAL reduction to give allylic alcohol **98**. Asymmetric epoxidation of the latter led to the intermediate epoxide which spontaneously cyclized to the bis-THF diol **99**. Selective silylation of the primary hydroxyl (TBDPSCl), tosylation of the free hydroxyl group and desilylation with TBAF gave the cyclized epoxide alcohol **100**. The latter was reacted with lithium trimethylsilylacetylide before removal of the TMS protecting group with potassium carbonate to give **101**. A Mitsunobu reaction on epoxide alcohol **100** before treatment with lithium trimethylsilylacetylide allowed the authors to obtain the epimeric compound **102**, a key intermediate in the total synthesis of asimicin (a C-24 epimer of rolliniastatin-2). Intermediates **101** or **102** was then coupled with vinyl iodide **111** (Fig. VI.12.) through a palladium catalyzed reaction leading to **103** or **104**, respectively. After hydrogenation with Wilkinson's catalyst and removal of the protecting groups with TBAF, (+)-asimicin and (+)-bullatacin were obtained (Fig. VI.11.).

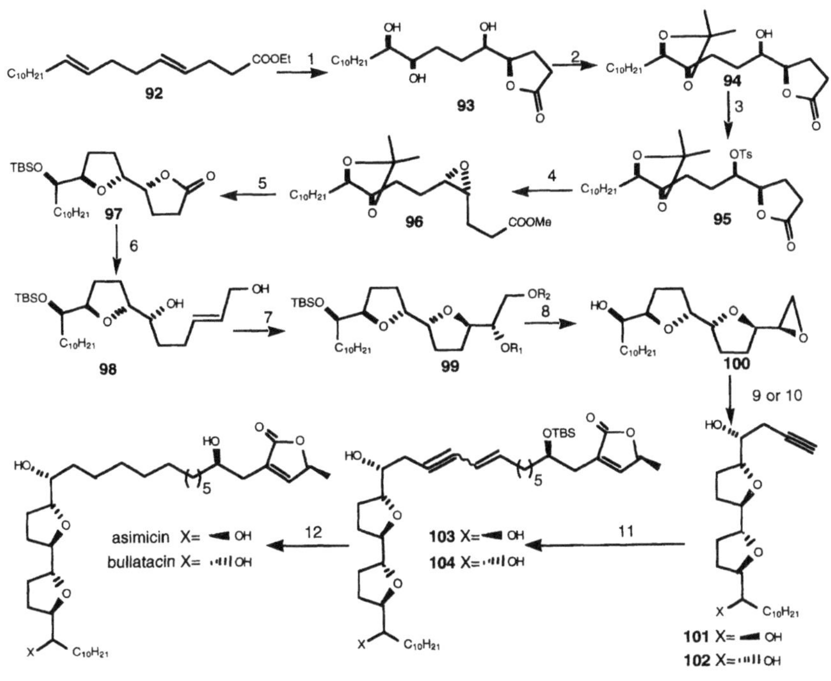

Fig. VI.11. *Reagents:* 1) AD-mix-β; 2) DMP, acetone, TsOH, 72% for the two steps; 3) TsCl, Et_3N, DMAP, CH_2Cl_2; 4) K_2CO_3, MeOH, 91% for the last two steps; 5) (i) $BF_3 \cdot OEt_2$, CH_2Cl_2, 63%; (ii) TBSCl, imid., DMF; 6) (i) DibalH; (ii) $Ph_3P = CHCO_2Et$; (iii) DibalH, 66% for the last 4 steps; 7) (i) (+)-DET, TBHP, 50 mol. % Ti(O-*i*-Pr)$_4$, 87%; (ii) TBDPSCl, imid., DMF, 86%; (iii) TsCl, 98%; 8) TBAF, THF, R.T., 88%; 9) (i) TMSC≡CLi, $BF_3 \cdot OEt_2$; (ii) K_2CO_3, 70% for the last two step; 10 (i) *p*-NO_2PhCO_2H, (ii) TMSC≡CLi, $BF_3 \cdot OEt_2$; (iii) NaOMe, MeOH, THF, 34% for the last three steps; 11) **111**, Pd(PPh$_3$)$_2$Cl$_2$, CuI, NEt_3, 79% for **103** and 82% for **104**; 12) (i) Rh(PPh$_3$)$_3$Cl, PhH, H_2 1 atm; (ii) AcCl in MeOH, Et_2O, 75%

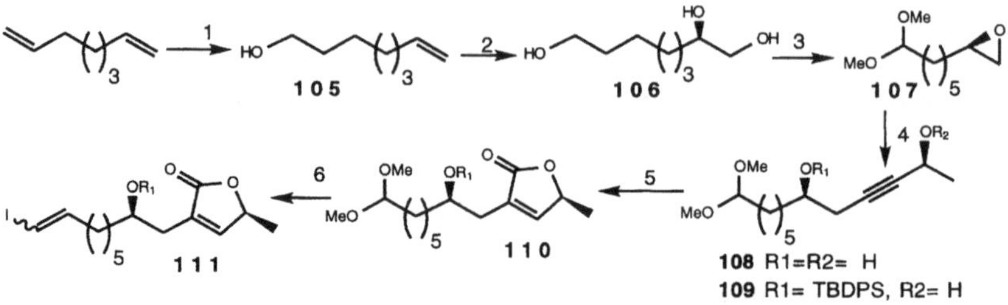

Fig. VI.12. *Reagents*: 1) (i) 9-BBN, (ii) H_2O_2, NaOH; 2) (i) AD-Mix-β; (ii) recrist. (x2), 64% for the last steps; 3; (i) $CH(OMe)_3$, PPTS; (ii) TMSCl; (iii) Swern; (iv) TsOH, MeOH, (v) K_2CO_3, MeOH, 86%; 4) (i) (*S*)-LiC≡CCH(OTBS)CH_3, $BF_3 \cdot OEt_2$; (ii) PPTS, MeOH, 88% for the last two steps; (iii) TBDPSCl; (iv) PPTS MeOH, 80% for the last two steps; 5) (i) Redal, (ii) I_2, (iii) $(Ph_3P)_2PdCl_2$, NH_2NH_2, K_2CO_3, THF, CO 45 psi, 83% for the last three steps; 6) (i) TFA, H_2O, $CHCl_3$; (ii) $CrCl_2$, CHI_3, dioxane, THF, 72% for the last 2 steps

The lactone fragment was prepared by the Sharpless asymmetric dihydroxylation of unsaturated alcohol **105**, leading to homochiral triol **106**. In a one-pot 4 steps sequence, triol **106** was transformed into epoxide **107** which upon addition of the homochiral lithium acetylide (derived from optically pure 3-butyn-2-ol) gave alcohol **108**. Silylation of the C-4 hydroxyl (TBDPSCl), followed by selective removal of the TBS group (PPTS, MeOH) produced the homopropargyl alcohol **109**. Redal reduction and iodine treatment followed by carbonylation under Stille conditions led to lactone **110**. Hydrolysis of the acetal followed by addition of iodoform in the presence of chrome dichloride produced vinyl iodide **111** (Fig. VI.12.).

Synthesis of (+)-bullatacin (rolliniastatin-2) (153)

Sazaki *et al.*'s synthesis of (+)-bullatacin started with diethyl isopropylidene D-tartrate, which was reduced with DIBAL to the dialdehyde before Wittig-Horner homologation and direct hydrogenation of the double bond to give **112**. The latter was again reduced at low temperature by DIBAL and homologated through a Wittig process leading to the di-unsaturated diester **113**, which was converted to diol **114** (by DIBAL reduction at −78 °C). **114** was submitted to a double asymmetric epoxidation with (+)-diethyl-tartrate leading to the corresponding bis-epoxide which was protected as the bis-*p*-nitrobenzoyl ester **115**. Acetonide removal and subsequent rearrangement of the epoxides to the bis-THF unit was performed by treatment with $BF_3 \cdot OEt_2$ to give **116** possessing a C2 symmetry axis. Monotosylation of the latter followed by basic hydrolysis

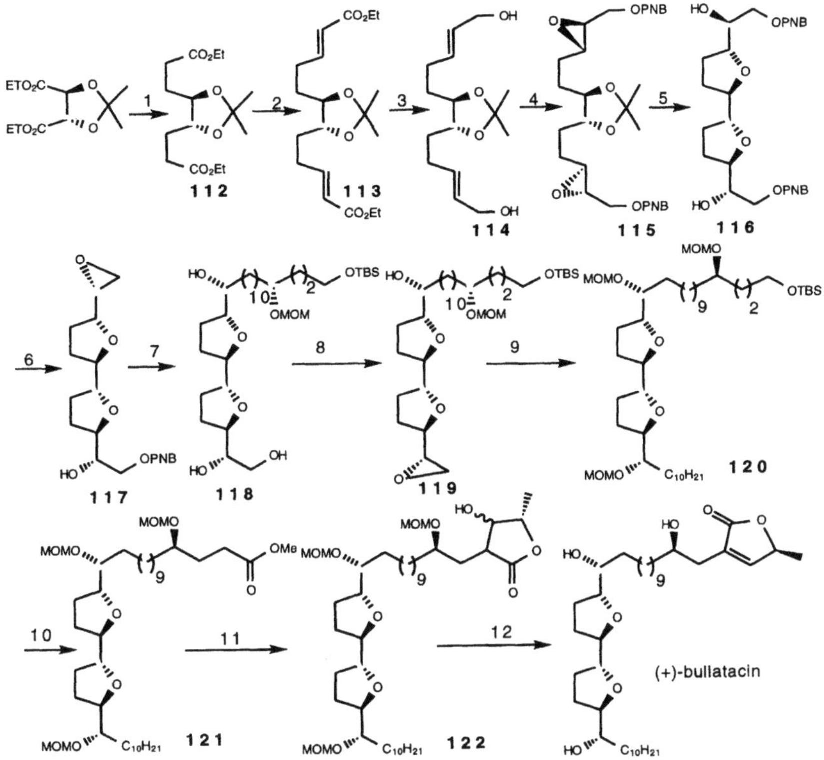

Fig. VI.13. *Reagents:* 1) (i) Dibal, toluene, −78 °C; (ii) $(EtO)_2P(O)CH_2CO_2Et$, NaH, DME, −78 °C to R.T., (iii) H_2, Pd/C, EtOH, 86% for the three steps; 2) (i) Dibal, toluene, −78 °C; (ii) $(EtO)_2P(O)CH_2CO_2Et$, NaH, DME, −78 °C to R.T., 78% for the two steps; 3) Dibal, CH_2Cl_2, toluene, −78 °C, 86%; 4) (i) (+)-DIPT, $Ti(O\textit{i}\text{-}Pr)_4$, TBHP, CH_2Cl_2, −30 °C to −20 °C; (ii) PNBCl, TEA, 0 °C, 96% for the two steps; 5) $BF_3 \cdot OEt_2$, MeOH—CH_2Cl_2, H_2O, 0 °C, 92%; 6) (i) MsCl, TEA, THF, 0 °C, 86%; (ii) $n\text{-}Bu_4NOH$, THF, 0 °C, 99%; 7) (i) **124**, *n*-BuLi, DME, R.T., 83%; (ii) Na-Hg, EtOH, R.T., 80%; 8) (i) TsCl, pyr., −20 °C; (ii) K_2CO_3, EtOH—H_2O, R.T., 57% for the two steps; 9) (i) CuBr, $CH_3(CH_2)_8MgBr$, THF, 0 °C; (ii) MOMCl, $i\text{-}Pr_2NEt$, CH_2Cl_2, 0 °C to R.T., 64% for the two steps; 10) (i) $n\text{-}Bu_4NF$, THF, 0 °C; (ii) CrO_3—H_2SO_4, acetone, −20 °C; (iii) CH_2N_2, Et_2O—AcOEt, 0 °C, 71% for the three steps; 11) (i) LDA; (ii) (*S*)-tetrahydropyranoyl lactaldehyde, THF, −78 °C; (iii) CSA, MeOH—H_2O, R.T.; 12) (i) BzCl, pyr., 0 °C to R.T.; (ii) NH_3 in MeOH, R.T., 48% for the last four steps, (iii) $BF_3 \cdot OEt_2$, DMS, R.T., 99%

of the *p*-nitrobenzoyl esters led to formation of the monoepoxide **117** (Fig. VI.13.).

The half of the molecule eventually containing the γ-lactone was introduced by reaction of the epoxide with the α-sulfonyl carbanion of phenyl sulfone **124** prepared in 12 steps and 18% overall yield from diol

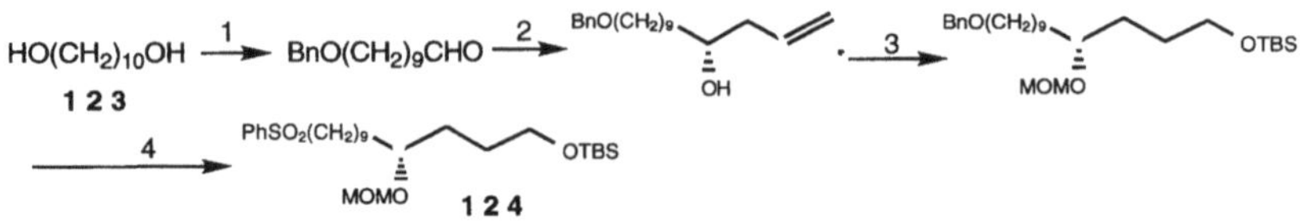

Fig. VI.14. *Reagents:* 1) (i) NaH, BnCl, DMF, 51%; (ii) PDC, Celite, CH_2Cl_2, 82%; 2) (i) allyl-B(^{d}Ipc)$_2$, Et_2O, −78 °C; (ii) NaOH, H_2O_2, −78 °C to R.T., 66% for the two steps; 3) (i) MOMCl, *i*-Pr_2NEt, CH_2Cl_2, 0 °C to R.T., 98%; (ii) BH_3·THF, THF, −20 °C to R.T.; (iii) NaOH, H_2O_2, 0 °C to R.T., 78%, (iv) TBDMSCl, TEA, imidazole, CH_2Cl_2, 0 °C to R.T., 99%; 4) (i) Li, liquid NH_3, −78 °C, 99%; (ii) TsCl, pyr., 0 °C, 97%; (iii) PhSH, NaH, 0 °C to R.T., 92%, (iv) MMPP, EtOH—H_2O, R.T., 95%

123 (Fig. VI.14.). The coupling reaction was best carried out by treating a mixture of 1 eq. of **117** with 2 eq. of **124** with 4 eq. of *n*-BuLi in DME at room temperature to give the coupled product which was desulfonated by treatment with sodium amalgam leading to triol **118** (Fig. VI.13.). Selective tosylation of the primary hydroxyl group and subsequent basic treatment led to epoxide **119**. Opening of the latter by the desired cuprate, followed by MOM protection of the secondary hydroxyl groups gave **120**. Selective deprotection of the primary hydroxyl group (TBAF, THF), Jones' oxidation and esterification led to methyl ester **121**. The lactone ring was then built up through the usual method: Enolization (LDA, −78 °C), aldolization ((*S*)-methyl lactaldehyde) and acidic hydrolysis (CSA, MeOH–H_2O), leading to alcohol **122**. After benzoylation of **122** and subsequent elimination by treatment with NH_3, deprotection of the protecting groups was performed with BF_3·OEt_2 to afford (+)-bullatacin. The synthesis was achieved in 26 steps and 11% overall yield.

VI.2.3. Synthesis of Rolliniastatin-1

Koert *et al.* reported the total synthesis of (+)-rolliniastatin-1 from L-glutamic acid, based on a sequential approach, using as a key step a diastereoselective copper catalyzed Grignard addition with functionalized aldehydes (*150*) (Fig. VI.15.).

Nitrile **127** (prepared from L-glutamic acid in 5 steps and 61% overall yield) was converted to the corresponding methyl ester by a known procedure, followed by reduction to the corresponding alcohol. Benzylation and desilylation of the latter afforded **128**, which by Swern oxidation led to the corresponding aldehyde. Copper catalyzed addition of the required Grignard reagent afforded the *threo* acetonide alcohol **129** with a 97:3 *threo*/*erythro* ratio. Removal of the acetonide, mesylation of the

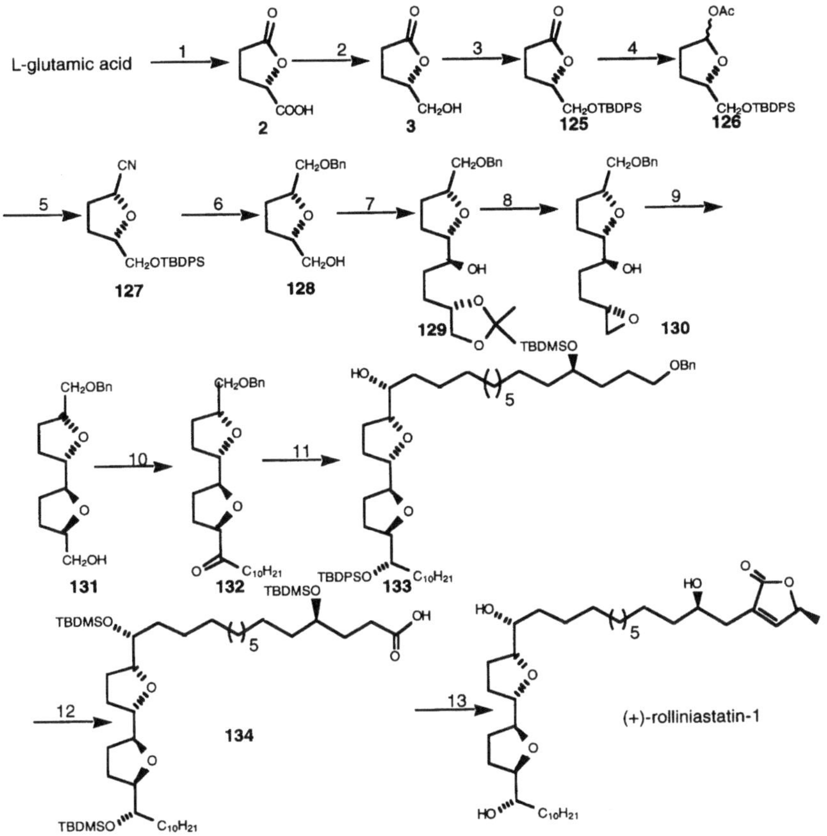

Fig. VI.15. *Reagents:* 1) $NaNO_2$, H_2SO_4, 70%; 2) $BH_3{\cdot}SMe_2$, THF, 98%; 3) TBDPSCl, imidazole, DMF, 92%; 4) (i) DIBAL; (ii) Ac_2O, Et_3N; 5) TMSCN, $BF_3{\cdot}OEt_2$, Et_2O, 0 °C (d.e. = 50:50); 6) (i) NaOMe, MeOH; (ii) $LiAlH_4$, THF; (iii) BnBr, NaH; (iv) TBAF, THF; 7) (i) $(COCl)_2$, DMSO, Et_3N, CH_2Cl_2, −78 °C; (ii) $BrMgCH_2CH_2CH^*$ $(OR)CH_2OR$, $CuBr{\cdot}SMe_2$, −78 °C, Et_2O; 8) (i) HOAc, THF/H_2O; (ii) MesitylSO_2Cl, pyridine, 0 °C; (iii) K_2CO_3, MeOH; 9) HOAc, CH_2Cl_2, R.T.; 10) (i) $(COCl)_2$, DMSO, Et_3N; (ii) $C_{10}H_{21}MgBr$; (iii) $(COCl)_2$, DMSO, Et_3N; 11) (i) $Zn(BH_4)_2$, Et_2O, −78 °C; (ii) TBDMSCl; (iii) H_2, Pd—C, THF; (iv) $(COCl)_2$, DMSO, Et_3N; (v) $CuBr{\cdot}SMe_2$, Et_2O, −78 °C, 14-benzyloxy-11-*tert*-butyldimethylsilyloxy-1-ylmagnesium bromide; 12) (i) TBAF, THF; (ii) TBDMSCl; (iii) H_2, Pd-C, THF; (iv) $(COCl)_2$, DMSO, Et_3N; (v) $KMnO_4$, $NaHPO_4$, *tert*-BuOH, H_2O; 13) (i) LDA; (ii) (*S*)-propylene oxide, then *p*-TsOH; (iii) MMPP, THF, MeOH, then toluene; (iv) HF, MeCN, THF, R.T

primary hydroxyl and base treatment gave epoxide **130**. Acid treatment of **130** led to the bis-THF compound **131** through an intramolecular epoxide opening. Swern oxidation of the free primary hydroxyl group and subsequent addition of the desired Grignard reagent followed by another Swern oxidation led to the expected ketone **132** which was reduced with $Zn(BH_4)_2$ to yield the *erythro* alcohol as the major isomer (*erythro*/*threo* ratio = 82:18). After protection of the free hydroxyl as a silyl ether by treatment with TBDPSCl in the presence of imidazole and debenzylation of the primary hydroxyl, the alcohol so obtained was oxidized under Swern conditions to afford the corresponding aldehyde. Addition of the required Grignard reagent in the presence of CuBr, gave *threo* alcohol **133** in a 95:5 *threo*/*erythro* ratio. Deprotection of the hydroxyl groups followed by reprotection as *tert*-butyldimethylsilyl ethers yielded the expected protected tetraol. Debenzylation of the primary hydroxyl and oxidation *via* the non-isolated corresponding aldehyde led to the desired carboxylic acid **134**. The latter was treated with 2 eq. of LDA, followed by PhSSPh and then with (*S*)-propylene oxide to give after acid treatment the desired lactone. Oxidation and thermal elimination of the sulfoxide allowed the introduction of the double bond in the lactone ring. Deprotection of the silyl ethers with 5% HF in CH_3CN/THF afforded rolliniastatin-1. The synthesis was achieved in 30 steps and in 2.1% overall yield.

VI.3. Synthesis of Acetogenins of Type E

VI.3.1. Synthesis of Reticulatamol and Reticulatamone

Total syntheses of natural reticulatamone and reticulatamol (as a C-15 diastereomeric mixture) in 5 and 6 steps, respectively, have been recently reported in the literature (*145*). Both strategies used are based on a Michael type radical addition of iodo compound **137** to enone **135**. This required the preparation of the enone which was achieved in two steps from octadecanal **136** through Grignard addition of vinylmagnesium bromide followed by oxidation with manganese dioxide. The iodo compound **137** was prepared from methyl 12-bromododecanoate **138** in 4 steps; enolization by LDA and alkylation with phenylselenium chloride followed by another enolization and alkylation with (*S*)-propylene oxide afforded after acidic treatment the desired lactone. Oxidation by H_2O_2 under acid conditions led to the unsaturated lactone which after bromine displacement with NaI in acetone under reflux led to iodo compound **137**. Treatment of a stoichiometric mixture of **135** and **137** with 2 eq. of n-Bu_3SnH and a catalytic amount of AIBN in toluene under reflux led,

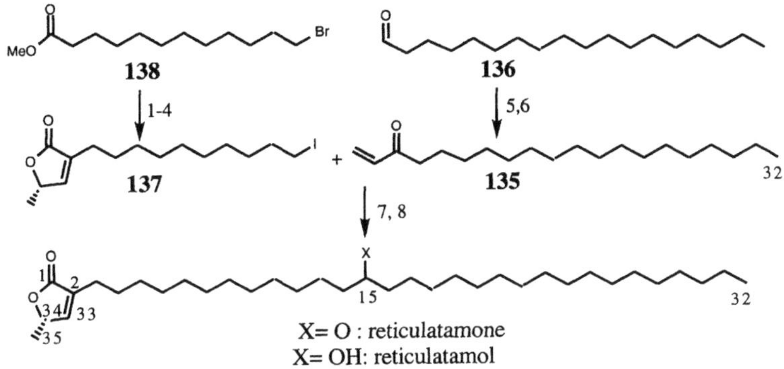

Fig. VI.16. *Reagents:* 1) (i) LDA, −78 °C; (ii) PhSeCl; 2) (i) LDA, −78 °C; (ii) (*S*)-propylene oxide; 3) H_2O_2, AcOH, 55% for the last 3 steps; 4) NaI, acetone, 100%; 5) vinyl MgBr; 6) MnO_2, 85% for the last 2 steps; 7) *n*-Bu_3SnH, AIBN cat., reflux toluene, 56%; 8) *n*-Bu_3SnH/ SiO_2, CH_2Cl_2, r.t., 48 h, 92%

after purification by flash chromatography, to (+)-reticulatamone in 57% yield. Reduction of the latter with *n*-$Bu_3SnH \cdot SiO_2$ in CH_2Cl_2 at room temperature for 48 h led to reticulatamol as an epimeric mixture at C-15 in 92% yield (*145*, *213*). The final reduction step is highly chemoselective, since neither the carbonyl of the lactone nor the conjugated double bond was affected by the reaction conditions (Fig. VI.16.).

VI.3.2. Synthesis of (+)- and/or (−)-Muricatacin

Many syntheses of the title compounds have been reported and are discussed in what follows.

Figadère's synthesis (*55*)

(+)-Muricatacin has been synthesized by FIGADERE *et al.* in 4 steps and 48% overall yield from a very inexpensive starting material, L-glutamic acid (Fig. VI.17.). The key steps of the synthesis are a nitrous deamination of the α-amino acid with retention of the configuration of the stereogenic centre, and a very diastereoselective reduction of α-butyrolactonic ketone **11** with L-Selectride™. The use of D-glutamic acid allowed the preparation of (−)-muricatacin as well.

Tochtermann's synthesis (*47*, *50*)

TOCHTERMANN *et al.*'s strategy was based on the chemical resolution of a racemic α-butyrolactonic ketone **140** followed by diastereoselective

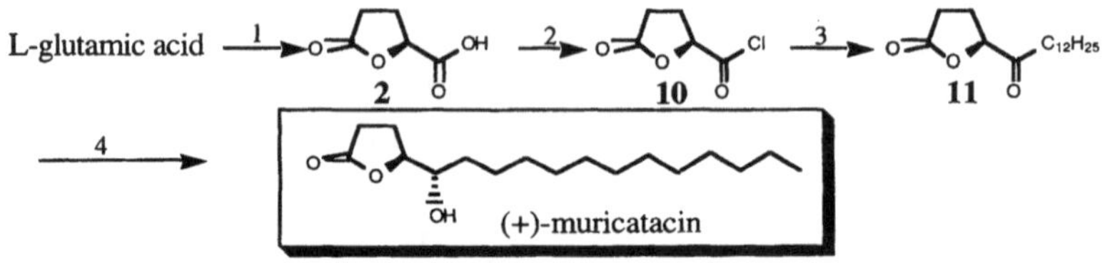

Fig. VI.17. *Reagents:* 1) $NaNO_2$, HCl, 70%; 2) $(COCl)_2$, DMF cat., CH_2Cl_2, 92%; 3) dodecylmagnesium bromide, $-78\,°C$, THF, 85%; 4) L-Selectride™, $-78\,°C$, THF, 88% (*syn*/*anti* = 98/2)

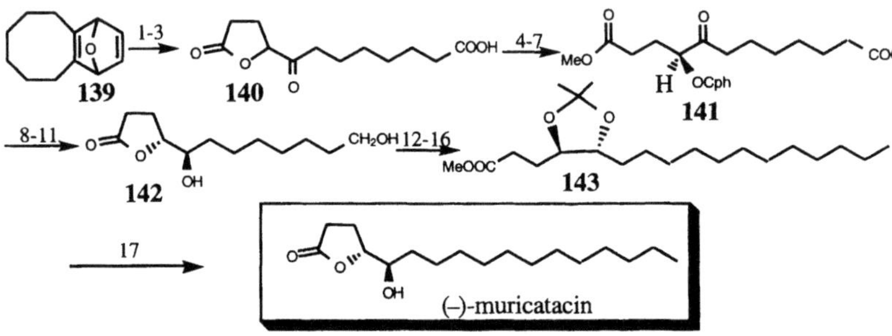

Fig. VI.18. *Reagents:* 1) H_2, Pt; 2) O_3; 3) O_2; 4) 2N NaOH, r.t.; 5) $AgNO_3$, H_2O, r.t.; 6) CH_3I, Et_2O; 7) (i) (−)-camphanoyl acid chloride, pyridine; (ii) separation; 8) L-selectride™, $-78\,°C$, THF; 9) Amberlyst-H 15, CH_2Cl_2; 10) (i) $NaOCH_3$, CH_3OH, reflux, 2 h; (ii) NaOH, H_2O, CH_3OH, reflux, 5 h; (iii) HCl; 11) $BH_3\cdot SMe_2$, THF, r.t.; 12) 2,2-dimethoxypropane, CH_3OH, Amberlyst-H 15, 60 h, r.t.; 13) PCC, CH_2Cl_2, 2.5 h, r.t.; 14) $Ph_3P^+C_5H_{11}Br^-$, KO*t*—Bu, Et_2O, reflux, 1 h; 15) H_2, Pt; 16) THF, H_2O, HCl, reflux, 1 h; 17) *id.* 16)

reduction of the obtained homochiral **140** leading to **141**, which after further chemical transformations led to the desired (−)-muricatacin (Fig. VI.18.). The synthesis has been achieved in 17 steps each transformation ranging from 77 to 97% yield.

Marshall's synthesis (*48*, *53*)

MARSHALL *et al.* separately prepared (+)- and (−)-muricatacin through 1,2-addition of homochiral silyloxy allylic stannanes **147** on α,β-unsaturated aldehydes. Acylstannane **145** was reduced with either (*R*)- or (*S*)-BINAL-H to the corresponding hydroxystannanes **146a** or **146b**, both with 95% ee, respectively. Rearrangement of the protected alcohols **146a, b** (as silyl ethers) by treatment with $BF_3\cdot OEt_2$ at low temperature led to allylic stannanes **147a** and **147b**. The latter two compounds, when

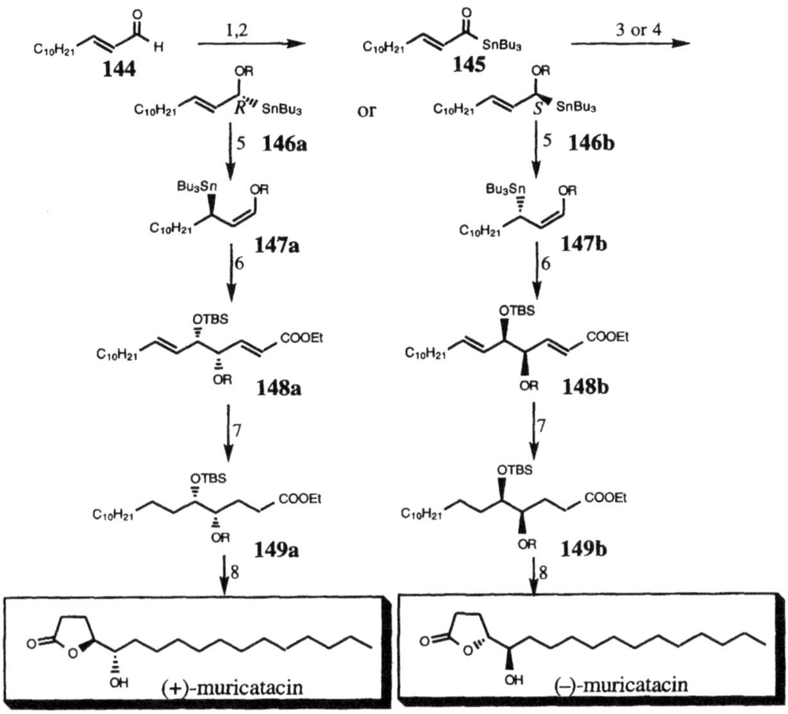

Fig. VI.19. *Reagents:* 1) *n*-Bu_3SnLi; 2) DIAD; 3) (*S*)- or (*R*)- BINAL-H, 95% e.e.; 4) TBDMSCl, imidazole, CH_2Cl_2, 48% for the last 4 steps; 5) $BF_3 \cdot OEt_2$, –78 °C, 95% e.e.; 6) $BF_3 \cdot OEt_2$, CH_2Cl_2, CHOCH=$CHCO_2Et$, 75–80%; 7) H_2, Pd/C, 86–88%; 8) HF, THF, H_2O, 83–87%

treated with the appropriate enal in the presence of $BF_3.OEt_2$ at –78 °C, led to adducts **148a** and **148b**, which after hydrogenation followed by HF cleavage, gave rise to (+)- and (–)-muricatacin. The syntheses were achieved in 8 steps and in 27.6% overall yield (Fig. VI.19.).

Sharpless' synthesis (*49*)

SHARPLESS' strategy is the most efficient preparation of (+)- and (–)-muricatacin reported so far. Asymmetric dihydroxylation (AD-mix.-α or β) or the γ,δ-unsaturated ester **152** (prepared from allylic alcohol **151** by a Claisen rearrangement), afforded directly the desired (+)- or (–)-muricatacin with 95 and 96% ee, respectively, and in 74% overall yield from tridecanal **150** (Fig. VI.20.).

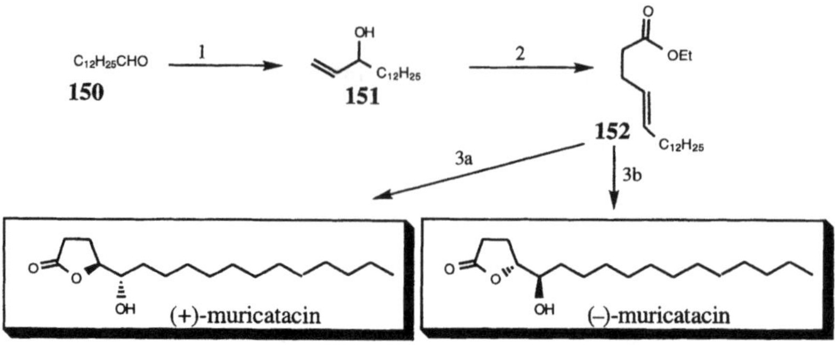

Fig. VI.20. *Reagents:* 1) vinylMgBr, 96%; 2) $CH_3C(OEt)_3$, $EtCO_2H$ cat., reflux, 92%; 3a) AD-mix-α., 82%, 95% e.e.; 3b) AD-mix-β., 84%, 96% e.e

Kang's synthesis (*52*)

D-glucose was chosen by KANG *et al.* as starting material which was converted to the known aldehyde **153**. Wittig homologation, followed by hydrogenation gave the product **154**. Protection of the free hydroxyl as a benzyl ether, removal of the acetonide, followed by oxidative cleavage, gave the acyclic aldehyde **155**. Reaction of the latter with the anion of triethylphosphonoacetate provided the unsaturated ester **156**. Hydrogenation and treatment with aqueous trifluoroacetic acid afforded (−)-muricatacin. The synthesis was achieved in 8 steps from the known aldehyde **153** and in 15.8% overall yield (Fig. VI.21.).

Bessodes' synthesis (*56*)

For the preparation of (+)-muricatacin, BESSODES and coworkers used as a key step the kinetic resolution by means of Sharpless' asymmetric

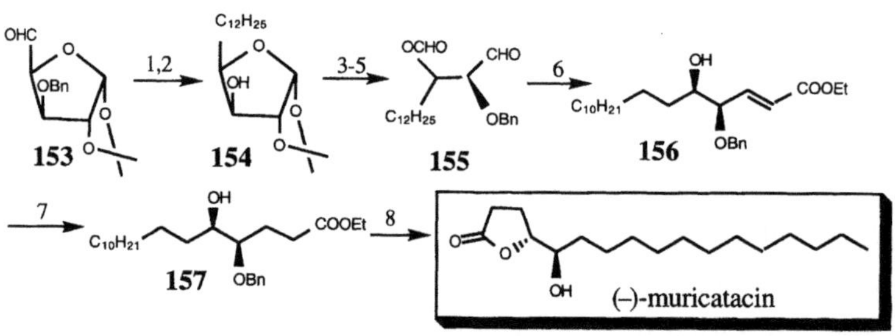

Fig. VI.21. *Reagents:* 1) $CH_3(CH_2)_{10}PPh_3Br$, *n*-BuLi, THF, r.t., 12 h, 86%; 2) H_2, Pd/C, EtOAc, r.t., 10 h, 91%; 3) NaH, BnCl, THF, r.t., 5 h; 4) 2N HCl, DME, r.t., 48 h; 5) $NaIO_4$, MeOH, r.t., 1 h, 53% for the last 3 steps; 6) NaH, $(EtO)_2POCH_2CO_2Et$, THF, r.t., 3 h, 67%; 7) H_2, Pd/C, EtOAc, r.t., 24 h, 71%; 8) TFA, H_2O (4/1), r.t., 3 h, 80%

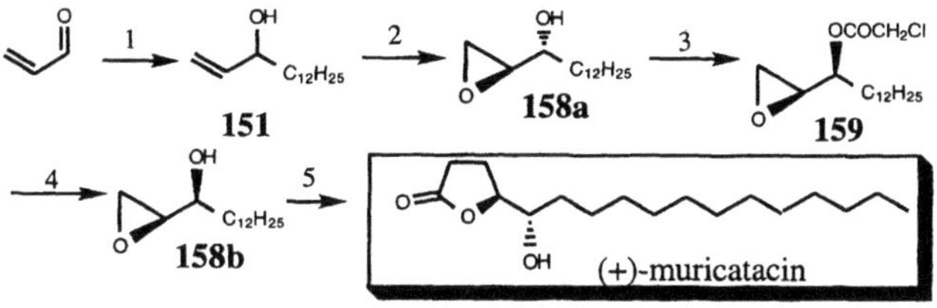

Fig. VI.22. *Reagents:* 1) $C_{12}H_{25}MgBr$; 2) $Ti(O\text{-}i\text{-}Pr)_4$, D-diisopropyl tartrate, $t\text{-}BuO_2H$, CH_2Cl_2, r.t., $-20\,°C$, 95% (resolution yield); 3) DEAD, PPh_3, $ClCH_2CO_2H$, benzene, reflux, 90%; 4) MeONa 1%, MeOH, 100%; 5) (i) $LiCH_2CO_2Li$, THF, reflux; (ii) H_3O^+, 65%

epoxidation of the allylic alcohol **151**, obtained by Grignard reaction of dodecylmagnesium bromide on acrolein, by treatment with *tert*-BuO_2H in the presence of $Ti(i\text{-}PrO)_4$ and D-diisopropyl tartrate under kinetic conditions to give the epoxy alcohol **158a**. After a Mitsunobu reaction (DEAD, PPh_3, $ClCH_2COOH$), chloroacetate **159** was obtained with inversion of configuration of the stereogenic centre bearing the hydroxyl group. Subsequent deacylation with sodium methanolate followed by treatment with lithium acetate and then acidic work-up, led to (+)-muricatacin. The synthesis was achieved in 5 steps from dodecyl bromide and in 26.7% overall yield (Fig. VI.22.).

In the same year, LIU *et al.* (*218*) reported the total synthesis of racemic muricatacin using a similar strategy, but performing the epoxidation step without asymmetric inducation and using the lithium enolate of acetonitrile for opening of the epoxide (+/−)-**158h** before acidic treatment leading to (+/−)-muricatacin.

Tanaka's synthesis (*51*)

TANAKA's group also prepared (−)-muricatacin from dodecylbromide in 7 steps and in 27% overall yield, using as a key step the Sharpless asymmetric epoxidation of allylic alcohol **46**, leading to the hydroxy epoxide **47** as already shown in Fig. VI. 6. Subsequent elongation of the alkyl chain with the lithium enolate of *t*-butyl acetate and acidic treatment, gave (−)-muricatacin.

Depezay's synthesis (*219*, *54*)

A synthesis by DEPEZAY *et al.* starts from (2*R*, 3*R*)-3,4-epoxy-1,2-O-methylidene butane 1,2-diol **161** prepared from D-isoascorbic acid **160** in 6 steps and 40% overall yield (*220*). Nucleophilic opening of the epoxide with undecylmagnesium bromide in the presence of Li_2CuCl_4 led to the

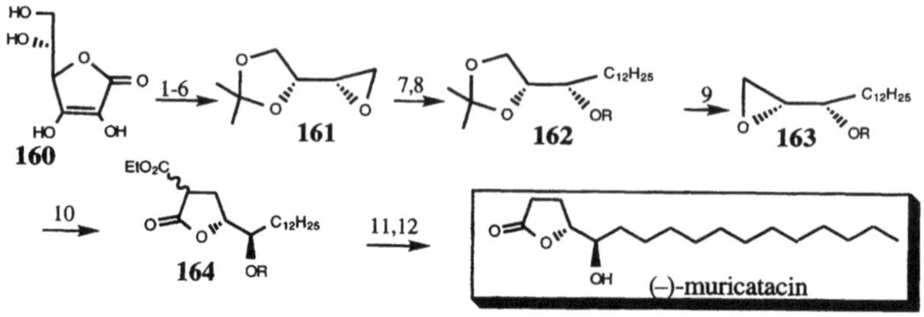

Fig. VI.23. *Reagents:* 1–6) ref. (*220*); 7) $C_{11}H_{23}MgBr$, Li_2CuCl_4, THF, −35 °C, 80%; 8) NaH, DMF, imidazole, 20 °C, MPMCl, *n*-Bu_4NI, 20 °C, 93%; 9) (i) AcOH/H_2O 4/1, 20 °C; (ii) Ph_3P, DEAD, 125 °C, 70–80% for the last 2 steps; 10) $CH_2(CO_2Et)_2$, EtONa, EtOH, 60 °C; 11) $MgCl_2 \cdot 6H_2O$, $CH_3CON(CH_3)_2$; 12) DDQ, CH_2Cl_2/H_2O 18/1, 20 °C, 37% for the last 3 steps

corresponding alcohol which was protected as its 4-methoxybenzyl ether **162**. Removal of the acetonide (AcOH,H_2O), followed by an intramolecular Mitsunobu reaction (PPh_3, DEAD, 125 °C), gave epoxide **163**. Addition of diethyl malonate anion on epoxide **163** led to a mixture of α-carbethoxy-γ-butyrolactones **164** which upon treatment with magnesium chloride hexahydrate in dimethylacetamide followed by deprotection of the alcohol through a DDQ oxidation led to the desired (−)-muricatacin. The synthesis was achieved in 12 steps from D-isoascorbic acid **160** and in 8.8% overall yield (Fig. VI.23.).

Bonini's synthesis (*221*)

The strategy of BONINI and coworkers for the synthesis of (−)-muricatacin is based on an original rearrangement-opening of 2,3-epoxy alcohol **167**, obtained by asymmetric epoxidation of the corresponding allylic alcohol **166**, by treatment LiI in DME at 70 °C. Coupling of the iodohydrin **168** so obtained with malonate afforded (−)-muricatacin **19** (Fig. VI.24.). The synthesis was achieved in 5 steps and 25% overall yield.

Somfai's synthesis (*57*)

A synthesis of (+)-muricatacin by SOMFAI team started with diethyl D-tartrate which in 4 steps and 83% overall yield led to D-threitol derivative **169**. Subsequent tosylation followed by displacement with a cuprate reagent gave **170**. Deprotection of the acetal gave **171** which after selective tosylation of the primary hydroxyl and basic treatment led to epoxide

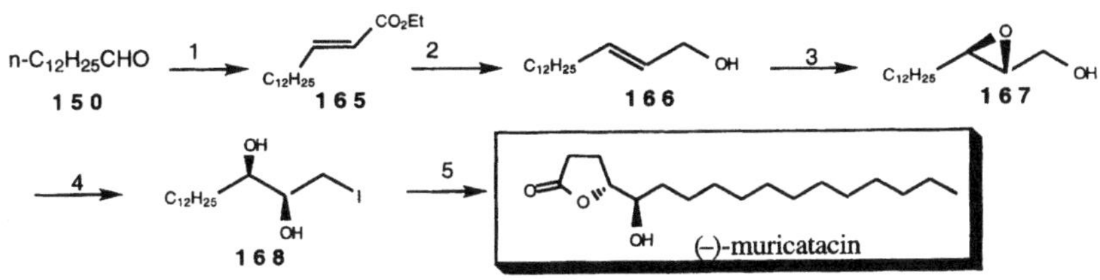

Fig. VI.24. *Reagents:* 1) $(C_2H_5O)_2POCH_2CO_2Et$, LiOH, THF, 90%; 2) DIBAL, 91%; 3) (−)-DET, Ti(O*i*-Pr)$_4$, TBHP, CH_2Cl_2, 89%; 4) LiI, DME, 70 °C, 70%; 5) $CH_2(CO_2Et)_2$, EtONa, EtOH, reflux, H^+, 50%

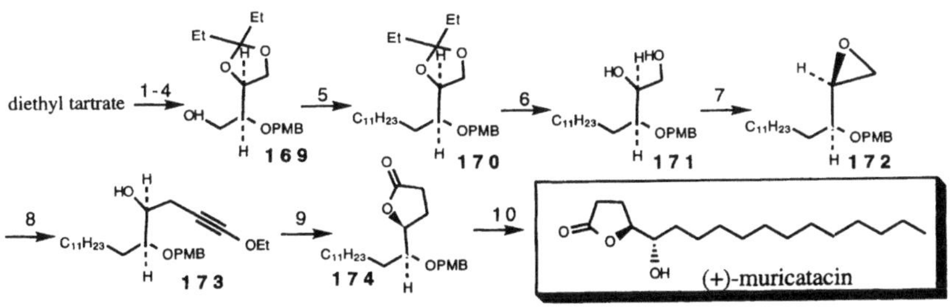

Fig. VI.25. *Reagents:* 1–4) ref. (57), 83%; 5) (i) TsCl, pyr., 0 °C, (ii) $C_{11}H_{23}MgBr$, CuI, THF, −30 °C, 82% for the last two steps; 6) 2% aq. H_2SO_4, MeOH, 91%; 7) (i) TsCl, pyr., 0 °C; (ii) K_2CO_3, MeOH, 83% for the last two steps; 8) LiC≡COEt, $BF_3 \cdot OEt_2$, THF, −78 °C, 79%; 9) xylenes, reflux, 79%; 10) DDQ, H_2O, CH_2Cl_2, 89%

172, which was then opened up by lithium ethoxyacetylide to deliver **173**. Subsequent heating of the latter in refluxing xylene led to the protected muricatacin **174** through a retro-ene reaction. Oxidative removal of the *p*-methoxybenzyl group with DDQ, water, gave (+)-muricatacin. The synthesis was performed in 11 steps and 28% overall yield (Fig. VI.25.).

Quayle's synthesis (*58*)

A synthesis by QUAYLE *et al.* started with the known enantiopure bis-epoxide **175**, obtained from L-(+)-tartaric acid and treated *in situ* with lithium acetylide-ethylenediamine complex to afford diol **176**. Treatment of the latter with $Cr(CO)_5$. THF generated the stable carbene complex **177** which by oxidation with CAN led to lactone **178**. Coupling reaction of this acetylene with the corresponding acetylenic iodide using Wityak's procedure afforded the conjugated acetylene **179** which was then hydro-

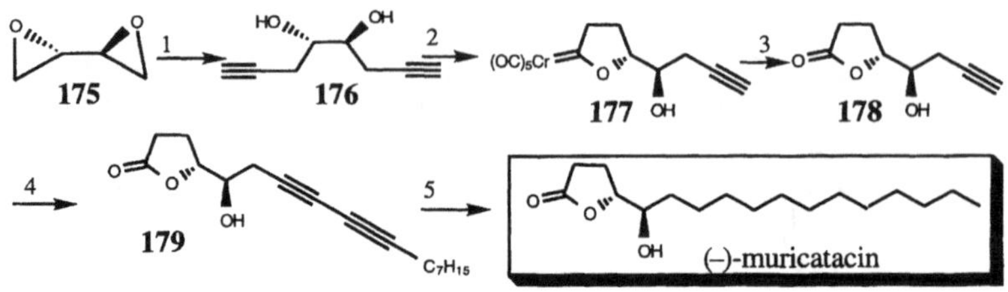

Fig. VI.26. *Reagents:* 1) lithium acetylide-EDA complex, 4 eq., DMSO, 0 °C, 87%; 2) $Cr(CO)_5 \cdot THF$, 2 eq., THF, 20 °C, 3) CAN, 1 eq., acetone, 20 °C, 68% from **191**; 4) $IC{\equiv}CC_7H_{15}$, $(Ph_3)_2PPdCl_2$, 3 mol. %, CuI, 3 mol. %, DIPA, 1.8 eq., THF, 45 °C, 1.25 hrs., 82%; 5) H_2, Pd/C, EtOAc, 20 °C, 16 hrs., 94%

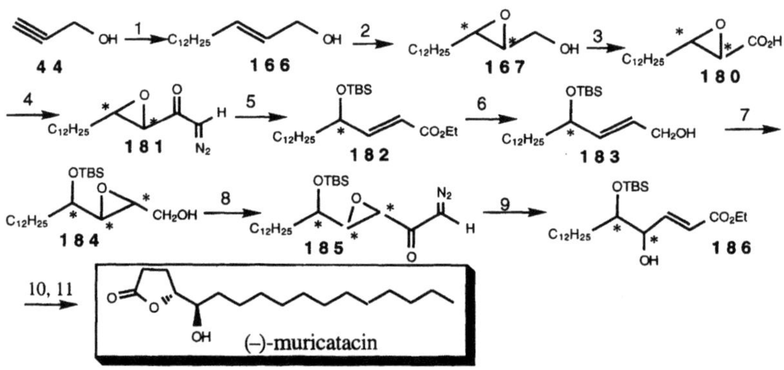

Fig. VI.27. *Reagents:* 1) (i) $LiNH_2$, NH_3(l), $C_{12}H_{25}Br$, 80%; (ii) $LiAlH_4$, Et_2O/THF 7/8, 93%; 2) $Ti(OiPr)_4$, TBHP, M.S., −20 °C, D-DET, 85%, e.e. 91%; 3) (i) DMSO, $(ClCO)_2$, DIEPA, −78 °C, 88%; (ii) $NaClO_2$, $NaHPO_4$, 90%; 4) $ClCO_2iBu$, Et_3N, CH_2N_2, 60%; 5) hν, EtOH, TBSCl, 55%; 6) Dibal-H, 81%; 7) $Ti(OiPr)_4$, TBHP, M.S., −20 °C, L-DET, 92%, d.e. 95%; 8) RuO_4, $ClCO_2iBu$, Et_3N, CH_2N_2, 52%; 9) hν, EtOH, 50%; 10) $Ni(OAc)_2$, $NaBH_4$, pTsOH, 50%; 11) TBAF, 75%

genated to afford (−)-muricatacin (Figure VI.26.). The synthesis was achieved in 8 steps and 49% overall yield.

Zwanenburg's synthesis (222)

A synthesis by the group of Zwanenburg in 13 steps and 1.5% used as key steps the Sharpless asymmetric epoxidation of an allylic alcohol and the photo-induced rearrangement of α,β-epoxy diazomethyl ketones in an alcoholic solvent (Fig. VI.27.).

VI. 4. Hemisynthesis of Acetogenins of Annonaceae

VI. 4.1. Hemisynthesis of Solamin and Reticulatacin

The first report of the hemisynthesis of a naturally-occurring acetogenin of Annonaceae was published by ROBLOT *et al.* (*108*). They have shown that treatment of a natural precursor, diepomuricanin-A, by NaOH followed by acetylation with Ac_2O-pyridine afforded solamin diacetate **187** in 60% yield. Unfortunately the composition of the expected diastereomeric mixture was not discussed. Indeed opening of the bis-epoxide system at either C-15 or C-20 would lead to two different compounds **187** and **188** with the same relative configuration but of opposite absolute configuration across the THF-ring (Fig. VI.28.).

In the same report (*108*) the authors showed that treatment of epomuricenin A with *m*-CPBA according to HOYE (*223*) allowed the partial synthesis of diepomuricanin-A. Again, no details were given concerning the expected epimeric mixture at C-19/20 of the products so obtained due to oxidation on either "a" or "b" face (Fig. VI.29.).

Hemisynthesis of reticulatacin was reported by TAM *et al.* (*110*) from dieporeticanin, but by treatment with 70% perchloric acid. Here again, two compounds—reticulatacin and its tetraepimer **189**—were expected with identical relative configuration, but with opposite absolute configuration across the THF ring because of epoxide ring opening either at C-17 or C-22 (Fig. VI.30.).

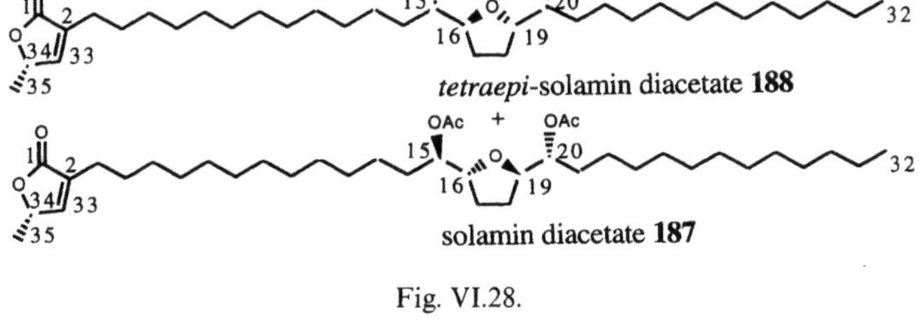

Fig. VI.28.

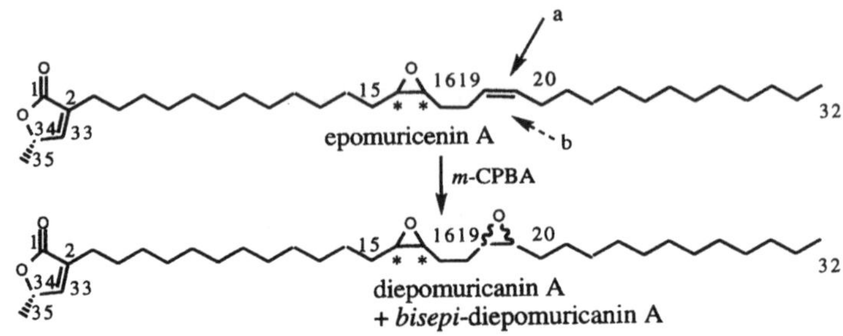

Fig. VI.29.

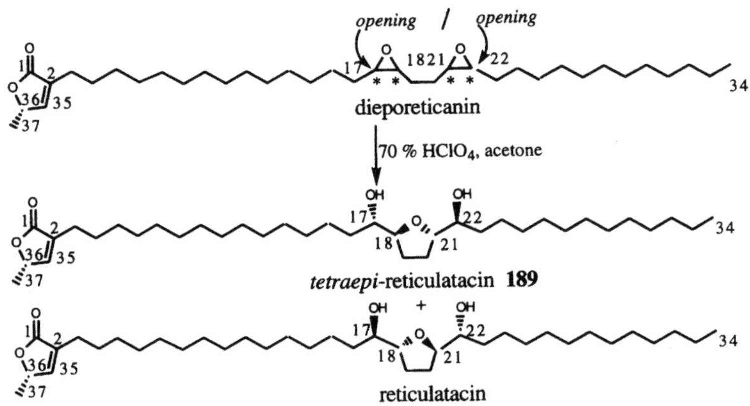

Fig. VI.30.

VI.4.2. Hemisynthesis of corossolone

TAM *et al.* also reported the hemisynthesis of trieporeticanin (*110*) and its bis-epimer (at C-23/24), by treatment of a natural precursor, dieporeticenin, with *m*-CPBA (Fig. VI.31.).

GROMEK *et al.* (*94*) reported the hemisynthesis of corossolone as part of a diastereomeric mixture at C-15/16/19/20, from corepoxylone by treatment with 70% perchloric acid in acetone. It was shown by HPLC that two products, corossolone and its tetraepimer **190**, were thus obtained, having identical spectroscopic data (^{1}H NMR, ^{13}C NMR, MS). These two compounds arise from opening of the oxiranes at either C-15 or at C-20 (Fig. VI.32.).

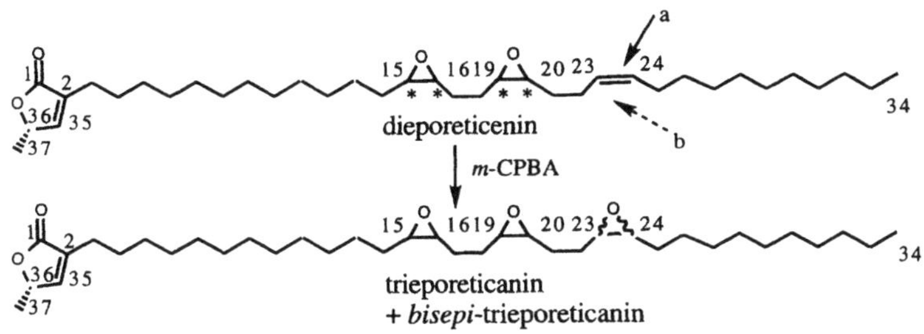

Fig. VI.31.

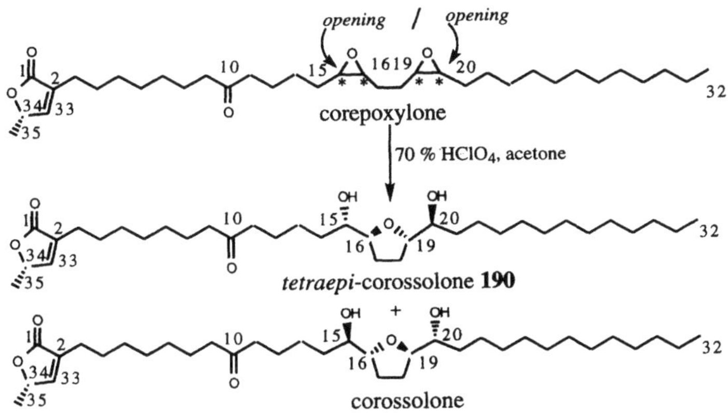

Fig. VI.32.

VI.4.3. Hemisynthesis of isodesacetyluvaricin

In 1993, SAHPAZ *et al.* (*165*) reported the hemisynthesis of isodesacetyluvaricin starting from tripoxyrollin. Treatment with perchloric acid (*94*), led to complex mixture in which isodesacetyluvaricin could be characterized as well as other minor unnatural acetogenins (Fig. VI.33.).

VI.4.4. Hemisynthesis of gigantecin

GU *et al.* reported the hemisynthesis of gigantecin (*100*) starting from gigantetronenin. The strategy used was based on the same sequence as that used earlier by ROBLOT (*108*) and GROMEK (*94*), namely epoxidation of the isolated double bond of gigantetronenin with *m*-CPBA leading to a 1:1 mixture of two epoxides (due to oxidation on both faces of the double bond), followed by treatment with perchloric acid to yield a mixture of the

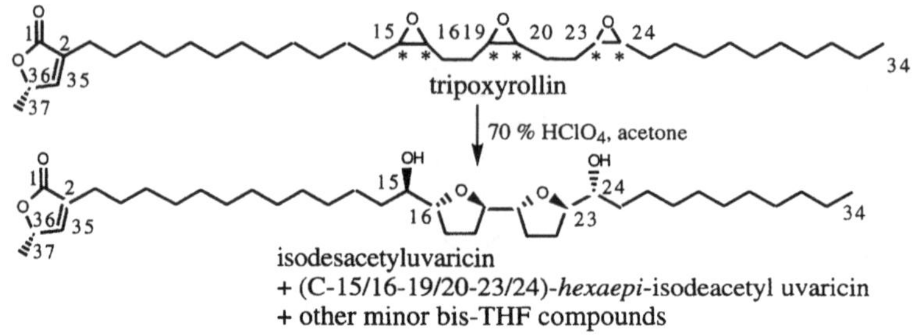

Fig. VI.33.

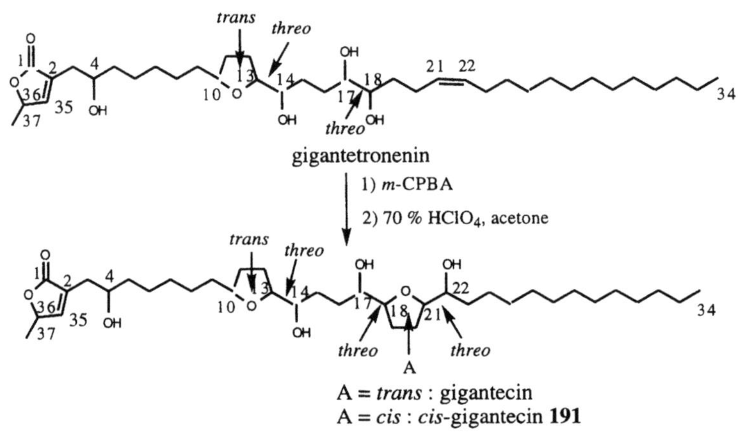

Fig. VI.34.

trans compound (38%), namely gigantecin, and the *cis* compound (37%), namely C-18/21 *cis* gigantecin **191** (Fig. VI.34.).

Interestingly, coriacin, a positional isomer of gigantetronenin, when exposed to *m*-CPBA, led to gigantecin (Fig. VI. 35.) *without formation of any other isomer* and without isolation of the intermediate epoxide(s) (*95*)!

On the other hand gonionenin, isolated from *Goniothalamus giganteus* (*100*), when exposed to the classical reaction conditions (*m*-CPBA, and then perchloric acid), gave a 1:1 mixture of unknown acetogenins, *trans*-cyclogonionenin and *cis*-cyclogonionenin (Fig. VI.36.).

More recently goniodenin when treated with *m*-CPBA and then *p*-TsOH ($HClO_4$ leads to the destruction of the intermediate epoxides), afforded cyclogoniodenin T and cyclogoniodenin C (*75*), the first hemisynthetic tris-THF acetogenins (of type D) and both epimers of goniocin, the single natural acetogenin of type D, known so far (Fig. VI.37.).

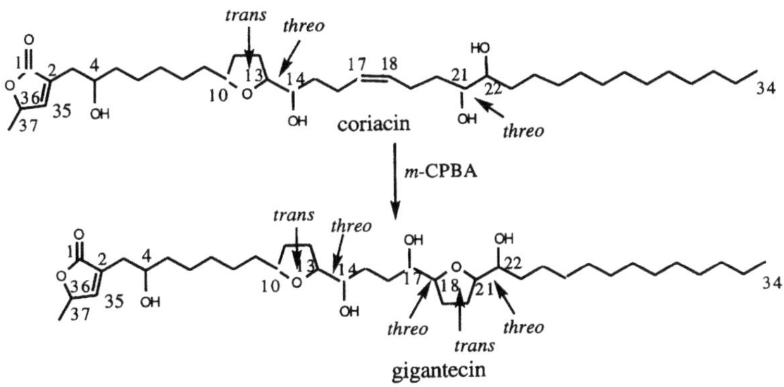

Fig. VI.35.

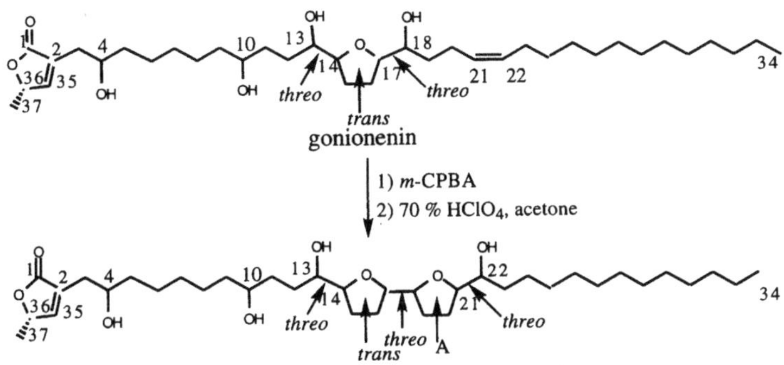

A = *trans* : *trans*-cyclogonionenin

Fig. VI.36.

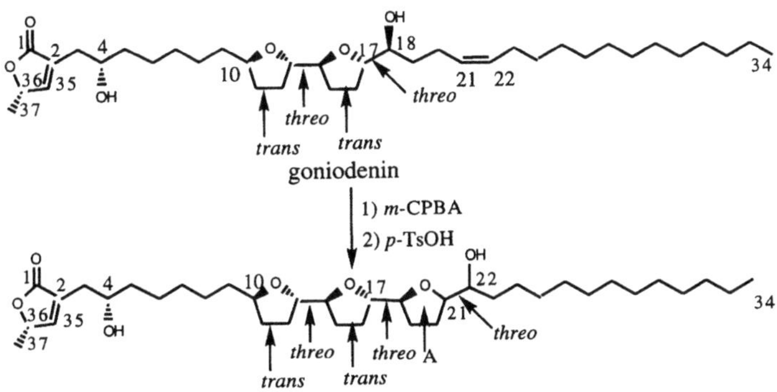

A = *trans* : cyclogoniodenin T
A = *cis* : cyclogoniodenin C

Fig. VI.37.

VII. Biological Activities

The first isolated acetogenin, uvaricin, was stated to be a new antitumor agent (*1*). Since then many studies have appeared describing cytotoxic, antitumor, antiparasitic, pesticidal, antimicrobial, antifungal and immunosuppressive properties of various acetogenins.

VII.1. Cytotoxic Activity

Uvaricin, described as a new antitumor agent, in fact exhibited significant *in vivo* cytotoxic activity against lymphocytic leukaemia 3PS with a T/C of 160% at a dose of 1.4 mg/k. All acetogenins possess more or less significant cytotoxicity determined by the brine shrimp assay (BST) (*170*) and/or against VERO and KB cells, ranging between 10^{-1} and 10^{-12} μg/ml.

Table VII.1. lists monotetrahydrofuran acetogenins from *Annona muricata* whose tetrahydrofuran pattern possesses the same relative stereochemistry (*threo-trans-threo*). It is apparent that activities are more pronounced for the alcohols than for the corresponding ketone derivatives as shown by the pairs annonacin/annonacinone and corossolin/corossolone. The presence and location of a hydroxyl on the chain seems to be important. Corossolin, with hydroxyl at C-10, exhibits activity at a concentration of 3×10^{-3} while murisolin, with one hydroxyl group at C-4, possesses activity at 10^{-1}, and annonacin, with two hydroxyl groups at C-4 and C-10, is more active, 10^{-4}. Acetylation of the hydroxyl groups reduces the cytotoxic activity. Lack of rings or the presence of one epoxide

Table VII. 1. *In vitro Cytotoxic Activity of Acetogenins from Annona muricata* (*ED_{50}; $\mu g. ml^{-1}$*) (*65, 108*)

	K.B. cells	VERO cells
Annonacin	1×10^{-4}	1×10^{-2}
Annonacinone	1×10^{-3}	1×10^{-1}
Murisolin	1×10^{-2}	1×10^{-1}
Corossolin	3×10^{-3}	3×10^{-2}
Corossolone	1×10^{-1}	3×10^{-1}
Solamin	3×10^{-1}	1
Epomuricenin	2	>10
Diepomuricanin	3×10^{-2}	2.4
Vinblastine	1×10^{-2}	>3

(epomuricenin) lowers the activity, whereas the presence of two epoxides separated by two methylenes enhances the activity (diepomuricanin). Presence of a γ-lactone seems to be essential for activity, since opening the lactone ring dramatically lowers the activity whereas reduction of the double bond has very little influence and slightly lowers the cytotoxicity (*65*).

In the bistetrahydrofuran acetogenins, the same structure/activity relationships are observed. For the same level of hydroxylation, acetogenins of type B are more cytotoxic than their type A counterparts. Acetogenins of type C have an intermediate activity.

In the biogenetic precursors of THF acetogenins (type E), activity is notable in the bisepoxy and amplified in the trisepoxy derivatives the latter are sometimes even slightly more active than the corresponding THF acetogenins.

Isoacetogenins, derived from the corresponding 4-hydroxyacetogenins by translactonization increase these activities, but also affect cell specificity. For example, the results of cytotoxic activity assay against KB, HeLa and Vero cell lines are given in Table VII.2.

It is difficult to draw conclusions about the significance of the absolute configuration of the various different asymmetric carbons, since this has not so far been determined for any acetogenin except (+)- and (−)-muricatacin (*55*). Relative stereochemistry, at least, is clearly important for instance the activity (*5*) described for asimicin and isomers is from 10^{-3} to 10^{-5} whereas activity from bullatacin with the relative configuration *threo-trans-threo-trans-erythro* has been claimed at a level of 10^{-13} (*83*).

These diverse studies lead to the structure/cytotoxicity relationships outlined in Scheme VII.1.

In addition to BST and assays on VERO and KB cells, studies of acetogenin activity have been carried out on cell line panels of human tumors of the lung, colon, breast (*115*), pancreas, prostate, kidney (*135*), melanoma, ovary (*93*) etc. In a three-day national cancer institute tumor cell panel using 60 human tumor cell lines, the average of ED50 growth inhibition ranged from 1 to 10 μM for various representative acetogenins (*6*) (Table VII.3.).

It appears that a selectivity for one or the other cell line depends on acetogenin structure. CASSADY *et al.* (*224*) have reported that isoannonacin and isoannonacinone, both belonging to the iso series, are 10,000 times less active against leukaemia cells and 1000 times more active against colon tumor cells than the parent acetogenins, annonacin and annonacinone. Other examples are given by MCLAUGHLIN *et al.* for gigantecin, which is active at 2.19×10^{-7}, 4.11×10^{-9} and 2.68×10^{-4} respectively against lung, breast and colon tumor cell lines (*115*).

Table VII. 2. *Cytotoxic Activity of Acetogenins/Isoacetogenins* (IC_{50} *in* $\mu g/ml$)

Acetogenins/Isoacetogenins	KB cells	HeLa cells	VERO cells
Cherimolin/Isocherimolin-1	$1 \times 10^{-1}/3 \times 10^{-4}$	$1 \times 10^{-4}/3 \times 10^{-5}$	ND
Molvizarin/Isomolvizarin	$1 \times 10^{-5}/3 \times 10^{-4}$	$1 \times 10^{-5}/1 \times 10^{-4}$	$1 \times 10^{-3}/1 \times 10^{-4}$
Rolliniastatin-2/Isorolliniastatin-2	$3 \times 10^{-3}/1 \times 10^{-5}$	$2 \times 10^{-2}/1 \times 10^{-5}$	$1 \times 10^{-1}/1 \times 10^{-3}$
Rolliniastatin-1/Isorolliniastatin-1	$1 \times 10^{-5}/3 \times 10^{-5}$	$3 \times 10^{-6}/1 \times 10^{-5}$	$1.5 \times 10^{-4}/1 \times 10^{-4}$

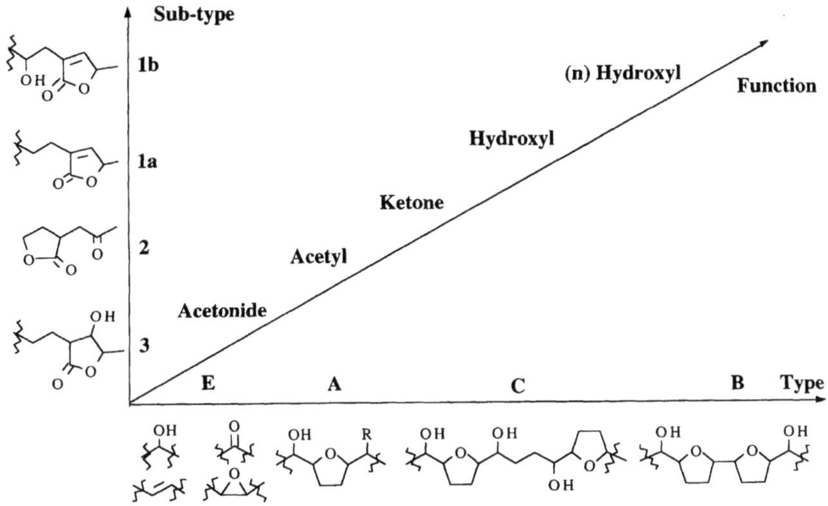

Scheme VII.1. Structure-cytotoxicity relationships

Table VII. 3. *Partial Results for Three Day NCI Tumor Cell Panel* (GI_{50} *in* μM *Units*) MGI_{50} *is the Average* GI_{50} *for 60 Cancerous Cell Lines*

Acetogenin	CNS tumor SF268	Ovarian cancer	All cell lines MGI_{50}
Annonacin	1.15	1.72	0.81
Annonacinone	1.08	2.56	1.55
Isoannonacin	1.08	3.06	1.35
Isoannonacinone	1.19	7.67	3.02
Gigantetrocin	0.55	2.11	0.87
Bullatacin	0.31	2.27	1.91
Bullatacinone	1.62	12.8	3.16
Bullatalicin	1.15	15.6	6.03
Bullatalicinone	2.40	14.4	3.39
Asimicin	1.11	1.73	1.32

30-, 31- and 32-hydroxybullatacinones show selective cytotoxicity for colon cancer cell lines (ED 50 (μg/ml) about 1×10^{-12}) *(122)*, whereas 10-, 12- and 29-hydroxybullatacinones show selective cytotoxicity for breast cancer cell lines (ED 50 (μg/ml) 1. $\times 10^{-4}$, 10^{-6}) *(121)*.

Table VII.4. contains some examples showing huge differences in activity depending on the tumor cell line (asimicin and venezenin). It is worth noting that isomers of different relative configuration across the

Table VII. 4. *Cytotoxic Activity of Acetogenins against Different Tumor Cell Lines*

Compound	Type	Brine shrimp test (BST) LC_{50} (μg/ml)	Human lung carcinoma ED_{50} (μg/ml)	Human breast carcinoma ED_{50} (μg/ml)	Human colon adenocarcinoma ED_{50} (μg/ml)
Asimicin	B1b	2.5×10^{-1}	8.46×10^{-4}	8.52×10^{-1}	$<10^{-15}$
Bullacin[a]	B1a	7.0×10^{-2}	1.79×10^{-5}	1.0×10^{-5}	5.23×10^{-3}
Gigantetrocin-A[b]	A1b	1.3	8.10×10^{-3}	5.30×10^{-1}	$<10^{-8}$
Gigantetrocin-B[b]	A1b	2.6	2.50×10^{-1}	6.30×10^{-1}	4.10×10^{-5}
Longicin[c]	A1b	0.11	1.77×10^{-6}	>1	2.40×10^{-5}
Gigantecin[d]	C1b	3.4×10^{-2}	2.19×10^{-7}	4.11×10^{-9}	2.68×10^{-4}
Gonitothalamicin[e]	A1b	0.24	8.00×10^{-3}	5.70×10^{-2}	1.10×10^{-3}
Goniothalamicinone[c]	A2	24	2.06×10^{-5}	9.67×10^{-1}	4.05×10^{-4}
Parviflorin[a]	B1b	8.8×10^{-2}	1.27×10^{-15}	1.72	5.49×10^{-1}
Venezenin[f]	E1b	6.87×10^{-1}	1.08×10^{-2}	$<10^{-2}$	1.58
Adriamycin		8.0×10^{-2}	1.53×10^{-4}	1.38×10^{-2}	5.58×10^{-4}

[a] *(79)*, [b] *(119)*, [c] *(42)*, [d] *(115)*, [e] *(43)*, [f] *(30)*

Table VII.5. *Cytotoxic Activity on Human Cancerous Cells: IC_{50} in μg/ml*

Compound	Fogarty	PC3	MCF7S	SKOV3	HT29
Annonacin	>10	6.1	4	10	>10
Isoannonacin	>10	4.2	1	10	>10
Rolliniastatin-1	>10	9	2	>10	>10
Squamocin	2.2	2	1.7	1.7	3.9

THF pattern (gigantetrocin-A and gigantetrocin-B) do not have the same activities. The same comment can be made for goniothalamicin and its translactonized derivative goniothalamicinone. It is interesting that reference of a tetrahydrofuran ring is not essential, since venezenin shows promising activity (*30*).

In others experiments (*225*), several acetogenins were tested on some cancerous cell lines which are generally refractory to chemotherapy such as glioma (Fogarty), prostatic (PC3), ovarian (SKOV3), mammalian (MCF7) and colon (HT 29), see Table VII.5.

As shown in Table VII.5., annonacin exhibited interesting activity against PC3 and MCF7S, isoannonacin was five times more efficient against mammary cancer MCF7S, while rolliniastatin-1 or squamocin displayed similar activities against these cells. In this series, squamocin was active against all five cell lines with IC_{50} ranging from 2.7 to 6 μM. It should be pointed out that acetogenins are just as active against parental sensitive cell lines as against the corresponding resistant lines, particularly those homologues that expressed an Multi Drug Resistence (MDR) phenotype. It seems therefore that acetogenins failed to be recognized by the glycoprotein Gp170 over expressed in these MDR cell lines.

VII.2. Antitumor Activity

If antitumor activity is defined as activity against a tumor in an animal model without undue toxicity towards the host (*226*), there are only a few published results relating to this criterion. For instance, bullatacin (= rolliniastatin-2) was evaluated *in vivo* in a murine ovarian teratocarcinoma, but toxicity was too great to prove antitumor activity (*227*). *In vivo* experiments were performed with uvaricin isolated from the roots of *Uvaria acuminata* (*1*) which exhibited antitumor activity against 3PS leukaemia in mice with a T/C of 157% at 1.4 mg/kg per day. *In vivo* data obtained against mice bearing implanted L1210 murine leukaemia showed increased life span of 38% for rolliniastatin-2 (bullatacin) at 50 μg/kg per day and 44% at 400 μg/kg per day isorolliniastatin-2 (bullatacinone) in a nine-day experiment (*228*). Antitumor efficacies of these acetogenins in athymic mice bearing subcutaneous implanted A2780 human ovarian carcinoma xenografts show 68% tumor growth inhibition at 0.1 mg/kg per day for bullatacin compared with 78% for cisplatin at 5 mg/kg per day used as a standard (*228*). Moreover, interesting antitumor activity of several acetogenins was demonstrated in nude mice with transplanted tumor with annonacin giving the best results (*26*).

From these results, it appears that *in vivo* activity is strongly limited by toxicity of acetogenins administered by the intra-peritoneal route, generally with at least 1 mg/kg in acute toxicity. We submitted rolliniastatin-1 and isorolliniastatin-1 to a toxicological trial on the mouse by the subcutaneous route. The results show that isorolliniastatin-1 with a LD_{50} of 200 mg/kg was considerably less toxic than rolliniastatin-1 (LD_{50} of 5 mg/kg). In general, isoacetogenins are less toxic than the corresponding acetogenin, even though the cytotoxic activity on cell lines is similar; therefore isoacetogenins, easy to prepare from the corresponding acetogenins, should be considered as interesting candidates to broaden the therapeutic index of these products.

VII.3. Antiparasitic Activity

Some acetogenins exhibit antiparasitic activity. A number of results have been published relating to activity a 4-aminoquinolin-resistant strain of *Plasmodium falciparum* (*5*). Uleicins from *Rollinia ulei* (*228*) possess significant *in vitro* activity against *Leishmania donovani* with a good therapeutic index. Acetogenins of *Annona senegalensis* were tested against *Leishmania major*, *L. donovani* and *Trypanosoma brucei* (*69*). The results indicated that senegalene, squamocin and molvizarin were effective against *Leishmania* with a minimum effective concentration ranging from 25 to 50 μg/ml, while senegalene was the most effective against *T. brucei* (MEC: 50 μg/ml).

Similarly, acetogenins isolated from *Annona muricata* and from *Annona cherimolia* possess filaricidal activity against *Molinema dessetae* (*230*) (Table VII.6.).

In the monotetrahydrofuran series, the most potent acetogenin is annonacin: acetylation reduces its activity. Oxidation of the hydroxyl at C-10 to a carbonyl slightly increases the activity. The absence of a hydroxyl at C-4 appreciably reduces potency. Among the bistetrahydrofuran series, the activity of cherimolin is comparable to that of annonacinone.

References, pp. 273–288

Table VII. 6. *Activity of Acetogenins on Infective Larvae of Molinema dessetae*

Compound	LD_{50} (μg/ml) day 1	LD_{50} (μg/ml) day 7
Annonacin	0.66	0.08
Annonacinone	0.52	0.28
Corossolin	5.12	0.41
Murisolin	1.5	0.30
Cherimolin	0.67	0.04
Otivarin	6.66	0.25
Tetraacetylannonacin	10.2	1.20
Diethylcarbamazine	> 500	330
Ivermectine	1.3	0.27

VII.4. Pesticidal Activity

The pesticidal activity of acetogenins has been described reportedly, particularly that of annonin (*132*, *231*) bullatacin (*83*), asimicin (*5*, *232a*), squamocin (*138*), goniothalamicin (*115*) and sylvaticin (*163*). Asimicin has been shown to be toxic for melon aphids, *Aphis gossypii* (100% mortality at 500 ppm), Mexican bean beetles, *Epilachna varvestis* (100% mortality at 50 ppm) and mosquito larvae, *Aedes aegypti* (100% mortality at 1 ppm) (*232*). Squamocin and neoannonin showed strong ovicidal and larvicidal activity in *Drosophila* feeding tests and oils of some *Annona* seeds are traditionally used to get rid of lice in the scalp (*138*) as, for example, ground seeds of *Annona reticulata* in Vietnam (*233*). It is also noteworthy that, in some countries of South America, ground bark or seeds of some species of Annonaceae are spread on soils as pesticides.

VII.5. Other Activities

Antimicrobial and antifungal activities were reported against from positive bacteria (*Staphylococcus, Bacillus...*) (*137*) from negative bacteria (*Escherichia, Proteus...*) (*4*, *137*) and fungi (*Candida, Trichophyton...*) (*4*, *137*).

An interesting immunosuppressive activity was evidenced in the mixed lymphocyte reaction (M.L.R.) of mouse system cells for acetogenins isolated from *Annona muricata* (*234*). IC50 of annonacin is about 3 nM on this model. By comparison, cyclosporin used as a reference in this test is active at 10 nM.

VII.6. Mechanism of Action

Some mechanisms for the biological activity of acetogenins have been proposed; in particular it has been suggested that they act as ionophores and that they affect mitochondrial respiration and the cytoplasmic membrane.

Acetogenins have been shown to form supramolecular complexes with metal cations as observed by mass spectrometry (*107*, *179*) and by nmr (*235*). They form a ligand/metal complex with calcium with high selectivity (*236*). Such ligands may be useful as neutral calcium ionophores and might explain the observed antimicrobial activities and, by playing an important role in cellular function, other biological effects such as pesticidal and antitumoral activities (*237*).

A number of tests failed to reveal a more classic target for antimitotic (*238*) and antitumor activity based on tubulin or DNA-linked mechanisms.

Research on the mechanism responsible for the insecticidal properties of *Annona squamosa* (*132*) showed that acetogenins in this species have an inhibitory effect on mitochondrial respiration, the molecular target being NADH-cytochrome C oxidoreductase with a specific action at the level of complex I in insect mitochondria. Only the flow of electrons through site I was strongly affected by acetogenins. Similar results were observed for mammalian respiration, particularly on mitochondria isolated from bovine heart and also on yeast mitochondria extracted *Neurospora crassa* cells (*244*). As an example, bullatacin (= rollinistatin-2) probably inhibits cancer cell growth through the inhibition of mitochondrial electron transport systems, thus reducing the ATP level and, because the cancer cells probably require more energy than normal cells, the antitumor activity can be explained in this way (*82*). This mechanism was confirmed with other insect or mammalian models (*239*). Inhibition of mitochondrial NADH oxidase could lead to a lowering of the cellular ATP pool which in turn could alter cellular viability. Moreover, the target may be linked to cell growth control through intracellular pH regulation and activation of protooncogenes into oncogenes (*241*). Recently, a new mechanism was proposed (*240*). It has been shown that acetogenins are active on both sensitive cancerous cell lines and their multidrug resistance phenotype (MDR). These cell lines contain NADH oxidase in their cytoplasmic membranes. Therefore, because of the external localisation of the target, one may expect cytotoxic activity to be retained in MDR cell lines with alteration in drug transport mechanisms (*242*, *243*). Furthermore, expression of the plasma-170 glycoprotein which is believed to be strongly correlated to the MDR phenotype, requires ATP, and acetogenins, by

Table VII.7. *Inhibition of NADH Oxidase of HeLa and Rat Liver Plasma Membrane Vesicles by Annonaceous Acetogenins*

Acetogenin	Rat liver ED_{50} in μM	HeLa ED_{50} in μM
Annonacin A	>10	1
Asimicin	>10	0.005
Bullatacin	>10	0.005–0.01
Bullatacinone	>10	0.1–1

depleting intracellular ATP levels, offer an exciting potential for the inhibition of MDR resistant cancer cell lines (*82*). Moreover, recent molecular biochemical results indicate that tumor cells overexpress NADH oxido-reductase in plasma membrane more strongly than normal cell lines. From all these considerations, it seems that this enzyme could be a specific target for new antitumor drugs acting as antiproliferatives. Indeed, acetogenins such as bullatacin show activity on NADH oxido-reductase in plasma membrane vesicles isolated from HeLa cells derived from human cervical carcinoma derivation, but not in plasma membrane vesicles isolated from rat livers (*240*) Table VII.7.

The results in Table VII.7. suggest that this specific target could be a new and promising approach for acetogenin antitumor activity in both sensitive and MDR cell lines.

Acknowledgements

We thank all members of the Laboratory of Pharmacognosy, URA CNRS 1843, whose names appear in the references. Moreover, we owe special gratitude to Olivier Laprévote, Marc Schwaller, Michel Lebœuf and Jacqueline Vaquette for their help in the realization of this manuscript.

References

1. Jolad, S.D., J.J. Hoffmann, K.H. Schram, J.R. Cole, M.S. Tempesta, G.R. Kriek, and R.B. Bates: Uvaricin, a New Antitumor Agent from *Uvaria acuminata* (Annonaceae). J. Org. Chem., **47**, 3151–3153 (1982).

2. Dabrah, T.T., and A.T. Sneden: Rollinone, a New Cytotoxic Acetogenin from *Rollinia papilionella*. J. Nat. Prod., **47**, 652–657 (1984).

3. Dabrah, T.T., and A.T. Sneden: Rollinicin and Isorollinicin, Cytotoxic Acetogenins from *Rollinia papilionella*. Phytochemistry, **23**, 2013–2016 (1984).

4. Cortes, D., J.L. Rios, A. Víllar, and S. Valverde: Cherimoline et dihydrocherimoline, deux nouvelles γ-lactones, bis-tétrahydrofuraniques possédant une activité antimicrobienne. Tetrahedron Lett., **25**, 3199–3202 (1984).
5. Rupprecht, J.K., Y.-H. Hui, and J.L. McLaughlin: Annonaceous Acetogenins: a Review. J. Nat. Prod., **53**, 237–278 (1990).
6. Fang, X.-P., M.J. Rieser, Z.-M. Gu, G.-X. Zhao, and J.L. McLaughlin: Annonaceous Acetogenins: an Updated Review. Phytochem. Anal. (1993), **4**, 27–48; Annonaceous Acetogenins: an Updated Review, Appendices. Phytochem. Anal., **4**, 49–67 (1993).
7. Xu, L., C.-J. Chang, J.-G. Yu, and J.M. Cassady: Chemistry and Selective Cytotoxicity of Annonacin-10-one, Isoannonacin, and Isoannonacin-10-one. Novel Polyketides from *Annona densicoma* (Annonaceae). J. Org. Chem., **54**, 5418–5421 (1989).
8. Cavé, A.: Acetogenins from Annonaceae. In: Phytochemistry of Plants Used in Traditional Medicine (Hostettmann, K., A. Marston, M. Maillard, and M. Hamburger, eds.), pp 228–248. Oxford: Clarendon Press, 1995.
9. Duret, P., A. Laurens, R. Hocquemiller, D. Cortes, and A. Cavé: Isoacetogenins, Artifacts Issued from Translactonization from Annonaceous Acetogenins. Heterocycles, **39**, 741–749 (1994).
10. Nonfon, M., F. Lieb, H. Moeschler, and D. Wendisch: Four Annonins from *Annona squamosa*, Phytochemistry, **29**, 1951–1954 (1990).
11. Fang, X.-P., Z.-M. Gu, M.J. Rieser, Y.-H. Hui, J.L. McLaughlin, M. Nonfon, F. Lieb, H.-F. Moeschler, and D. Wendisch: Structural Revisions of Some Non-adjacent Bis-Tetrahydrofuran Annonaceous Acetogenins. J. Nat. Prod., **56**, 1095–1100 (1993).
12. Cortes, D., B. Figadère, and A. Cavé: Bis-Tetrahydrofuran Acetogenins from Annonaceae. Phytochemistry, **32**, 1467–1473 (1993).
13. Yu, J.G., X.Z. Luo, L. Sun, C.Y. Liu, S.L. Hong, and L.B. Ma: Squamostatin-B, a New Polyketide from *Annona squamosa* (Annonaceae). Chin. Chem. Lett., **4**, 423–426 (1993).
14. Yu, J.G., X.Z. Luo, C.Y. Liu, L. Sun, S.L. Hong, and L.B. Ma: Studies on the Chemical Constituents of *Annona squamosa* Seed. Acta Pharmaceutica Sinica, **29**, 443–448 (1994).
15. Fujimoto, Y., C. Murasaki, K. Kakinuma, T. Eguchi, N. Ikekawa, M. Furuya, K. Hirayama, T. Ikekawa, M. Sahai, Y.K. Gupta, and A.B. Ray: Squamostatin-A: Unprecedented Bis-Tetrahydrofuran Acetogenin from *Annona squamosa.* Tetrahedron Lett., **31**, 535–538 (1990).
16. Cortes, D., S.H. Myint, B. Dupont, and D. Davoust: Bioactive Acetogenins from Seeds of *Annona cherimolia.* Phytochemistry, **32**, 1475–1482 (1993).
17. Duret, P., D. Gromek, R. Hocquemiller, A. Cavé, and D. Cortes: Isolation and Structure of Three New Bis-Tetrahydrofuranic Acetogenins from the Roots of *Annona cherimolia.* J. Nat. Prod., **57**, 911–916 (1994).
18. Gu, Z.-M., L. Zeng, X.-P. Fang, T. Colman-Saizarbitoria, M. Huo, and J.L. McLaughlin: Determining Absolute Configurations of Stereocenters in Annonaceous Acetogenins through Formaldehyde Acetal Derivatives and Mosher Ester Methodology. J. Org. Chem., **59**, 5162–5172 (1994).
19. Nishioka, S., H. Araya, C. Murasaki, M. Sahai, and Y. Fujimoto: Determination of Absolute Stereochemistry at Carbinol Stereocenters of Tetrahydrofuranic Acetogenins by the Advanced Mosher Ester Method. Nat. Prod. Lett., **5**, 117–121 (1994).
20. Fujimoto, Y., C. Murasaki, H. Shimada, S. Nishioka, K. Kakinuma, S. Singh, M. Singh, K.K. Gupta, and M. Sahai: Annonaceous Acetogenins from the seeds of

Annona squamosa. Non-adjacent Bis-Tetrahydrofuranic Acetogenins. Chem. Pharm. Bull., **42**, 1175–1184 (1994).

21. SAHPAZ, S.: Étude chimique et biologique des acétogénines des graines d'*Annona senegalensis* et de *Rollinia membranacea* (Annonaceae). Doctorat de l'Université Paris-Sud, Châtenay-Malabry, 1995.
22. SAHPAZ, S., M.C. GONZÁLEZ, R. HOCQUEMILLER, M.C. ZAFRA-POLO, and D. CORTES: Annonasenegalin and Annogalene: Two Cytotoxic Monotetrahydrofuran Acetogenins from *Annona senegalensis* and *Annona cherimolia.* Phytochemistry, **42**, 103–107 (1996).
23. COLMAN-SAIZARBITORIA, T., Z.-M. GU, and J.L. MCLAUGHLIN: Two New Bioactive Monotetrahydrofuran Annonaceous Acetogenins from the Bark of *Xylopia aromatica.* J. Nat. Prod., **57**, 1661–1669 (1994).
24. ETCHEVERRY, S., S. SAHPAZ, D. FALL, A. LAURENS, and A. CAVÉ: Annoglaucin, an Acetogenin from *Annona glauca.* Phytochemistry, **38**, 1423–1426 (1995).
25. ZENG, L., F.-E. WU, and J.L. MCLAUGHLIN: Annohexocin, a Novel Mono-THF Acetogenin with Six Hydroxyls, from *Annona muricata* (Annonaceae). Bioorg. Med. Chem. Lett., **5**, 1865–1868 (1995).
26. JOSSANG, A., B. DUBAELE, A. CAVÉ, M.-H. BARTOLI, and H. BÉRIEL: Deux nouvelles acétogénines monotétrahydrofuraniques cytotoxiques: l'annomonicine et la montanacine. Tetrahedron Lett., **31**, 1861–1864 (1990).
27. CHANG, F.-R., Y.-C. WU, C.-Y. DUH, and S.-K. WANG: Studies on the Acetogenins of Formosan Annonaceous Plants, II. Cytotoxic Acetogenins from *Annona reticulata.* J. Nat. Prod., **56**, 1688–1694 (1993).
28. JOSSANG, A., B. DUBAELE, A. CAVÉ, M.-H. BARTOLI, and H. BÉRIEL: Annomontacine: une nouvelle acétogénine γ-lactone-monotétrahydrofuranique cytotoxique de l'*Annona montana.* J. Nat. Prod, **54**, 967–971 (1991).
29. FANG, X.-P., J.E. ANDERSON, D.L. SMITH, J.L. MCLAUGHLIN, and K.V. WOOD. Gigantetronenin and Gigantrionenin: Novel Cytotoxic Acetogenins from *Goniothalamus giganteus.* J. Nat. Prod., **55**, 1655–1663 (1992).
30. COLMAN-SAIZARBITORIA, T., Z.-M. GU, G.-X. ZHAO, L. ZENG, J.F. KOZLOWSKI, and J.L. MCLAUGHLIN: Venezenin: A New Bioactive Annonaceous Acetogenin from the Bark of *Xylopia aromatica.* J. Nat. Prod., **58**, 532–539 (1995).
31. WU, F.-E., Z.-M. GU, L. ZENG, G.-X. ZHAO, Y. ZHANG, J.L. MCLAUGHLIN, and S. SASTRODIHARDJO: Two New Cytotoxic Monotetrahydrofuran Annonaceous Acetogenins, Annomuricins A and B, from the Leaves of *Annona muricata.* J. Nat. Prod., **58**, 830–836 (1995).
32. WU, F.-E, L. ZENG, Z.-M. GU, G.-X. ZHAO, Y. ZHANG, J.T. SCHWEDLER, J.L. MCLAUGHLIN, and S. SASTRODIHARDJO: New Bioactive Monotetrahydrofuran Annonaceous Acetogenins, Annomuricin C and Muricatocin C from the Leaves of *Annona muricata.* J. Nat. Prod., **58**, 909–915 (1995).
33. WU, F.-E., G.-X. ZHAO, L. ZENG, Y. ZHANG, J.T. SCHWEDLER, J.L. MCLAUGHLIN, and S. SASTRODIHARDJO: Additional Bioactive Acetogenins, Annomutacin and (2,4-*trans* and *cis*)-10R-Annonacin-A-ones, from the Leaves of *Annona muricata.* J. Nat. Prod., **58**, 1430–1437 (1995).
34. ZHANG, L.-L., R.-Z. YANG, and S.-J. WU: Studies on the Chemical Composition of *Goniothalamus howii* (I). Acta Botanica Sinica (Zhiwu Xuebao), **35**, 390–396 (1993); Chem. Abstracts, **120**: 129491m (1994).
35. MCCLOUD, T.G., D.L. SMITH, C.-J. CHANG, and J.M. CASSADY: Annonacin, A novel, Biologically Active Polyketide from *Annona densicoma.* Experientia, **43**, 947–949 (1987).

36. Chen, W.-S., Z.-J. Yao, and Y.-L. Wu: Study on the Chemical Constituents of *Annona glabra* L. Youji Huaxue **15**, 85–88 (1995); Chem. Abstracts, **122**: 261023f (1995).
37. Lieb, F., M. Nonfon, U. Wachendorff-Neumann, and D. Wendisch: Annonacins and Annonastatin from *Annona squamosa.* Planta Med., **56**, 317–319 (1990).
38. Cortes, D., S.H. Myint, A. Laurens, R. Hocquemiller, M. Leboeuf, and A. Cavé: Corossolone et corossoline, deux nouvelles γ-lactones mono tétrahydrofuraniques cytotoxiques. Can. J. Chem., **69**, 8–11 (1991).
39. Rieser, M.J., J.F. Kozlowski, K.V. Wood, and J.L. McLaughlin: Muricatacin: A Simple Biologically Active Acetogenin Derivative from the Seeds of *Annona muricata* (Annonaceae). Tetrahedron Lett., **32**, 1137–1140 (1991).
40. Rieser, M.J., X.-P. Fang, J.K. Rupprecht, Y.-H. Hui, D.L. Smith, and J.L. McLaughlin: Bioactive Single-Ring Acetogenins from Seed Extracts of *Annona muricata.* Planta Med., **59**, 91–92 (1993).
41. Yang, R.-Z., S.-J. Wu, R.-S. Xu, G.-W. Qin, and D.-J. Fan: Annonaceous Acetogenins from *Annona muricata.* Acta Botanica Sinica (Zhiwu Xuebao), **36**, 805–808 (1994); Chem. Abstracts, **122**: 209773m (1995).
42. Ye, Q., L. Zeng, Y. Zhang, G.-X. Zhao, J.L. McLaughlin, M.H. Woo, and D.R. Evert: Longicin and Goniothamamicinone: Novel Bioactive Monotetrahydrofuran Acetogenins from *Asimina longifolia.* J. Nat. Prod., **58**, 1398–1406 (1995).
43. Ratnayake, S., Z.-M. Gu, L.R. Miesbauer, D.L. Smith, K.V. Wood, D.R. Evert, and J.L. McLaughlin: Parvifloracin and Parviflorin: Cytotoxic Bistetrahydrofuran Acetogenins with 35 Carbons from *Asimina parviflora* (Annonaceae). Can. J. Chem., **72**, 287–293 (1994).
44. Alkofahi, A., J.K. Rupprecht, D.L. Smith, C.-J. Chang, and J.L. McLaughlin: Goniothalamicin and Annonacin: Bioactive Acetogenins from *Goniothalamus giganteus* (Annonaceae). Experientia, **44**, 83–85 (1988).
45. Hoye, T.R., P.R. Hanson, L.E. Hasenwinkel, E.A. Ramirez, and Z. Zhuang. Stereostructural Studies on the 4-Hydroxylated Annonaceous Acetogenins: a Novel Use of Mosher Ester Data for Determining Relative Configuration [between C(4) and C(36)]. Tetrahedron Lett., **35**, 8529–8532 (1994).
46. Rieser, M.J., Y.-H. Hui, J.K. Rupprecht, J.F. Kozlowski, K.V. Wood, J.L. McLaughlin, P.R. Hanson, Z. Zhuang, and T.R. Hoye: Determination of Absolute Configuration of Stereogenic Carbinol Centers in Annonaceous Acetogenins by ^{1}H and ^{19}F-NMR Analysis of Mosher Ester Derivatives. J. Am. Chem. Soc., **114**, 10203–10213 (1992).
47. Scholz, G., and W. Tochtermann: Optisch aktive γ-lactone aus Cyclooctin und Furan-Synthese von (−)-Muricatacin. Tetrahedron Lett., **32**, 5535–5538 (1991).
48. Marshall, J.A., and G.S. Welmaker: Stereoselective Synthesis of the Cytotoxic Acetogenins (+)-and (−)-Muricatacin. Synlett, 537–538 (1992).
49. Wang, Z.-M., X.-L. Zhang, K.B. Sharpless, S.C. Sinha, A. Sinha-Bagchi, and E. Keinan: A General Approach to γ-lactones via Osmium-catalyzed Asymmetric Dihydroxylation. Synthesis of (−)- and (+)-Muricatacin. Tetrahedron Lett., **33**, 6407–6410 (1992).
50. Tochtermann, W., G. Scholz, G. Bunte, C. Wolff, E.-M. Peters, K. Peters, and H.G. von Schnering: Tetrahydrofuran and γ-Lactones, V. Optically Active δ-Hydroxy-γ-lactones from Cyclooctine and Furan. Synthesis of (−)-(*R*, *R*)- and (+)-(*S*, *S*)-Muricatacin and Related Compounds. Liebigs Ann. Chem., 1069–1080 (1992).
51. Makabe, H., A. Tanaka, and T. Oritani: Synthesis of (−)-Muricatacin. Biosci. Biotech. Biochem, **57**, 1028–1029 (1993).

52. KANG, S.-K., H.-S. CHO, H.-S. SIM, and B.-K. KIM: Synthesis of (4*R*, 5*S*)-(−)- and (4*S*, 5*S*)-(+)-L-Factors and Muricatacin from D-Glucose. J. Carbohyd. Chem., **11**, 807–812 (1992).
53. MARSHALL, J.A., and G.S. WELMAKER: Enantioselective Synthesis of (+)- and (−)-Muricatacin through $SE^{2'}$ Addition of Nonracemic γ-Silyloxy Allylic Stannanes to Aldehydes. J. Org. Chem., **59**, 4122–4125 (1994).
54. SANIÈRE, M., I. CHARVET, Y. LE MERRER, and J.-C. DEPEZAY: Enantiopure Hydroxylactones from *L*-Ascorbic and *D*-Isoascorbic Acids. Part I. Synthesis of (−)-Muricatacin. Tetrahedron, **51**, 1653–1662 (1995).
55. FIGADÈRE, B., J.-C. HARMANGE, A. LAURENS, and A. CAVÉ: Stereospecific Synthesis of (+)-Muricatacin: a Biologically Active Acetogenin Derivative. Tetrahedron Lett., **32**, 7539–7542 (1991).
56. SAÏAH, M., M. BESSODES, and K. ANTONAKIS: Regioselective Opening of Chiral Hydroxyepoxides: a Short Route to Muricatacin and Its Diastereomer *epi*-Muricatacin. Tetrahedron Lett., **34**, 1597–1598 (1993).
57. SOMFAI, P.: An Enantiospecific Total Synthesis of (+)-Muricatacin. J. Chem. Soc. Perkin Trans. I, 817–819 (1995).
58. QUAYLE, P., S. RAHMAN, and J. HERBERT: Transition Metal Promoted Acetylene Isomerisation Reactions in Organic Synthesis: A Synthesis of (+)-(4*S*, 5*S*)-Muricatacin. Tetrahedron Lett., **36**, 8087–8088 (1995).
59. WU, F.-E., L. ZENG, Z.-M. GU, G.-X. ZHAO, Y. ZHANG, J.T. SCHWEDLER, J.L. MCLAUGHLIN, and S. SASTRODIHARDJO: Muricatocins A and B, Two New Bioactive Monotetrahydrofuran Annonaceous Acetogenins from the Leaves of *Annona muricata*. J. Nat. Prod., **58**, 902–908 (1995).
60. ZHAO, G.-X., M.J., RIESER, Y.-H. HUI, L.R. MISBAUER, D.L. SMITH, and J.L. MCLAUGHLIN: Biologically Active Acetogenins from Stem Bark of *Asimina triloba*. Phytochemistry, **33**, 1065–1073 (1993).
61. YANG, R.-Z., X.-C. ZHENG, S.-J. WU, and G.-W. QIN: Annonsilin A, a Novel Seco-tris-tetrahydrofuranyl Annonaceous Acetogenin. Acta Botanica Sinica, **37**, 492–495 (1995).
62. WU, Y.-C., F.-R. CHANG, C.-Y. DUH, and S.-K. WANG: Annoreticuin and Isoannoreticuin: Two New Cytotoxic Acetogenins from *Annona reticulata*. Heterocycles, **34**, 667–674 (1992).
63. DURET, P., A.-I. WAECHTER, R. HOCQUEMILLER, A. CAVÉ, and T. BATTEN: Annotemoyin-1 and -2: Two Novel Mono-tetrahydrofuranic γ-Lactone Acetogenins from the Seeds of *Annona atemoya*. Nat. Prod. Lett., **8**, 89–95 (1996).
64. SAHAI, M., S. SINGH, M. SINGH, Y.K. GUPTA, S. AKASHI, R. YUJI, K. HIRAYAMA, H. ASAKI, H. ARAYA, N. HARA, T. EGUCHI, K. KAKINUMA, and Y. FUJIMOTO: Annonaceous Acetogenins from the Seeds of *Annona squamosa*. Adjacent Bis-Tetrahydrofuranic Acetogenins. Chem. Pharm. Bull., **42**, 1163–1174 (1994).
65. MYINT, S.H., D. CORTES, A. LAURENS, R. HOCQUEMILLER, M. LEBOEUF, A. CAVÉ, J. COTTE, and A.-M. QUÉRO: Solamin, a Cytotoxic Monotetrahydrofuranic γ-Lactone Acetogenin from *Annona muricata* Seeds. Phytochemistry, **30**, 3335–3338 (1991).
66. RÍOS, J.L., D. CORTES, and S. VALVERDE: Acetogenins, Aporphinoids, and Azaantraquinone from *Annona cherimolia* Seeds. Planta Med., **55**, 321–323 (1989).
67. OHSAWA, K., S. ATSUZAWA, T. MITSUI, and I. YAMAMOTO: Isolation and Insecticidal Activity of Three Acetogenins from Seeds of Pond Apple, *Annona glabra*. J. Pesticide Sci., **16**, 93–96 (1991).
68. SAHPAZ, S., A. LAURENS, R. HOCQUEMILLER, A. CAVÉ, and D. CORTES: Senegalene, une nouvelle acétogénine oléfinique mono-tétrahydrofuranique des graines d'*Annona senegalensis*. Can. J. Chem., **72**, 1533–1536 (1994).

69. Sahpaz, S., C. Bories, P.M. Loiseau, D. Cortes, R. Hocquemiller, A. Laurens, and A. Cavé: Cytotoxic and Antiparasitic Activity from *Annona senegalensis* Seeds. Planta Med., **60**, 538–540 (1994).
70. Ratnayake, S., X.-P. Fang, J.E. Anderson, J.L. McLaughlin, and D.R. Evert: Bioactive Constituents from the Twigs of *Asimina parviflora*. J. Nat. Prod., **55**, 1462–1467 (1992).
71. Rupprecht, J.K., C.-J. Chang, J.M. Cassady, J.L. McLaughlin, K.L. Mikolajczak, and D. Weisleder: Asimicin, a New Cytotoxic and Pesticidal Acetogenin from the Pawpaw, *Asimina triloba* (Annonaceae). Heterocycles, **24**, 1197–1201 (1986).
72. Zhao, G., Y. Hui, J.K. Rupprecht, J.L. McLaughlin, and K.V. Wood: Additional Bioactive Compounds and Trilobacin, a Novel Highly Cytotoxic Acetogenin, from the Bark of *Asimina triloba*. J. Nat. Prod., **55**, 347–356 (1992).
73. Woo, M.H., L. Zeng, and J.L. McLaughlin: Asitribin and Asiminenins A and B, Novel Bioactive Annonaceous Acetogenins from the Seeds of *Asimina triloba*. Heterocycles, **41**, 1731–1742 (1995).
74. Hoye, T.R., and L. Tan: Total Synthesis of the Potent Antitumor, Bistetrahydrofuranyl Annonaceous Acetogenins (+)-Asimicin and (+)-Bullatacin. Tetrahedron Lett., **36**, 1981–1984 (1995).
75. Zhang, Y., L. Zeng, M-H. Woo, Z.-M. Gu, Q. Ye, F.-E. Wu, and J.L. McLaughlin: Goniodenin, a New Bioactive Annonaceous Acetogenin from *Goniothalamus giganteus* and Its Conversion to tri-THF Acetogenins. Heterocycles, **41**, 1743–1755 (1995).
76. Zhao, G.-X., L.R. Miesbauer, D.L. Smith, and J.L. McLaughlin: Asimin, Asiminacin, and Asiminecin: Novel Highly Cytotoxic Asimicin Isomers from *Asimina triloba*. J. Med. Chem., **37**, 1971–1976 (1994).
77. Chen, W.-S., Z.-J. Yao, Y.-B. Zhang, Y.-Z. Xu, and Y.-L. Wu: Atemoyacin-A: a New Bis-Tetrahydrofuranyl Annonaceous Acetogenin from *Annona atemoya*. Chin. J. Chem., **13**, 263–266 (1995).
78. Duret, P., R. Hocquemiller, A. Laurens, and A. Cavé: Atemoyin, a New Bis-Tetrahydrofuran Acetogenin from the Seeds of *Annona atemoya*. Nat. Prod. Lett., **5**, 295–302 (1995).
79. Gu, Z.-M., X.-P. Fang, L. Zeng, K.V. Wood, and J.L. McLaughlin: Bullacin: a New Cytotoxic Annonaceous Acetogenin from *Annona bullata*. Heterocycles, **36**, 2221–2228 (1993).
80. Gu, Z.-M., X.-P. Fang, L. Zeng, J.F. Kozlowski, and J.L. McLaughlin: Novel Cytotoxic Annonaceous Acetogenins: (2,4-*cis* and *trans*)-Bulladecinones from *Annona bullata* (Annonaceae). Bioorg. Med. Chem. Lett., **4**, 473–478 (1994).
81. Gu, Z.-M., L. Zeng, J.T. Schwedler, K.V. Wood, and J.L. McLaughlin: New Bioactive Adjacent Bis-THF Annonaceous Acetogenins from *Annona bullata*. Phytochemistry, **40**, 467–477 (1995).
82. Zhao, G.-X., J.H. Ng, J.F. Kozlowski, D.L. Smith, and J.L. McLaughlin: Bullatin and Bullanin: Two Novel, Highly Cytotoxic Acetogenins from *Asimina triloba*. Heterocycles, **38**, 1897–1908 (1994).
83. Hui, Y.-H., J.K. Rupprecht, Y.M. Liu, J.E. Anderson, D.L. Smith, C.-J. Chang, and J.L. McLaughlin: Bullatacin and Bullatacinone: Two Highly Potent Bioactive Acetogenins from *Annona bullata*. J. Nat. Prod., **52**, 463–477 (1989).
84. Li, X.-H., Y.-H. Hui, J.K. Rupprecht, Y.-M. Liu, K.V. Wood, D.L. Smith, C.-J. Chang, and J.L. McLaughlin: Bullatacin, Bullatacinone, and Squamone, a New Bioactive Acetogenin from the Bark of *Annona squamosa*. J. Nat. Prod., **53**, 81–86 (1990).

85. COPLEANU, F., K. OHTANI, M. HAMBURGER, M.P. GUPTA, P. SOLIS, and K. HOSTETTMANN: Novel Acetogenins from the Leaves of *Annona purpurea*. Helv. Chim. Acta., **76**, 1379–1388 (1993).
86. HUI, Y.-H., J.K. RUPPRECHT, J.E. ANDERSON, Y.-M. LIU, D.L. SMITH, C.-J. CHANG, and J.L. MCLAUGHLIN: Bullatacin, a Novel Bioactive Acetogenin from *Annona bullata* (Annonaceae). Tetrahedron, **45**, 6941–6948 (1989).
87. SHI, G., L. ZENG, Z.-M. GU, J.M. MACDOUGAL, and J.L. MCLAUGHLIN: Absolute Stereochemistries of Sylvaticin and 12,15-*cis*-Sylvaticin, Bioactive C-20,23-*cis* Non-adjacent Bis-Tetrahydrofuran Annonaceous Acetogenins, from *Rollinia mucosa*. Heterocycles, **41**, 1785–1796 (1995).
88. GU, Z.-M., L. ZENG, and J.L. MCLAUGHLIN: Isolation and Structural Elucidation of Bioactive C-12,15-*cis* Non-adjacent *bis*-THF Annonaceous Acetogenins. Heterocycles, **41**, 229–236 (1995).
89. GU, Z.-M., X.-P. FANG, M.J. RIESER, Y.-H. HUI, L.R. MIESBAUER, D.L. SMITH, K.V. WOOD, and J.L. MCLAUGHLIN: New Cytotoxic Annonaceous Acetogenins: Bullatanocin and *cis*- and *trans*-Bullatanocinone, from *Annona bullata* (Annonaceae). Tetrahedron, **49**, 747–754 (1993).
90. HUI, Y.-H., J.K. RUPPRECHT, J.E. ANDERSON, K.V. WOOD, and J.L. MCLAUGHLIN: Bullatalicinone, a New Potent Bioactive Acetogenin, and Squamocin from *Annona bullata* (Annonaceae). Phytother. Res., **5**, 124–129 (1991).
91. PINHEIRO SANTOS L., M.A.D. BOAVENTURA, and A. BRAGA DE OLIVEIRA: Crassiflorina, uma acetogenina tetrahydrofurânica citotóxica de *Annona crassiflora* (Araticum). Quim. Nova, **17**, 387–390 (1994).
92. LI, C.-M., H.-D. SUN, H.-L. ZHENG, and G.-D. TAO: Annonaceous Acetogenins from *Annona glabra*. Yunnan Zhiwu Yanjiu, **17**, 221–224 (1995); Chem. Abstracts, **123**: 222803m (1995).
93. HUI, Y.-H., K.V. WOOD, and J.L. MCLAUGHLIN: Bullatencin, 4-Deoxyasimicin, and the Uvariamicins: Additional Bioactive Annonaceous Acetogenins from *Annona bullata* Rich (Annonaceae). Natural Toxins, **1**, 4–14 (1992).
94. GROMEK, D., B. FIGADÈRE, R. HOCQUEMILLER, A. CAVÉ, and D. CORTES: Corepoxylone, a Possible Precursor of Mono-tetrahydrofuran γ-lactone Acetogenins. Biomimetic Synthesis of Corossolone. Tetrahedron, **49**, 5247–5252 (1993).
95. MENESES DA SILVA, E.L., F. ROBLOT, O. LAPRÉVOTE, P. VARENNE, and A. CAVÉ: Coriacin and 4-Deoxycoriacin, Two New Mono-THF Acetogenins from the Roots of *Annona coriacea*. Nat. Prod. Lett., **7**, 235–242 (1995).
96. MENESES DE SILVA, E.L., F. ROBLOT, J. MAHUTEAU, and A. CAVÉ: Coriadienin, the First Annonaceous Acetogenin with Two Double Bonds Isolated from *Annona coriacea* Mart. J. Nat. Prod., **59**, 528–530 (1996).
97. YAO, Z.-J., and Y.-L. WU: Synthetic Studies Toward Mono-THF Annonaceous Acetogenins: A Diastereoselective and Convergent Approach to Corossolone and (10*RS*)-Corossolin. J. Org. Chem., **60**, 1170–1176 (1995).
98. YAO, Z.-J., and Y.-L. WU: Total Synthesis of (10ζ, 15*R*, 16*S*, 19*S*, 34*R*)-Corossolin. Tetrahedron Lett., **35**, 157–160 (1994).
99. GU, Z.-M., X.-P. FANG, L. ZENG, and J.L. MCLAUGHLIN: Goniocin from *Goniothalamus giganteus*: The First tri-THF Annonaceous Acetogenin. Tetrahedron Lett., **35**, 5367–5368 (1994).
100. GU, Z.-M., X.-P. FANG, L. ZENG, R. SONG, J.H. NG, K.V. WOOD, D.L. SMITH, and J.L. MCLAUGHLIN: Gonionenin: A New Cytotoxic Annonaceous Acetogenin from *Goniothalamus giganteus* and the Conversion of Mono- THF Acetogenins to Bis-THF Acetogenins. J. Org. Chem., **59**, 3472–3479 (1994).

101. YU, J.G., D.K. HO, J.M. CASSADY, L. XU, and C.-J. CHANG: Cytotoxic Polyketides from *Annona densicoma* (Annonaceae): 10,13-*trans*-13,14-*erythro*-Densicomacin, 10,13-*trans*-13,14-*threo*-Densicomacin, and 8-Hydroxyannonacin. J. Org. Chem., **57**, 6198–6202 (1992).
102. HU, X.-E., and J.M. CASSADY: Determination of Absolute Configuration of Vicinal Diols by ^{1}H NMR Analysis of Mosher Ester Derivatives. Absolute Configuration of Polyketide (+)-Densicomacin. 207th A.C.S. Meeting, San Diego, 13–18/03/1993.
103. FANG, X.-P., J.E. ANDERSON, D.L. SMITH, K.V. WOOD, and J.L. MCLAUGHLIN: Giganenin, a Highly Potent Monotetrahydrofuran Acetogenin and 4-Deoxygigantecin from *Goniothalamus giganteus*. Heterocycles, **34**, 1075–1083 (1992).
104. ZHENG, X.-C., R.-Z. YANG, G.-W. QIN, R.-S. XU, and D.-J. FAN: Three Novel Chemical Compounds of Annonaceous Acetogenins from the Seeds of *Annona squamosa*. Acta Botanica Sinica, **37**, 238–243 (1995).
105. JOLAD, S.D., J.J. HOFFMANN, J.R. COLE, C.E. BARRY, R.B. BATES, G.S. LINZ, and W.A. KONIG: Desacetyluvaricin from *Uvaria acuminata*. Configuration of Uvaricin at C-36. J. Nat. Prod., **48**, 644–645 (1985).
106. LAPRÉVOTE, O., F. ROBLOT, R. HOCQUEMILLER, and A. CAVÉ: Structural Elucidation of Two New Acetogenins, Epoxyrollins A and B, by Tandem Mass Spectrometry. Tetrahedron Lett., **31**, 2283–2286 (1990).
107. LAPRÉVOTE, O., C. GIRARD, B.C. DAS, T. LAUGEL, F. ROBLOT, M. LEBOEUF, and A. CAVÉ: Location of Epoxy Rings in a Long Chain Acetogenin by Fast Atom Bombardment and Linked Scan (*B*/*E*) Mass Spectrometry of Lithium Cationized Molecules. Rapid Commun. Mass Spectrom., **6**, 352–355 (1992).
108. ROBLOT, F., T. LAUGEL, M. LEBOEUF, A. CAVÉ, and O. LAPRÉVOTE: Two Acetogenins from *Annona muricata* Seeds. Phytochemistry, **34**, 281–285 (1993).
109. HISHAM, A., U. SREEKALA, L. PIETERS, T. DE BRUYNE, H. VAN DEN HEUVEL, and M. CLAEYS: Epoxymurins A and B, two Biogenetic Precursors of Annonaceous Acetogenins from *Annona muricata*. Tetrahedron, **49**, 6913–6920 (1993).
110. VU THI TAM, PHAN QUAN CHI HIEU, B. CHAPPE, F. ROBLOT, O. LAPRÉVOTE, B. FIGADÈRE, and A. CAVÉ: Four New Acetogenins from the Seeds of *Annona reticulata*. Nat. Prod. Lett., **4**, 255–262 (1994).
111. SAHPAZ, S., R. HOCQUEMILLER, A. CAVÉ, J. SAEZ, and D. CORTES: Diepoxyrollin and Diepomuricanin-B, Two New Bis-Epoxyacetogenins from *Rollinia membranacea* Seeds. J. Nat. Prod., in press.
112. WU, Y.-C., F.-R. CHANG, K.-S. CHEN, S.-C. LIANG, and M.-R. LEE: Diepoxymontin, a Novel Acetogenin from *Annona montana*. Heterocycles, **38**, 1475–1478 (1994).
113. FANG, X.-P., R. SONG, Z.-M. GU, M.J. RIESER, L.R. MIESBAUER, D.L. SMITH, and J.L. MCLAUGHLIN: A New Type of Cytotoxic Annonaceous Acetogenin: Giganin from *Goniothalamus giganteus*. Bioorg. Med. Chem. Lett., **3**, 1153–1156 (1993).
114. YU, J.-G., X.E. HU, D.K. HO, M.F. BEAN, R.E. STEPHENS, J.M. CASSADY, L.S. BRINEN, and J. CLARDY: Absolute Stereochemistry of (+)-Gigantecin from *Annona coriacea* (Annonaceae). J. Org. Chem., **59**, 1598–1599 (1994).
115. ALKOFAHI, A., J.K. RUPPRECHT, Y.-M. LIU, C.-J. CHANG, D.L. SMITH, and J.L. MCLAUGHLIN: Gigantecin: A Novel Antimitotic and Cytotoxic Acetogenin, with Nonadjacent Tetrahydrofuran Rings, from *Goniothalamus giganteus* (Annonaceae). Experientia, **46**, 539–541 (1990).
116. FANG, X.-P., J.K. RUPPRECHT, A. ALKOFAHI, Y.-H. HUI, Y.-M. LIU, D.L. SMITH, K.V. WOOD, and J.L. MCLAUGHLIN: Gigantetrocin and Gigantriocin: Two Novel Bioactive Annonaceous Acetogenins from *Goniothalamus giganteus*. Heterocycles, **32**, 11–17 (1991).

117. Yang, R.-Z., L.-L. Zhang, and S.-J. Wu: The Chemical Constituents of *Goniothalamus howii* (II). Acta Botanica Sinica (Zhiwu Xuebao), **36**, 561–567 (1994); Chem. Abstracts **122**: 310664b (1995).
118. Li, C.M., Q. Mu, X.J. Hao, H.D. Sun, H.L. Zheng, and Y.C. Wu: Three New Bioactive Annonaceous Acetogenins from *Annona muricata*. Chin. Chem. Lett., **5**, 747–750 (1994).
119. Rieser, M.J., X.-P. Fang, J.E. Anderson, L.R. Miesbauer, D.L. Smith, and J.L. McLaughlin: Muricatetrocins A and B and Gigantetrocin B: Three New Cytotoxic Monotetrahydrofuran-ring Acetogenins from *Annona muricata*. Helv. Chim. Acta, **76**, 2433–2444 (1993).
120. Waechter, A.-I., R. Hocquemiller, A. Laurens, and A. Cavé: Glaucanisin, a New Acetogenin from *Annona glauca*. Nat. Prod. Lett., **6**, 133–138 (1995).
121. Gu, Z.-M., X.-P. Fang, Y.-H. Hui, and J.L. McLaughlin: 10-, 12-, and 29-Hydroxybullatacinones: New Cytotoxic Annonaceous Acetogenins from *Annona bullata* Rich (Annonaceae). Natural Toxins, **2**, 49–55 (1994).
122. Gu, Z.-M., X.-P. Fang, L.R. Miesbauer, D.L. Smith, and J.L. McLaughlin: 30-, 31-, and 32-Hydroxybullatacinones: Bioactive Terminally Hydroxylated Annonaceous Acetogenins from *Annona bullata*. J. Nat. Prod., **56**, 870–876 (1993).
123. Pan, X.P., and D.Q. Yu: Uvarigranin: A New Acetogenin from *Uvaria grandiflora* Roxb. Chin. Chem. Lett., **6**, 473–476 (1995).
124. Hisham, A., L.A.C. Pieters, M. Claeys, H. Van den Heuvel, E. Esmans, R. Dommisse, and A.J. Vlietinck: Acetogenins from Root Bark of *Uvaria narum*. Phytochemistry, **30**, 2373–2377 (1991).
125. Myint, S.H., A. Laurens, R. Hocquemiller, A. Cavé, D. Davoust, and D. Cortes: Murisolin: A New Cytotoxic Mono-tetrahydrofuran-γ-lactone from *Annona muricata*. Heterocycles, **31**, 861–867 (1990).
126. Yang, R.-Z., S.-J. Wu, R.-S. Xu, and G.-W. Qin: Annonaceous Acetogenins from *Annona muricata* 2. Acta Botanica Yunnanica (Yunnan Zhiwu Yanjiu), **16**, 187–190 (1994); Chem. Abstracts **121**: 297130g (1994).
127. Cortes, D., S.H. Myint, M. Leboeuf, and A. Cavé: A New Type of Cytotoxic Acetogenins: The Tetrahydrofuranic β-Hydroxymethyl γ-lactones. Tetrahedron Lett., **32**, 6133–6134 (1991).
128. Saez, J., S. Sahpaz, L. Villaescusa, R. Hocquemiller, A. Cavé, and D. Cortes: Rioclarine et membranacine, deux nouvelles acétogénines bis-tétrahydrofuraniques des graines de *Rollinia membranacea*. J. Nat. Prod., **56**, 351–356 (1993).
129. Chen, W.-S., Z.-J. Yao, and Y.-L. Wu: Study on the Chemical Constituents of *Annona atemoya* Hort and the Isolation and Structure of Atemoyacin-B. Huaxue Xuebao, **53**, 516–520 (1995); Chem. Abstracts **123**: 79506j (1995).
130. Cortes, D., S.H. Myint, and R. Hocquemiller: Molvizarin and Motrilin: Two Novel Cytotoxic Bis-Tetrahydrofuranic γ-lactone Acetogenins from *Annona cherimolia*. Tetrahedron, **47**, 8195–8202 (1991).
131. Hisham, A., C. Sunitha, U. Sreekala, L. Pieters, T. De Bruyne, H. van den Heuvel, and M. Claeys: Reticulacinone, an Acetogenin from *Annona reticulata*. Phytochemistry, **35**, 1325–1329 (1994).
132. Londershausen, M., W. Leicht, F. Lieb, H. Moeschler, and H. Weiss: Molecular Mode of Action of Annonins. Pestic. Sci., **33**, 427–438 (1991).
133. Shi, G., D. Alfonso, M.O. Fatope, L. Zeng, Z.-M. Gu, G.-X. Zhao, K. He, J.M. MacDougal, and J.L. McLaughlin: Mucocin: A New Annonaceous Acetogenin Bearing a Tetrahydropyran Ring. J. Am. Chem. Soc., **117**, 10409–10410 (1995).
134. Gui, H.Q., J.G. Yu, and Z.L. Yu: Muricatalin, a New Polyketide from *Annona muricata* (Annonaceae). Chin. Chem. Lett., **6**, 45–48 (1995).

135. ZENG, L., F.-E. WU, Z.-M. GU, and J.L. MCLAUGHLIN: Murihexocins A and B, Two Novel Mono-THF Acetogenins with Six Hydroxyls, from *Annona muricata* (Annonaceae). Tetrahedron Lett., **36**, 5291–5294 (1995).
136. WOO, M.H., L. ZENG, Q. YE, Z.-M. GU, G.-X. ZHAO, and J.L. MCLAUGHLIN: 16,19-*cis*-Murisolin and Murisolin A, Two Novel Bioactive Mono-tetrahydrofuran Annonaceous Acetogenins from *Asimina triloba* Seeds. Bioorg. Med. Chem. Lett., **5**, 1135–1140 (1995).
137. PADMAJA, V., V. THANKAMANY, and A. HISHAM: Antibacterial, Antifungal and Anthelmintic Activities of Root Barks of *Uvaria hookeri* and *Uvaria narum*. J. Ethnopharmacol., **40**, 181–186 (1993).
138. KAWAZU, K., J.P. ALCANTARA, and A. KOBAYASHI: Isolation and Structure of Neoannonin, a Novel Insecticidal Compound from the Seeds of *Annona squamosa*. Agric. Biol. Chem., **53**, 2719–2722 (1989).
139. HISHAM, A., L.A.C. PIETERS, M. CLAEYS, E. ESMANS, R. DOMMISSE, and A.J. VLIETINCK: Squamocin-28-one and Panalicin, Two Acetogenins from *Uvaria narum*. Phytochemistry, **30**, 545–548 (1991).
140. ZHENG, X.-C., R.-Z. YANG, R.-S. XU, and G.-W. QIN: Plagionicin A, with C-5-OH, a New Monotetrahydrofuran Acetogenin. Acta Botanica Sinica, **36**, 557–560 (1994).
141. HISHAM, A., L.A.C. PIETERS, M. CLAEYS, E. ESMANS, R. DOMMISSE, and A.J. VLIETINCK: Uvariamicin-I, II and III: Three Novel Acetogenins from *Uvaria narum*. Tetrahedron Lett., **31**, 4649–4652 (1990).
142. SAAD, J.M., Y.-H. HUI, J.-K. RUPPRECHT, J.E. ANDERSON, J.F. KOZLOWSKI, G.-X. ZHAO, K.V. WOOD, and J.L. MCLAUGHLIN: Reticulatacin: A New Bioactive Acetogenin from *Annona reticulata* (Annonaceae). Tetrahedron, **47**, 2751–2756 (1991).
143. VU THI TAM, PHAN QUAN CHI HIEU, B. CHAPPE, F. ROBLOT, B. FIGADÈRE, and A. CAVÉ: Reticulatain-1 and -2 with Reticulatamone: Three New Polyketides from the Seeds of *Annona reticulata*. Bull. Soc. Chim. Fr., **132**, 324–329 (1995).
144. MAKABE, H., A. TANAKA, and T. ORITANI: Total Synthesis of Solamin and Reticulatacin. J. Chem. Soc. Perkin Trans. I, 1975–1981 (1994).
145. VU THI TAM, C. CHABOCHE, B. FIGADÈRE, B. CHAPPE, BUI CHI HIEU, and A. CAVÉ: First Synthesis of a New Acetogenin of Anonaceae, Reticulatamol: Activated Tin Hydride with Enhanced Reducing Ability. Tetrahedron Lett., **35**, 883–886 (1994).
146. ABREO, M.J., and A.T. SNEDEN: 4-Hydroxy-25-desoxyneorollinicin, a New Bistetrahydrofuranoid Acetogenin from *Rollinia papilionella*. J. Nat. Prod., **52**, 822–828 (1989).
147. VU THI TAM, BUI CHI HIEU, and B. CHAPPE: Squamocin and Rolliniastatin-1 from the Seeds of *Annona reticulata*. Planta Med., **59**, 576 (1993).
148. PETTIT, G.R., G.M. CRAGG, J. POLONSKY, D.L. HERALD, A. GOSWAMI, C.R. SMITH, C. MORETTI, J.M. SCHMIDT, and D. WEISLEDER: Isolation and Structure of Rolliniastatin-1 from the South American Tree *Rollinia mucosa*. Can. J. Chem., **65**, 1433–1435 (1987).
149. BARNES, J.N., B.T. SCHANEBERG, and A.T. SNEDEN: Bistetrahydrofuranoid Acetogenins from *Rollinia sericea*. Planta Med., **61**, 486–487 (1995).
150. KOERT, U.: Total Synthesis of (+)-Rolliniastatin-1. Tetrahedron Lett., **35**, 2517–2520 (1994).
151. ETSE, J.T., and P.G. WATERMAN: Chemistry in the Annonaceae, XXII. 14-Hydroxy-25-desoxyrollinicin from the Stem Bark of *Annona reticulata*. J. Nat. Prod., **49**, 684–686 (1986).
152. PETTIT, G.R., R. RIESEN, J.E. LEET, J. POLONSKY, C.R. SMITH, J.M. SCHMIDT, C. DUFRESNE, D. SCHAUFELBERGER, and C. MORETTI: Isolation and Structure of Rolliniastatin-2: A New cell Growth Inhibitory Acetogenin from *Rollinia mucosa*. Heterocycles, **28**, 213–217 (1989).

153. NAITO, H., E. KAWAHARA, K. MARUTA, M. MAEDA, and S. SASAKI: The First Total Synthesis of (+)-Bullatacin, a Potent Antitumor Annonaceous Acetogenin, and (+)-15,24-*bisepi*-Bullatacin. J. Org. Chem., **60**, 4419–4427 (1995).
154. ABREO, M.J., and A.T. SNEDEN: Rollinone, a Revision and Extension of Structure. J. Nat. Prod., **53**, 983–985 (1990).
155. SINHA, S.C., and E. KEINAN: Total Synthesis of Naturally Occurring Acetogenins: Solamin and Reticulatacin. J. Am. Chem. Soc., **115**, 4891–4892 (1993).
156. TROST, B.M., and Z. SHI: A Concise Convergent Strategy to Acetogenins. (+)-Solamin and Analogues. J. Am. Chem. Soc., **116**, 7459–7460 (1994).
157. FUJIMOTO, Y., T. EGUCHI, K. KAKINUMA, N. IKEKAWA, M. SAHAI, and Y.K. GUPTA: Squamocin, a New Cytotoxic Bis-Tetrahydrofuran Containing Acetogenin from *Annona squamosa*. Chem. Pharm. Bull., **36**, 4802–4806 (1988).
158. BORN, L., F. LIEB, J.P. LORENTZEN, H. MOESCHLER, M. NONFON, R. SÖLLNER, and D. WENDISCH: The Relative Configuration of Acetogenins Isolated from *Annona squamosa*: Annonin I (Squamocin) and Annonin VI. Planta Med., **56**, 312–316 (1990).
159. HIRAYAMA, K., S. AKASHI, R. YUJI, U. NIITSU, and Y. FUJIMOTO: Structural Studies of Polyhydroxybis(tetrahydrofuran)acetogenins from *Annona squamosa* Using the Combination of Chemical Derivatization and Precursor-ion Scanning Mass Spectrometry. Org. Mass Spectrom., **28**, 1516–1524 (1993).
160. ARAYA, H., N. HARA, Y. FUJIMOTO, and M. SAHAI: Squamostanal-A, Apparently Derived from Tetrahydrofuranic Acetogenin, from *Annona squamosa*. Biosci. Biotech. Biochem., **58**, 1146–1147 (1994).
161. ARAYA, H., N. HARA, Y. FUJIMOTO, A. SRIVASTAVA, and M. SAHAI: Squamosten-A, a Novel Mono-tetrahydrofuranic Acetogenin with a Double Bond in the Hydrocarbon Chain, from *Annona squamosa* L. Chem. Pharm. Bull., **42**, 388–391 (1994).
162. LAPRÉVOTE, O., F. ROBLOT, R. HOCQUEMILLER, A. CAVÉ, B. CHARLES, and J.-C. TABET: Structural Elucidation of Five Stereoisomeric Acetogenins, Uleicins A-E, by Tandem Mass Spectrometry. Phytochemistry, **30**, 2721–2727 (1991).
163. MIKOLAJCZAK, K.J., R.V. MADRIGAL, J.K. RUPPRECHT, Y.-H. HUI, Y.-M. LIU, D.L. SMITH, and J.L. MCLAUGHLIN: Sylvaticin: A New Cytotoxic and Insecticidal Acetogenin from *Rollinia sylvatica* (Annonaceae). Experientia, **46**, 324–327 (1990).
164. ZHAO, G.-X., Z.-M. GU, L. ZENG, J.-F. CHAO, J.F. KOSLOWSKI, K.V. WOOD, and J.L. MCLAUGHLIN: The Absolute Configuration of Trilobacin and Trilobin, a Novel Highly Potent Acetogenin from the Stem Bark of *Asimina triloba* (Annonaceae). Tetrahedron, **51**, 7149–7160 (1995).
165. SAHPAZ, S., B. FIGADÈRE, J. SAEZ, R. HOCQUEMILLER, A. CAVÉ, and D. CORTES: Tripoxyrollin, a New Epoxy-acetogenin from the Seeds of *Rollinia membranacea*. Nat. Prod. Lett., **2**, 301–308 (1993).
166. LAPRÉVOTE, O.: Alcaloïdes, lignanes et acétogenines de quatre Annonaceae, *Unonopsis spectabilis*, *Rollinia mucosa*, *Rollinia exsucca* et *R. ulei*. Thèse de Doctorat de l'Université Paris-Sud, Châtenay-Malabry, 1989.
167. COLMAN-SAIZARBITORIA, T., J. ZAMBRANO, N.R. FERRIGNI, Z.-M. GU, J.-H. NG, D.L. SMITH, and J.L. MCLAUGHLIN: Bioactive Annonaceous Acetogenins from the Bark of *Xylopia aromatica*. J. Nat. Prod., **57**, 486–493 (1994).
168. ETSE, J.T., A.I. GRAY, and P.G. WATERMAN: Chemistry in the Annonaceae, XXIV. Kaurane and Kaur-16-ene-Diterpenes from the Stem Bark of *Annona reticulata*. J. Nat. Prod., **50**, 979–983 (1987).
169. YANG, R.-Z., X.-C. ZHENG, G.-W. QIN, and R.-S. XU: Squamosinin A: A Novel Para-Tris-Tetrahydrofuranyl Annonaceous Acetogenin. Acta Botanica Sinica, **36**, 809–812 (1994).

170. MCLAUGHLIN, J.L., C.-J. CHANG, and D.L. SMITH: "Bench-Top" Bioassays for the Discovery of Bioactive Natural Products: An Update. *In*: Studies in Natural Products Chemistry, vol 9, Atta-ur-Rahman ed., Elsevier, Amsterdam, pp 383–409 (1991).

171. DREUX, M., and M. LAFOSSE: Mesure de la diffusion de la lumière sur des microparticules en phase gazeuse. Utilisations actuelles. Analusis, **20**, 587–595 (1992).

172. GROMEK, D., R. HOCQUEMILLER, and A. CAVÉ: Qualitative and Quantitative Evaluation of Annonaceous Acetogenins by High Performance Liquid Chromatography. Phytochem. Anal., **5**, 133–140 (1994).

173. GYPSER, A., C. BÜLOW, and H.-D. SCHARF: Determination of the Absolute Configuration of Annonin I, a Bioactive Natural Acetogenin from *Annona squamosa*. Tetrahedron, **51**, 1921–1930 (1995).

174. YANG, R.-Z., and S.-J. WU: Annonaceous Acetogenins from *Annona muricata* 3. Acta Botanica Yunnanica (Yunnan Zhiwu Yanjiu), **16**, 309–310 (1994); Chem. Abstracts, **122**: 27791h (1995).

175. CAVÉ, A., D. CORTES, B. FIGADÈRE, R. HOCQUEMILLER, O. LAPRÉVOTE, A. LAURENS, and M. LEBOEUF: Recent Advances in the Acetogenins of Annonaceae. *In*: Phytochemical Potential of Tropical Plants (K.R. DOWNUM, J.T. ROMEO, H.E. STAFFORD, eds.) pp 167–202. New York: Plenum Press. 1993.

175a. CORTES, D., S.H. MYINT, J.-C. HARMANGE, S. SAHPAZ, and B. FIGADÈRE: Catalytic Hydrogenation of Annonaceous Acetogenins. Tetrahedron Lett., **33**, 5225–5226 (1992).

176. LAPRÉVOTE, O., C. GIRARD, B.C. DAS, D. CORTES, and A. CAVÉ: Formation of Gas-Phase Lithium Complexes from Acetogenins and their Analysis by Fast Atom Bombardment Mass Spectrometry. Tetrahedron Lett., **33**, 5237–5240 (1992).

177. LAPRÉVOTE, O., and B.C. DAS: Structural Elucidation of Acetogenins from Annonaceae by Fast Atom Bombardment Mass Spectrometry. Tetrahedron, **50**, 8479–8490 (1994).

178. SMITH, D.L., Y.-M. LIU, and K.V. WOOD: Structure Elucidation of Natural Products by Mass Spectrometry. *In*: Modern Phytochemical Methods (N.H. FISCHER, M.B. ISMAN, H.A. STAFFORD, eds.), pp 251–269. Plenum Press. 1991.

179. LAPRÉVOTE, O., C. GIRARD, B.C. DAS, A. LAURENS, and A. CAVÉ: Desorption of Lithium Complexes of Acetogenins by Fast Atom Bombardment: Application to Semi-quantitative Analysis of Crude Plant Extracts. Analusis, **21**, 207–210 (1993).

180. BROWN, P., J. KOSSANYI, and C. DJERASSI: Mass Spectrometry in Structural and Stereochemical Problems, CXVIII, Aliphatic Epoxides, Tetrahedron, Suppl. 8, part I, 241–267 (1966).

181. KEOUGH, T., E.D. MIHELICH, and D.J. EICKHOFF: Differentiation of Monoepoxide Isomers of Polyunsaturated Fatty Acids and Fatty Acid Esters by Low-energy Charge Exchange Mass Spectrometry. Anal. Chem., **56**, 1849–1852 (1984).

182. TUMLINSON, J.H., R.R. HEATH, and R.E. DOOLITTLE: Application of Chemical Ionization Mass Spectrometry of Epoxides to the Determination of Olefin Position in Aliphatic Chains. Anal. Chem., **46**, 1309–1312 (1974).

183. TABET, J.C., and J. EINHORN: The Use of Constant Neutral Spectra to Determine the Origin of the Acylium Ions in the Cl/NO^+ Spectra of Aliphatic Epoxides. Org. Mass spectrom., **20**, 310 (1985).

184. RAMIREZ, E.A., and T.R. HOYE: Determination of Relative and Absolute Configuration in the Annonaceous Acetogenins. *In*: Studies in Natural Products Chemistry, vol. 17 (Atta-ur-Rahman ed.), pp 251–282. Elsevier Science, B.V. 1995.

185. HOYE, T.R., P.R. HANSON, L.E. HASENWINKEL, E.A. RAMIREZ, and Z. ZHUANG: Stereostructural Studies on the 4-Hydroxylated Annonaceous Acetogenins: Synthesis of Model Butenolides of Known Relative and Absolute Configuration Involving an Intriguing Translactonization Reaction. Tetrahedron Lett., **35**, 8525–8528 (1994).

186. HOYE, T.R., and Z.-P. ZHUANG: Validation of the ^{1}H NMR Chemical Shift Method for Determination of Stereochemistry in the Bis-(tetrahydrofuranyl) Moiety of Uvaricin-Related Acetogenins from Annonaceae, Rolliniastatin-1 (and Asimicin). J. Org. Chem., **53**, 5578–5580 (1988).

187. GALE, J.B., J.-G. YU, X.E. HU, A. KHARE, D.K. HO, and J.M. CASSADY: Stereochemistry of Mono-tetrahydrofuranyl Moiety in Cytotoxic Polyketides. Part A: Synthesis of Model Compounds. Tetrahedron Lett., **34**, 5847–5850 (1993).

188. GALE, J.B., J.-G. YU, A. KHARE, X.E. HU, D.K. HO, and J.M. CASSADY: Stereochemistry of Mono-tetrahydrofuranyl Moiety in Cytotoxic Polyketides. Part B: Application of Proton Chemical Shift Patterns. Tetrahedron Lett., **34**, 5851–5854 (1993).

189. HARMANGE, J.-C., B. FIGADÈRE, and A. CAVÉ: Stereocontrolled Synthesis of 2,5-Linked Monotetrahydrofuran Units of Acetogenins. Tetrahedron Lett., **33**, 5749–5752 (1992).

189a. FIGADÈRE, B., J.-C. HARMANGE, L.X. HAI, and A. CAVÉ: Synthesis of 2,33-Dihydro-4-oxo-murisolin: Conjugate Addition of Primary Alkyl Iodides to α, β-Unsaturated Ketones. Tetrahedron Lett., **33**, 5189–5192 (1992).

190. DALE, J.A., and H.S. MOSHER: Nuclear Magnetic Resonance Enantiomer Reagents. Configurational Correlations *via* Nuclear Magnetic Resonance Chemical Shifts of Diastereomeric Mandelate, O-Methylmandelate, and α-Methoxy-α-trifluoromethylphenylacetate (MTPA) Esters. J. Am. Chem. Soc., **95**, 512–520 (1973).

191. SULLIVAN, G.R., J.A. DALE, and H.S. MOSHER: Correlation of Configuration and ^{19}F Chemical Shifts of α-Methoxy-α-trifluoromethylphenylacetate Derivatives. J. Org. Chem., **38**, 2143 (1973).

192. YAMAGUCHI, S: Asymmetric Synthesis. *In*: (Morrison, J.D., ed.), vol.1, pp 125–152. New York: Academic Press 1983.

193. POTIN, D., F. DUMAS, and J. d'ANGELO: New Chiral Auxiliaries: Their Use in the Asymmetric Hydrogenation of β-Acetamidocrotonates. J. Am. Chem. Soc., **112**, 3483–3486 (1990).

194. SCHMITZ, F.J., and E.D. LORANCE: Chemistry of Coelantherates XXI. Lactones from the Gorgonian *Pterogorgia guadeloupensis*. J. Org. Chem., **36**, 719–721 (1971).

195. SCHULTZ, W.J., M.C. ETTER, A.V. POCIUS, and S. SMITH: New Family of Cation Binding Compounds *threo*-α-, ω-Poly(cycloalkane)diyl. J. Am. Chem. Soc., **102**, 7981–7982 (1980).

196. DJERASSI, C., and W.-K. LAM: Sponge Phospholipids. Acc. Chem. Res., **24**, 69–75 (1991).

197. CARBALLEIRA, N.M., and J.R. MEDINA: New δ-5,9-Fatty Acids in the Phospholipids of the Sea Anemone *Stoichactis helianthus* J. Nat. Prod., **57**, 1688–1695 (1994).

198. HUTCHINSON, J.: Evolution and Phylogeny of Flowering Plants. New York: Academic Press 1969.

199. TAKHTAJAN, A.: Flowering Plants, Origin and Dispersal. Edinburgh: Oliver and Boyd Ltd. 1969.

200. FIGADÈRE, B.: Syntheses of Acetogenins of Annonaceae: A New Class of Bioactive Polyketides. Acc. Chem. Res., **28**, 359–365 (1995).

201. FIGADÈRE, B., and A. CAVÉ: Total Stereoselective Synthesis of Acetogenins of Annonaceae: A New Class of Bioactive Polyketides. *In*: Studies in Natural Products Chemistry, vol. 18 (Atta-ur-Rahman, ed.) pp 193–227. Amsterdam: Elsevier, 1996.

202. HARMANGE, J.C., and B. FIGADÈRE: Synthetic Routes to 2,5-Disubstituted Tetrahydrofurans. Tetrahedron: Asymmetry, **4**, 1711–1754 (1993).

203. Figadère, B., C. Chaboche, J.-F. Peyrat, and A. Cavé: Stereocontrolled Synthesis of Key Intermediates in the Total Synthesis of Acetogenins of Annonaceae. Tetrahedron Lett., **34**, 8093–8096 (1993).
204. Hoppe, R., M. Flasche, and H.-D. Scharf: An Approach Towards 2,5-Disubstituted Tetrahydrofurans of Annonaceous Acetogenins. Tetrahedron Lett., **35**, 2873–2876 (1994).
205. Rama Rao, A.V., K.L.N. Reddy, and K. Ashok Reddy: Studies on Polyether Natural Products: Asymmetric Synthesis of Bis(tetrahydrofuran). Indian J. Chem., **32B**, 1203–1208 (1993).
206. Koert, U., H. Wagner, and U. Pidun: Stereoselective Additions of Chiral, Functionalized Organozinc Reagents to Achiral and Chiral Aldehydes: a Matched-mismatched Case in Organozinc Chemistry. Chem. Ber., **127**, 1447–1457 (1994).
207. Koert, U., H. Wagner, and M. Stein: An Enantiomerically Pure Epoxyorganolithium Reagent for the Synthesis of Oligo (tetrahydrofurans) by an Epoxide-Cascade Reaction. Tetrahedron Lett., **35**, 7629–7632 (1994).
208. Harmange, J.-C., B. Figadère, and R. Hocquemiller: Enantiospecific Preparation of the Lactone Fragment of Murisolin. Tetrahedron: Asymmetry, **2**, 347–350 (1991).
209. Hoye, T.R., P.E. Humpal, J.I. Jiménez, M.J. Mayer, L. Tan, and Z. Ye: An Efficient and Versatile Synthesis of the Butenolide Subunit of 4-Hydroxylated Annonaceous Acetogenins. Tetrahedron Lett., **35**, 7517–7520 (1994).
210. Figadère, B., X. Franck, and A. Cavé: Synthesis of C1-C11 Fragment of Annonacin: A Polyketide Acetogenin of Annonaceae. Tetrahedron Lett., **36**, 1637–1640 (1995).
211. Trost, B.M., and T.L. Calkins: Synthetic Strategies to Acetogenins. The Hydroxybutenolide Terminus. Tetrahedron Lett., **36**, 6021–6024 (1995).
212. Hoye, T.R., and P.R. Hanson: Assigning the Relative Stereochemistry Between C(2) and C(4) of the 2-Acetonyl-4-alkylbutyrolactone Substructures of the Appropriate Annonaceous Acetogenins. J. Org. Chem., **56**, 5092–5095 (1991).
213. Figadère, B., C. Chaboche, X. Franck, J.-F. Peyrat, and A. Cavé: Carbonyl Reduction of Functionalized Aldehydes and Ketones by Tri-*n*-butyltin Hydride and SiO_2. J. Org. Chem., **59**, 7138–7141 (1994).
214. Harmange, J.-C.: Synthèse totale énantiosélective d'acétogénines d'Annonaceae (γ-lactones monotétrahydrofuraniques). Thèse de Doctorat de l'Université Paris-Sud, Châtenay-Malabry, 1992.
215. Iwai K., H. Kosugi, H. Uda, and M. Kawai: New Methods for Synthesis of Various Types of Substituted 2(5H)-Furanones. Bull. Chem. Soc. Jpn, **50**, 242 (1977).
216. Hoye, T.R., P.R. Hanson, A.C. Kovelesky, T.D. Ocain, and Z. Zhuang: Synthesis of (+)-15,16,19,20,23,24)-*hexepi*-Uvaricine: A Bis(tetrahydrofuranyl) Annonaceous Acetogenin Analogue. J. Am. Chem. Soc., **113**, 9369–9371 (1991).
217. Hoye, T.R., and P.R. Hanson: Synthesis of (−)-Bullatacin: The Enantiomer of a Potent, Antitumor, 4-Hydroxylated, Annonaceous Acetogenin. Tetrahedron Lett., **34**, 5043–5046 (1993).
218. Liu, Z.Y., J.J. Zhang, and W. Chen: A Facile Synthesis of (±)-Muricatacin. Chin. Chem. Lett., **4**, 663–664 (1993).
219. Gravier-Pelletier, C., M. Sanière, I. Charvet, Y. Le Merrer, and J.-C. Depezay: Synthesis of (−)-Muricatacin and (−)-(5*R*, 6*S*)-6-Acetoxy-5-hexadecanolide, the Mosquito Oviposition Attractant Pheromone, from *D*-Isoascorbic Acid. Tetrahedron Lett., **35**, 115–118 (1994).
220. Gravier-Pelletier, C., J. Dumas, Y. Le Merrer, and J.-C. Depezay: A General Way from L- and D-Isoascorbic Acids to Homochiral α-Hydroxy, α,β-Dihydroxy and α,β-

epoxy Aldehydes Useful Building Blocks for the Synthesis of Linear Oxygenated Fatty Acids Metabolites. J. Carbohydr. Chem., **11**, 969–998 (1992).

221. Bonini, C., C. Federici, L. Rossi, and G. Righi: C-1 Reactivity of 2,3-Epoxyalcohols *via* Oxirane Opening with Metal Halides: Applications and Synthesis of Naturally Occurring 2,3-Octanediol, Muricatacin, 3-Octanol, and 4-Dodecanolide. J. Org. Chem., **60**, 4803–4812 (1995).

222. Van Aar, M.P.M., L. Thijs, and B. Zwanenburg: Synthesis of (4*R*, 5*R*)-Muricatacin and its (4*R*, 5*S*)-Analog by Sequential Use of the Photo-induced Rearrangement of Epoxy Diazomethyl Ketones. Tetrahedron, **51**, 11223–11234 (1995).

223. Hoye, T.R., and J.C. Suhadolnik: Stereocontrolled Synthesis of 2,5-Linked Bistetrahydrofurans *via* the Triepoxide Cascade Reaction. Tetrahedron, **42**, 2855–2862 (1986).

224. Cassady, J.M., W.M. Baird, and C.-J. Chang: Natural Products as a Source of Potential Cancer Chemotherapeutic and Chemopreventive Agents. J. Nat. Prod., **53**, 23–41 (1990).

225. Poupon, M.-F. (unpublished results).

226. Suffness, M., D.J. Newman, and K. Snader: Discovery and Development of Antineoplastic Agents from Natural Sources. Bioorganic Marine Chemistry **3**, pp 134. Berlin Heidelberg: Springer Verlag 1989.

227. Holschneider, C.H., M.T. Johnson, R.B. Knox, A. Rezai, W.J. Ryan, and F.J. Montz: Bullatacin – *in vivo* and *in vitro* Experience in an Ovarian Cancer Model. Cancer Chemother. Pharmacol., **34**, 166–170 (1994).

228. Ahammadsahib, K.I., R.M. Hollingworth, J.P. McGovren, Y.-H. Hui, and J.L. McLaughlin: Mode of Action of Bullatacin: A Potent Antitumor and Pesticidal Annonaceous Acetogenin. Life Science, **53**, 1113–1120 (1993).

229. Cavé, A., R. Hocquemiller, and O. Laprévote: Utilisation d'acétogénines en thérapeutique en tant que substances antiparasitaires. F. Patent 1048 N° 88 09 674 (1989).

230. Bories, C., P. Loiseau, D. Cortes, S.H. Myint, R. Hocquemiller, P. Gayral, A. Cavé, and A. Laurens: Antiparasitic Activity of *Annona muricata* and *Annona cherimolia* Seeds. Planta Med., **57**, 434–436 (1991).

231. Moeschler, H.F., W. Pfluger, and D. Wendisch. US Patent N° 4 689 232 issued August 25 (1987).

232. Lewis, M.A., J.T. Arnason, B.J.R. Philogene, J.K. Rupprecht, and J.L. McLaughlin: Inhibition of Respiration at Site 1 by Asimicin, an Insecticidal Acetogenin of the Paw paw, *Asimina triloba*, Annonaceae. Pesticide Biochem. Physiol., **45**, 15–23 (1993).

232a. Ratnayake, S., J.K. Rupprecht, W.M. Potter, and J.L. McLaughlin: Evaluation of Various Parts of the Paw paw Tree, *Asimina triloba* (Annonaceae), as Commercial Sources of the Pesticidal Annonaceous Acetogenins. J. Economic Entomol., **85**, 2353–2356 (1992).

233. Vu Thi Tam: Étude chimique et biologique des acétogénines des graines d'*Annona reticulata*, Annonaceae. Doctorat de l'Université Paris-Sud, Châtenay-Malabry (1995).

234. Laurens, A., P. Dutartre, R. Hocquemiller, and A. Cavé: Immunomodulating Activity of Annonacin Isolated from *Annona muricata* Seeds. Communication to the 18th International IUPAC Symposium on the Chemistry of Natural Products, Strasbourg, France, 30 August-4 September (1992).

235. Peyrat, J.-F., B. Figadère, A. Cavé, and J. Mahuteau: Study of the Binding Activity of Oligo-tetrahydrofuranic γ-lactones with Cations. Tetrahedron Lett., **36**, 7653–7656 (1995).

236. SASAKI, S., H. NAITO, K. MARUTA, E. KAWAHARA, and M. MAEDA: Novel Calcium Ionophores: Supramolecular Complexation by the Hydroxylated-Bistetrahydrofuran Skeleton of Potent Antitumor Annonaceous Acetogenins. Tetrahedron Lett., **35**, 3337–3340 (1994).
237. SASAKI, S., K. MARUTA, H. NAITO, H. SUGIHARA, K. HIRATANI, and M. MAEDA: New Calcium-selective Electrodes Based on Annonaceous Acetogenins and Their Analogs with Neighboring Bistetrahydrofuran. Tetrahedron Lett., **36**, 5571–5574 (1995).
238. PADMAJA, V., S.M. JESSY, C.R. SUDHAKARAN NAIR, G.R. NAIR, V. THANKAMANI, and A. HISHAM: Antimitotic Effects of *Uvaria narum* and *U. hookeri*. Fitoterapia, **65**, 77–81 (1994).
239. DEGLI ESPOSTI, M., A. GHELLI, M. RATTA, D. CORTES, and E. ESTORNELL: Natural Substances (Acetogenins) from the Family Annonaceae Are Powerful Inhibitors of Mitochondrial NADH Dehydrogenase (Complex I). Biochem. J., **301**, 161–167 (1994).
240. MORRÉ, D.J., R. DE CABO, C. FARLEY, N.H. OBERLIES, and J.L. MCLAUGHLIN: Mode of Action of Bullatacin, a Potent Antitumor Acetogenin: Inhibition of NADH Oxidase Activity of HELA and HL-60, but not Liver, Plasma Membranes. Life Sciences, **56**, 343–348 (1995).
241. FAULK, W.P., K. BARABAS, I.L. SUN, and F.L. CRANE: Transferrin-Adriamycin Conjugates which Inhibit Tumor Cell Proliferation without Interaction with DNA Inhibit Plasma Membrane Oxido-reductase and Proton Release in K562 Cells. Biochem. Int., **25**, 815–822 (1991).
242. GOTTESMAN, M.M.: How Cancer Cells Evade Chemotherapy: Sixteenth Richard and Hinda Rosenthal Foundation Award Lecture. Cancer Research, **53**, 747–754 (1993).
243. SIMON, S.M., and M. SCHINDLER: Cell Biological Mechanisms of Multidrug Resistance in Tumors. Proc. Natl. Acad. Sci. USA, **91**, 3497–3504 (1994).
244. FRIEDRICH, T., P. VAN HEEK, H. LEIF, T. OHNISHI, E. FORCHE, B. KUNZE, R. JANSEN, W. TROWITZSCH-KIENAST, G. HÖFLE, H. REICHENBACH, and H. WEISS: Two Binding Sites of Inhibitors in NADH: Ubiquinone Oxidoreductase (Complex I). Relationship of One Site with the Ubiquinone Binding Site of Bacterial Glucose: Ubiquinone Oxidoreductase. Eur. J. Biochem., **219**, 691–698 (1994)
245. GU, Z.-M., G.-X. ZHAO, N.H. OBERLIES, L. ZENG, and J.L. MCLAUGHLIN: Annonaceous Acetogenins. *In*: Phytochemistry of Medicinal Plants (J.T. ARNASON, R. MATA, and J.T. ROMEO, eds.) pp 249–310, New York; Plenum Press 1995.

(*Received January 16, 1996*)

Author Index

Abraham, R.J. 30, *77*
Abreo, M.J. *282, 283*
Aguilar, E. *78*
Ahammadsahib, K.I. *287*
Akashi, S. *277, 283*
Alcantara, J.P. *282*
Alfonso, D. *281*
Alkofahi, A. *276, 280*
Anderson, G.W. *77*
Anderson, J.E. *275, 278–282*
Antonakis, K. *277*
Aoyagi, R. *72*
Araya, H. *274, 277, 283*
Arnason, J.T. *287*
Arseniyadis, S. *76*
Asaki, H. *277*
Ashok Reddy, K. *286*
Atsuzawa, S. *277*

Baczynskyj, L. *72*
Bahr, J. *74*
Bai, R. *78, 79*
Baird, W.M. *287*
Ball, R.G. *76*
Baltzer, F. *71*
Barabas, K. *288*
Barkóczy, J. *76–78*
Barnes, J.N. *282*
Barry, C.E. *280*
Bartoli, M.-H *275*
Bartroli, J. *77*
Bates, R.B. *273, 280*
Bates, R.E. *73*
Batten, T. *277*
Bean, M.F. *280*
Bebbington, A. *71*
Benoiton, N.L. 30, *77*
Bériel, H. *275*
Bernier, J.-L. *74*
Bessodes, M. 254, *277*
Bhat, K.L. *76*
Biskupiak, J.E. *74*
Boaventura, M.A.D. *279*
Boettner, F.E. *72, 73*
Boger, D.L. *77*
Bonini, C. 256, *287*
Bontems, R.J. *72, 73*
Bories, C. *278, 287*
Born, L. 212, 216, 217, *283*
Bowman, R.E. *77*
Boyd, M.R. *72, 78*
Bozich, F.A. *75*
Braga de Oliveira, A. *279*
Brauer, S. *79*
Braun, M. *76*
Bredenkamp, M.W. *74, 75*
Brinen, L.S. *280*
Brown, P. *72, 284*
Bui Chi Hieu *282*
Bülow, C. *284*
Bunte, G. *276*
Burkett, D.D. *74–76*
Burkholder, L.M. *70*
Burkholder, P.R. *70*
Burow, D.F. *76*

Calkins, T.L. *286*
Callahan, F.M. *77*
Calventi, I. *70*
Carballeira, N.M. *285*
Cardellina II, J.H. *78*
Carmely, S. *78*
Cassady, J.M. 265, *274, 275, 278, 280, 285, 287*
Castro, B. *77, 78*
Cavé, A. *274–287*

Cecil, J.T. *71*
Cerny, R.L. *73*
Chaboche, C. *282, 286*
Chaet, A.B. *71*
Chaney, M.O. *78*
Chang, C.-J. *274–276, 278–280, 284, 287*
Chang, F.-R. *275, 277, 280*
Chanley, J.D. *71*
Chao, J.-F. *283*
Chappe, B. *280, 282*
Chapuis, J.-C. *72, 78*
Charles, B. *283*
Charvet, I. *277, 286*
Chen, F.M.F. *77*
Chen, K.-S. *280*
Chen, W. *286*
Chen, W.-S. *276, 278, 281*
Cho, H.-S. *277*
Choi, H.S. *79*
Ciereszko, L.S. *70, 71*
Claeys, M. *280–282*
Clardy, J. *280*
Clewlow, P.J. *74*
Coggins, J.J.R. *77*
Cole, J.R. *273, 280*
Colman-Saizarbitoria, T. *274, 275, 283*
Colson, P.-J. *76*
Cook, B. *77*
Copleanu, F. 172, *279*
Cortes, D. 82, *274–281, 283, 284, 287, 288*
Coste, J. *77, 78*
Costello, C.E. *74*
Cotte, J. *277*
Cragg, G.M. *74, 282*
Crane, F.L. *288*
Crimmins, M.T. *76*
Curran, T.T. *77*

Dabrah, T.T. 82, *273*
Dale, J.A. *285*
Danda, H. *76*
D'Angelo, J. *285*
Das, B.C. *280, 284*
Davoust, D. *274, 281*
Day, J.F. *71*
De Bruyne, T. *280, 281*
De Cabo, R. *288*
Degli Esposti, M. *288*
Depezay, J.-C. 255, *277, 286*
Devant, R.M. *76*
Diem, M.J. *76*
Djerassi, C. 207, *284, 285*
Dommisse, R. *281, 282*
Donati, G. *71*
Doolittle, R.E. *284*
Dormann, D.E. *78*
Doubek, D.L. *72, 73*
Dreux, M. *284*
Drexler, H.D. *79*
Dubaele, B. *275*
Dudeck, L.E. *71*
Dufour, M.-N. *77, 78*
Dufresne, C. *73, 282*
Duh, C.-Y. *275, 277*
Dumas, F. *285*
Dumas, J. *286*
Duncan, K. *69*
Dunn, A. *71*
Dupont, B. *274*
Duret, P. 194, *274, 277, 278*
Durkin, K.P.M. *78*
Durso, Jr., A.R. *74*
Dutartre, P. *287*

Eales, N.B. *71*
Eckart, K. *73*
Eguchi, T. *274, 277, 283*
Eickhoff, D.J. *284*
Einhorn, J. *284*
Endo, M. *74*
Engel, H. *71*
Esmans, E. *281, 282*
Estornell, E. *288*
Etcheverry, S. *275*
Etse, J.T. *282, 283*
Etter, M.C. *285*
Evans, D.A. 22, *76, 77*
Evert, D.R. *276, 278*
Ewing, W.R. *76*

Fall, D. *275*
Fan, D.-J. *276, 280*
Fang, X.-P. *274–276, 278–281*
Farley, C. *288*
Fatiadi, A.J. *75*
Fatope, M.O. *281*
Faulk, W.P. *288*

Federici, C. *287*
Ferrigni, N.R. *283*
Figadère, B. 251, *274, 277, 279, 280, 282–287*
Fischer, P. *74*
Flahive, E.J. *78*
Flasche, M. *286*
Forche, E. *288*
Franck, X. *286*
Franck-Neumann, M. *76*
Friedman, S.J. *79*
Friedrich, T. *288*
Friess, S.L. *71*
Fry, J.L. *76*
Fuji, Y. *72*
Fujimoto, Y. 215–218, *274, 277, 283*
Furuya, M. *274*

Gage, J.R. *76*
Gale, J.B. 213, 214, 216, 217, *285*
Galeotti, N. *78*
Gargus, J.L. *71*
Gayral, P. *287*
Gebhard, I. *75*
Gebreyesus, T. *78*
Geoffroy, P. *76*
Ghelli, A. *288*
Gignac, S.M. *79*
Girard, C. *280, 284*
Gleich, P. *75*
Gloer, J.B. *75*
González, M.C. *275*
Goswami, A. *282*
Goto, T. *70*
Gottesman, M.M. *288*
Gougoutas, J.Z. *70*
Gravier-Pelletier, C. *286*
Gray, A.I. *283*
Griesser, H. *74, 75*
Gromek, D. 260, 261, *274, 279, 284*
Groy, T.L. *77*
Grunicke, D. *79*
Gu, Z.-M. 218, 261, *274–283, 288*
Gui, H.Q. *281*
Gupta, K.K. *274*
Gupta, M.P. *279*
Gupta, Y.K. *274, 277, 283*
Gust, D.J. *72*
Gutierrez, C.G. *77*
Gypser, A. *284*

Hachiya, I. *76*
Hacker, M.P. *74*
Hagiwara, H. *76*
Hai, L.X. *285*
Halstead, B.W. *70*
Hamada, Y. *73–75*
Hamamoto, Y. *74*
Hamburger, M. *279*
Hamel, E. 68, 69, *78, 79*
Hamill, R.L. *78*
Hansen, M.M. *76*
Hanson, P.R. *276, 284, 286*
Hao, X.J. *281*
Hara, N. *277, 283*
Harmange, J.-C. 214, 216, 217, *277, 284–286*
Harris, B.D. *76*
Hartwell, J.L. *71*
Hasenwinkel, L.E. *276, 284*
Hayashi, K. *74, 75*
He, K. *281*
Heath, R.R. *284*
Heathcock, C.H. *76*
Heilbrunn, L.V. *71*
Henichart, J.-P. *74*
Herald, C.L. *71–73*
Herald, D.L. *73, 74, 76–78, 282*
Herbert, J. *277*
Hill, R.B. *72*
Hill, R.R. *77*
Hirata, Y. *70*
Hiratani, K. *288*
Hirayama, K. *274, 277, 283*
Hisham, A. *280–282, 288*
Ho, D.K. *280, 285*
Hocquemiller, R. *274–281, 283, 284, 286, 287*
Hoehn, M.M. *78*
Hoffmann, J.J. *273, 280*
Höfle, G. *288*
Hogan, F. *74–77*
Hogan-Pierson, F. *76*
Hollingworth, R.M. *287*
Holschneider, C.H. *287*
Holzapfel, C.W. 18, 35, *72–75*
Hong, S.L. *274*
Hong, Y.T. *79*
Hoppe, R. *286*
Hostettmann, K. *279*
Houssin, R. *74*

Hoye, T.R. 208, 210–212, 216, 217, 241, 243, 244, 259, *276, 278, 284–287*
Hu, X.E. *280, 285*
Hu, Z.-B. *79*
Huala, C.V. *71*
Hui, Y.-H. *274, 276–283, 287*
Humpal, P.E. *286*
Huo, M. *274*
Hutchinson, J. 227, *285*
Hyman, L.H. *71*

Iitaka, Y. *77*
Ikekawa, N. *274, 283*
Ikekawa, T. *274*
Ikota, I. *77*
Inoue, M. *72*
Irako, N. *75*
Ireland, C.M. *74*
Ishigaki, T. *79*
Ishiwata, H. *73, 79*
Iwai, K. *286*

Jackson, J.T. *77*
Jahnes, W.B. *71*
Jakowska, S. *70*
Jansen, R. *288*
Jayaram, S.K. *78*
Jeffs, G.E. *77*
Jessy, S.M. *288*
Jiménez, J.I. *286*
Johnson, M.T. *287*
Jolad, S.D. 81, *273, 280*
Jones, N.D. *78*
Jossang, A. *275*
Jouillié, M.M. *76*
Jouin, P. *74, 78*

Kakinuma, K. *274, 277, 283*
Kamano, Y. *72, 73*
Kanai, M. *74*
Kang, S.-K. 254, *277*
Kano, S. *74*
Kantoci, D. *73, 75–78*
Kashman, Y. *78*
Kato, S. *74, 75*
Kawahara, E. *283, 288*
Kawai, M. *286*
Kawazu, K. *282*
Keinan, E. 236, 244, *276, 283*
Kelly, R.C. *75*
Keough, T. *284*
Kepler, J.A. *79*
Khare, A. *285*
Kigoshi, H. *73, 79*
Kim, B.-K. *277*
Kim, I.K. *79*
Kimura, K. *76*
Kirst, H.A. *78*
Kishi, Y. *70*
Kishore, B. *75*
Kishore, V. *75*
Kitahara, K. *72*
Kizu, H. *72, 73*
Knox, R.B. *287*
Kobayashi, A. *282*
Kobayashi, S. *76*
Koert, U. 248, *282, 286*
Koga, K. *74*
Kohda, K. *73*
Kondo, Y. *75*
Konig, W.A. *280*
Kossanyi, J. *284*
Kosugi, H. *286*
Kovelesky, A.C. *286*
Kozlowski, J.F. *275, 276, 278, 282, 283*
Kriek, G.R. *273*
Kroner, M. *75*
Kunze, B. *288*
Kuwasawa, K. *72*
Kwon, N.Y. *79*

Lafosse, M. *284*
Lake, R.J. *75*
Lam, W.-K. *285*
Laprévote, O. *279, 280, 283, 284, 287*
Laugel, T. *280*
Laurens, A. *274–278, 281, 284, 287*
Leboeuf, M. *276, 277, 280, 281, 284*
Lee, B.S. *79*
Lee, C.F. *71*
Lee, M.-R. *280*
Leenay, T.L. *76*
Leet, J.E. *282*
Leicht, W. *281*
Leif, H. *288*
Le Merrer, Y. *277, 286*
Lewis, M.A. *287*
Li, C.M. *279, 281*
Li, C.P. *71*
Li, K.-M. *75*

Li, X.-H. *278*
Liang, S.-C. *280*
Licgter, W. *71*
Lieberknecht, A. *74*
Lieb, F. *274, 276, 281, 283*
Ligouri, V.R. *71*
Linz, G.S. *280*
Liu, C.Y. *274*
Liu, H.-J. *76*
Liu, Y.-M. *278–280, 283, 284*
Liu, Z.Y. 255, *286*
Lloyd-Williams, P. *74, 76*
Loiseau, P. *278, 287*
Londershausen, M. *281*
Lorance, E.D. *285*
Lorentzen, J.P. *283*
Lucas, A.H. *71*
Luduena, R.F. *78, 79*
Luo, X.Z. *274*

Ma, L.B. *274*
MacDougal, J.M. *279, 281*
Madrigal, R.V. *283*
Maeda, M. *283, 288*
Mahuteau, J. *279, 287*
Maibaum, J. *76*
Makabe, H. *276, 282*
Makino, N. *72*
Maleczka Jr., R.E. *75*
Marderosian, A.H.D. *71*
Marner, F.-J. *78*
Marshall, J.A. 252, *276, 277*
Maruta, K. *283, 288*
Matsui, K. *72*
Maugras, I. *74*
Mayer, M.J. *286*
McCloud, T.G. *275*
McDermott, J.R. *77*
McGovren, J.P. *287*
McLaughlin, J.L. 265, *274–284, 287, 288*
Medina, J.R. *285*
Meneses da Silva, E.L. *279*
Meyers, A.I. *78*
Michel, C. *71, 72*
Miesbauer, L.R. *276, 278–281*
Mihelich, E.D. *284*
Mikolaiczak, K.L. *278, 283*
Misbauer, L.R. *277*
Mitsui, T. *277*
Mizukawa, K. *74*

Moeschler, H. *274, 281, 283, 287*
Montz, F.J. *287*
Moore, R.E. *78*
Moretti, C. *282*
Morré, D.J. *288*
Mosher, H.S. 221, 222, *285*
Mu, Q. *281*
Mullaney, J.T. *78*
Munro, M.H.G. *75*
Murasaki, C. *274*
Musso, H. *76*
Mutou, T. *79*
Myint, S.H. *274, 276, 277, 281, 284, 287*

Nagarajan, S. *75*
Nair, G.R. *288*
Naito, H. *283, 288*
Nakagawa, M. *74*
Nakanishi, T. *74*
Nelson, L.A.K. *76*
Nelson, P.S. *73*
Nemoto, T. *73*
Newman, D.J. *287*
Newman, R.A. *74*
Ng, J.H. *278, 279, 283*
Niel, G. *74*
Nieman, R.A. *73*
Nigrelli, R.F. *70, 71*
Niitsu, U. *283*
Nisato, D. *78*
Nishioka, S. *274*
Nonfon, M. *274, 276, 283*

Oberhansli, W.E. *78*
Oberlies, N.H. *288*
Ocain, T.D. *286*
Occolowitz, J.L. *78*
Ode, R.H. *71, 72*
Ogawa, T. *79*
Ohnishi, T. *288*
Ohsawa, K. *277*
Ohtani, K. *279*
Oikawa, Y. *78*
Ojika, M. *73, 79*
Oritani, T. *276, 282*

Padmaja, V. *282, 288*
Pan, X.P. *281*
Pantaloni, A. *77*
Parikh, J.R. *75*

Patino, N. *78*
Peters, E.M. *276*
Peters, K. *276*
Pettit, G.R. *71–79, 282*
Pettit, W.E. *73*
Peyrat, J.-F. *286, 287*
Pfluger, W. *287*
Phan Quan Chi Hieu *280, 282*
Philogene, B.J.R. *287*
Pidun, U. *286*
Pieters, L.A.C. *280–282*
Pietsch, T. *79*
Pinheiro Santos, L. *279*
Pliny the Elder 2, 4, *71*
Pocius, A.V. *285*
Polonsky, J. *282*
Poncet, J. 24, 31, 35, 47, *74, 78*
Porfirio, B. *71*
Potin, D. *285*
Potter, W.M. *287*
Potzolli, B. *74*
Poupon, M.-F. *287*
Prasad, V. *79*
Prescott, B. *71*
Putt, S.R. *76*

Qin, G.-W. *276, 277, 280–283*
Quayle, P. 257, *277*
Quentmeier, H. *79*
Quero, A.-M. *277*

Radunz, H.-E. *76*
Rahman, S. *277*
Rama Rao, A.V. *286*
Ramirez, E.A. 208, *276, 284*
Ratnayake, S. *276, 278, 287*
Ratta, M. *288*
Ray, A.B. *274*
Reddy, K.L.N. *286*
Reichenbach, R. *288*
Rezai, A. *287*
Rich, D.H. *76*
Riesen, R. *282*
Rieser, M.J. 221, *274, 276, 277, 279–281*
Righi, G. *287*
Rinehart, K.L. *75*
Rios, J.L. *274, 277*
Roach, M.C. *78, 79*
Roblot, F. 259, 261, *279, 280, 282, 283*
Rossi, L. *287*
Roux, F. *74*
Rudloe, J.J. *72*
Ruggieri, C.D. *71*
Rupprecht, J.K. *274, 276, 278–280, 282, 283, 287*
Ryan, W.J. *287*

Saad, J.M. *282*
Saez, J. *280, 281, 283*
Sahai, M. *274, 277, 283*
Sahpaz, S. 261, *275, 277, 278, 280, 281, 283, 284*
Saïah, M. *277*
Sakai, R. *75*
Sakakura, A. *79*
Sanière, M. *277, 286*
Sasaki, S. 246, *283, 288*
Sastrodihardjo, S. *275, 277*
Schaneberg, B.T. *282*
Scharf, H.-D. *284, 286*
Schaufelberger, D. *282*
Scheuer, P.J. *72, 74*
Schidt, J.M. *72*
Schindler, M. *288*
Schmeer, M.R. *71*
Schmidt, J.M. *73, 78, 79, 282*
Schmidt, U. 17, *73–75*
Schmitz, F.J. *285*
Scholz, G. *276*
Schram, K.H. *273*
Schroll, R.M. *73*
Schultz, W.J. *285*
Schwedler, J.T. *275, 277, 278*
Segel, M.M. *71*
Sharpless, K.B. 253, *276*
Shi, G. 172, *279, 281*
Shi, Z. *283*
Shibata, M. *75*
Shibuya, S. *74*
Shih, T.L. *77*
Shimada, H. *274*
Shioiri, T. 17, 23, 30, 35, *73–75, 77, 78*
Shwarz, H. *73*
Sifford, D.H. *70*
Sigel, M.M. *71*
Sim, H.-S. *277*
Simon, S.M. *288*
Singh, M. *274, 277*
Singh, S. *274, 277*
Singh, S.B. *74–78*

Sinha, S.C. *276, 283*
Sinha-Bagchi, A. *276*
Sjerassi, C. *285*
Slack, J.A. *78*
Smith III, A.B. *76*
Smith, C.R. *282*
Smith, D.L. *275–281, 283, 284*
Smith, S. *285*
Snader, K. *287*
Sneden, A.T. 82, *273, 282, 283*
Snyman, R.M. *74*
Sobotka, H. *71*
Solis, P. *279*
Söllner, R. *283*
Somfai, P. 256, *277*
Sone, H. *73, 79*
Song, R. *279, 280*
Sreekala, U. *280, 281*
Srirangam, J.K. *75–78*
Srivastava, A. *283*
Standaert, F.G. *71*
Stein, M. *286*
Steinauer, R. *77*
Stempien, M.F. *71*
Stephens, R.E. *280*
Steube, K.G. *79*
Stroud, H.H. *77*
Sudhakaran Nair, C.R. *288*
Suenaga, K. *79*
Suffness, M. *287*
Sugano, K. *78*
Sugi, H. *72*
Sugihara, H. *288*
Suhadolnik, J.C. *287*
Sullins, D.W. *75*
Sullivan, G.R. *285*
Summerhays, L.R. *77*
Sun, H.-D. *279, 281*
Sun, I.L. *288*
Sun, L. *274*
Sunitha, C. *281*
Suzuki, S. *72*
Szymanski, E.F. *78*

Taba, K.M. *76*
Tabet, J.C. *283, 284*
Tachibana, S. *77*
Takahashi, S. *70*
Takhtajan, A. 227, *285*
Tan, L. *278, 286*
Tanaka, A. 237, 255, *276, 282*
Tao, G.-D. *279*
Taylor, G.F. *79*
Tempesta, M.S. *273*
Thankamani, V. *282, 288*
Thij, L. *287*
Thomas, W.A. *77*
Thornton, T.J. *78*
Titus, H.W. *71*
Tochtermann, W. 251, *276*
Todsen, W.L. *75*
Tomer, K.B. *72*
Tomioka, K. *74*
Torii, K. *77*
Trost, B.M. 239, *283, 286*
Trowitzsch-Kienast, W. *288*
Tucker, B.E. *73*
Tuinman, A.A. *72*
Tumlinson, J.H. *284*

Uda, H. *76, 286*
Utz, R. *73–75*

Valverde, S. *274, 277*
Van Aar, M.P.M. *287*
Van den Heuvel, H. *280, 281*
Van Draanen, N.A. *76*
Van Heek, P. *288*
Van Zyl, W.J. *72, 74, 75*
Varenne, P. *279*
Villaescusa, L. *281*
Víllar, A. *274*
Vlietinck, A.J. *281, 282*
Von Dreele, R.B. *71*
Von E. Doering, W. *75*
Von Hofheinz, W. *78*
Von Schnering, H.G. *276*
Vu Thi Tam 259, 260, *280, 282, 287*

Wachendorff-Neumann, U. *276*
Waechter, A.-I. *277, 281*
Wagner, H. *286*
Wagner, I. *76*
Wagner, K. *74*
Walker, M.A. *76*
Wang, S.-K. *275, 277*
Wang, Z.-M. *276*
Wardlaw, T.R. *76*
Wasylyk, J.M. *74*
Waterman, P.G. *282, 283*

Watson, M. *72*
Weinheimer, A.J. *70*
Weisleder, D. *278, 282*
Weiss, H. *281, 288*
Weller, D. *74*
Wellham, L.L. *71*
Welmaker, G.S. *276, 277*
Wendisch, D. *274, 276, 283, 287*
Wenger, R.M. *77*
Whitcomb, E.R. *71*
Whitson, D. *71*
Wicnienski, N. *75*
Williams, M.D. *75–79*
Wilson, W.L. *71*
Wolff, C. *276*
Woo, M.H. *276, 278, 282*
Wood, H.B. *71*
Wood, K.V. *275, 276, 278–280, 282*–284
Woodward, R.B. *70*
Wu, F.-E. *275, 277, 278, 282*
Wu, S.-J. *275–277, 281, 284*
Wu, Y.-C. *275, 277, 280, 281*
Wu, Y.-L. 234, *276, 278, 279, 281*
Wuts, P.G.M. *76*

Xu, L. *274, 280*
Xu, R.-S. *276, 280–283*
Xu, Y.-Z. *278*

Yamada, K. 11, 47, 48, *73, 79*
Yamada, S. *77*
Yamaguchi, S. 221, *285*
Yamamoto, I. *277*
Yamazaki, M. 11, *73*
Yang, R.-Z. *275–277, 280–284*
Yao, Z.-J. *276, 278, 279, 281*
Ye, Q. *276, 278, 282*
Ye, Z. *286*
Yonemitsu, O. *78*
Yoshida, T. *79*
Yosief, T. *78*
Yuasa, Y, *74*
Yu, D.Q. *281*
Yu, J.-G. *274, 280, 281, 285*
Yu, Z.L. *281*
Yuji, R. *277, 283*

Zafra-Polo, M.C. *275*
Zambrano, J. *283*
Zeng, L. *274–279, 281–283, 288*
Zhang, J.J. *286*
Zhang, L.-L. *275, 281*
Zhang, X.-L. *276*
Zhang, Y. *275–278*
Zhang, Y.-B. *278*
Zhao, G.-X. *274–278, 281–283, 288*
Zheng, H.-L. *279, 281*
Zheng, X.-C. *277, 280, 282, 283*
Zhuang, Z. *276, 284–286*
Zwanenburg, B. 258, *287*

Subject Index

Acetic acid 35, 36
Acetic anhydride 233
Acetogenins 81–84, 188–191, 193–195, 197, 199–202, 204–207, 209, 210, 212, 214, 217–221, 226–229, 231, 261, 264, 265, 269–272
Acetogenins Isoacetogenins 266
2,4-*cis*-Acctogenins 189
2,4-*trans*-Acetogenins 189
Acetone 39, 161, 233, 237, 239, 244, 247, 250, 251, 258, 260, 262, 263
12,13-Acetonide 164
Acetonide diacetate 118, 119
Acetonitrile 7, 36, 38, 39
cis-2-Acetonyl-4-butyl-γ-butyrolactone 194, 209, 210
trans-2-Acetonyl-4-butyl-γ-butyrolactone 194, 209, 210
Acetylatemoyin 168
N-Acetylphenylalanine methyl ester 57
Acetylsquamocin-N 168
Acrolein 255
Actin 63, 69
Acylstannane 252
Adriamycin 268
Aedes aegypti 271
2-Allenyl-1,3-dioxa-2-borolane-(4*S*, 5*S*)-dicarboylic acid bis(1′-methylethyl) ester 236
Allyl bromide 232
Allylic stannanes 252
Allylmagnesium chloride 235, 236
Almunequin 84, 89, 90, 170, 181
Althiomycin 45
Amberlyst-H 15 252
Amberlyst L H-15 242
2-(1-Aminoalkyl)thiazole-4-carboxylic acid 14
2-Aminoethanethiol 15
3-Amino-6-hydroxy-2-piperidone 52
Annogalene 84, 89, 91, 178, 184
Annoglaucin 84, 91, 182
Annohexocin 84, 91, 182
Annomonicin 84, 91, 182, 183
Annomontacin 84, 92, 182, 186, 188
Annomontacin-10-one 84, 92, 178
18,21-*cis*-Annomontacin-10-one 84, 93, 178
Annomuricin-A 84, 93, 182
Annomuricin-B 84, 93, 182
Annomuricin-C 84, 94, 150, 182
Annomutacin 84, 94, 182
Annona atemoya 100, 105, 120, 139, 144, 154, 157, 159, 162, 165, 175, 180
Annona bullata 105–112, 120, 133–137, 139, 157, 158, 162, 165, 175, 176, 180
Annona cherimolia 90, 96, 99, 100, 107, 109, 110, 126, 139, 140, 142–145, 156, 162, 165, 181, 270
Annonaceae 81, 82, 101, 180, 189, 227, 231, 259, 271
Annonacin 84, 86, 95, 96, 98, 138, 146, 179, 181, 182, 184–186, 188, 213, 214, 264, 265, 267, 269–271
Annonacin-10-one 84, 97, 138, 181, 182
Annonacin-1 84
Annonacin-2 84, 96, 184
Annonacin-A 84, 88, 96, 154, 181, 182, 184, 185, 264, 265, 267, 270, 271, 273
Annonacin-A-one 84, 87, 96, 97, 182, 185
10-*R*-Annnonacin-A-one 97, 182
10-*S*-Annonacin-A-one 97
Annonacinone 97
Annona coriacea 113, 114, 119, 125, 128, 181, 229
Annona crassiflora 95, 97, 110, 117, 118, 126, 131, 132, 137, 138, 181

Annona glabra 95, 100, 110, 115, 120, 164, 165, 171, 182
Annona glauca 91, 129, 162, 182
Annona montana 91, 92, 95, 97, 122, 131, 145, 182
Annona muricata 91, 93–97, 113–115, 118, 120, 123, 124, 126–128, 131, 137, 138, 141, 146–152, 154, 155, 164, 182, 228, 264, 270, 271
Annona purpurea 107, 110, 158, 161, 162, 172, 183
Annona reticulata 91, 98, 99, 120–122, 138, 144, 158–162, 164, 165, 169, 174, 175, 183, 229
Annona senegalensis 95, 96, 99, 100, 126, 144, 162, 164, 165, 178, 184, 227, 270
Annona sp. 271
Annona squamosa 90, 95, 96, 98, 100, 102, 105, 107, 110, 120, 139, 145, 154, 157, 159, 162, 165–170, 171, 184, 272
Annonareticin 84
Annonastatin 100
Annonastatin 84, 85, 184
Annonin-I 88, 165, 212, 220
Annonin-III 87, 98, 145
Annonin-IV 85, 110, 113
Annonin VI 84, 162, 184
Annonin VIII 85, 107, 113, 184
Annonin XIV 85, 184
Annonin XVI 85, 89, 170, 184
Annonsilin A 85, 98, 184
Annoreticuin 85, 98, 183
Annoreticuin-9-one 85, 99, 169, 183
Annosenegalin 85, 99, 181, 184
Annotemoyin-1 85, 100, 180
Annotemoyin-2 85, 100, 180
Antibiotic activity 2
Anticancer activity 2–4, 43, 63, 64, 70
Antifungal activity 264, 271
Antimicrobial activity 82, 264, 271, 272
Antimitotic activity 272
Antineoplastic activity 3–5, 7, 11, 18, 39, 48, 57, 63, 64
Antiparasitic activity 82, 264, 271
Antitumor activity 2, 81, 82, 188, 264, 269, 270, 272, 273
Aphis gossypii 271
Aplyronine A 47, 48
Aplysia kurodai 48
Aplysia sp. 4
Aplysiidae 3
Aplysiomorpha 3
Ascidiacyclamide 14
Asimicin 84, 85, 88, 100, 181, 182, 184, 185, 188, 224, 245, 265, 267, 268, 271, 273
Asimilobin 85, 186
Asimilobine 101
Asiminacin 85, 88, 102, 167, 185
Asimina longifolia 95, 126, 127, 131, 132, 137, 143, 147, 178, 185
Asimina parviflora 95, 100, 131, 144, 156, 157, 162, 185
Asimina triloba 96, 97, 100, 102–104, 106, 107, 112, 127, 137, 145, 152, 153, 157, 162, 165, 173, 185
Asiminecin 85, 103, 185
Asiminenin-A 85, 103, 185
Asiminenin-B 85, 104, 185
Asimin 85, 102, 185
Asitribin 85, 104, 185
Atemoyacin-A 85, 157, 167
Atemoyacin-B 85, 88, 154
Atemoyin 85, 88, 105, 168, 180, 202,
Auristatin PE 64, 67, 70
Auristatins 64
2,2′-Azobis-(2-methylpropionitrile) 28

Bacillus sp. 271
Benoiton procedure 25, 30
Benzene 27
N-(Benzyloxycarbonyl)-L-isoleucine 25
Biological activity 4, 14, 264, 272
Bis(2-oxo-3-oxazolidinyl)phosphinic chloride 35
Bistetrahydrofuran acetogenins 265
Boc-dolaisoleuine 31
Boc-dolaphenine 16–18, 35
Boc-dolaproine 20–24, 31, 43
N-Boc-dolaproine 23
N-Boc-dolaproinyl-dolaphenine 32, 33, 35, 40
Boc-dolaproinyl-(6*R*)-*iso*-dolaphenine 40
Boc-(9*R*, 10*S*, 11*S*)-*iso*-dolaproinyl-dolaphenine 32
Boc-(9*S*, 10*R*, 11*S*)-*iso*-dolaproinyl-dolaphenine 32
Boc-(9*S*, 10*S*, 11*S*)-*iso*-dolaproinyl-dolaphenine 32
Boc-(*S*)-isoleucinol 30
Boc-isostatine 30

Boc-phenylalanine 15, 17
Boc-phenylalaninol 15
Boc-(*S*)-proline 19, 22
Bonellia fulginosa 2
Borane 15, 19, 25
Brain glioma 60
15-O-*p*-Bromophenylurethane 162
Bromotris(dimethylamino)phos-
 phonium hexafluorophosphate 33, 43
Bryostatins 3
Bugula neritina 3
Bullacin 85, 105, 180, 268
Bulladecinone 85, 106, 180
Bullanin 85, 106, 185
Bullatacin 84, 85, 88, 106, 107, 113, 162, 163, 180, 183–185, 222, 225, 226, 244–248, 265, 267, 269–273
(+)-Bullatacin 244–248
Bullatacinone 85, 87, 107, 163, 180, 181, 184, 186, 193, 267, 270, 273
Bullatalicin 85, 89, 107, 109, 113, 170, 180, 184, 185, 187, 267
12,15-*cis*-Bullatalicin 108
C-12,15-*cis*-Bullatalicin 85, 180
Bullatalicinone 85, 87, 109, 139, 180, 267
2,4-*cis*-Bullatalicinone 109
12,15-*cis*-Bullatalicinone 108, 111, 112
C-12,15-*cis*-Bullatalicinone 180
2,4-*trans*-Bullatalicinone 109
Bullatanocin 84, 85, 88, 89, 110, 113, 171, 180, 184, 226
12,15-*cis*-Bullatanocin 111
C-12,15-*cis*-Bullatanocin 85, 180
Bullatanocinone 85, 111, 180
cis-Bullatanocinone 111
2,4-*cis*-Bullatanocinone 109, 112
12,15-*cis*-Bullatanocinone 112
C-12,15-*cis*-Bullatanocinone 85, 180
trans-Bullatanocinone 111
2,4-*trans*-Bullatanocinone 112
Bullatencin 85, 112, 180
Bullatin 85, 112, 186
Butanol 188
tert-Butanol 233
N-*tert*-Butoxycarbonyl-dolaproine 22
N-(*tert*-Butoxycaronyl)-(*S*)-prolinal 19
tert-Butyl acetate 25
tert-Butyl alcohol 13
Butyllithium 22
n-Butyl lithium 236
tert-Butyldimethylsilyl chloride 233
(*S*,*R*)-*tert*-Butyldimethylsilyl-Hiva-
 dolapyrrolidone 47
(*S*,*S*)-*tert*-Butyldimethylsilyl-Hiva-
 dolapyrrolidone 47

Cahn-Ingold-Prelog priority
 order 221, 222
(−)-Camphanoyl acid chloride 252
α-Carbethoxy-γ-butyrolactones 256
N-Carbobenzyloxy-*N*-methyl-
 isoleucinal 27
Carbon tetrachloride 5
3-(12-Carboxydodecanyl)-5-methylfuran-2-
 (5H)-one 166
Carboxylic acid 47
Cathepsin D 18
Cherimolin 183, 270, 271
Cherimolin-1 85, 89, 107, 113, 170, 181
Cherimolin-2 85, 88, 89, 110, 113, 171, 181, 182
Chloroacetate 255
Chloroform 5, 14, 39, 55
Chlorotrimethylsilane 219
Chrome dichloride 246
Cisplatin 270
Citric acid 244
Claisen rearrangement 253
Colchicine 68
Colon carcinoma 60
Corepoxylone 85, 113, 182, 231, 260
Coriacin 85, 113, 126, 181, 229, 230, 263
Coriadienin 85, 114, 181, 229, 230
Corossolin 85, 86, 114, 115, 182, 264, 271
16,19,20,34-*epi*-Corossolin 115
10*RS*-Corossolin 114, 115
(10ξ,15*R*,16*S*,19*S*,20*S*,24*R*)-Corossolin 115
Corossolone 85, 113, 115, 182, 189, 191, 192, 196, 234–236, 260, 261, 264
Cotton effect 107, 109, 155, 163, 165, 168, 174, 220
Crassiflorin 85, 110, 113, 181
Cyclogoniodenin C 85, 116, 117, 130, 131, 262, 263
Cyclogoniodenin T 85, 86, 116, 117, 130, 131, 262, 263
Cyclogonionenin 230
Cysteamine 15
Cyclohexene 25, 35
Cyclohexylisoprophlamine 238

Cyclosporin 271
S-Cysteine methyl ester 15
Cytostatic activity 3, 57, 60, 61
Cytotoxic activity 3, 14, 47, 82, 264, 265, 267–270, 272

α,β-Dehydro-2-aminobutanoic acid 52
Dehydrodolastatin-13 55
Densicomacin-1 86, 117–119
13,14-*erythro*-Densicomacin 86, 117, 118, 181
13,14-*threo*-Densicomacin 86, 118, 119, 127, 181
Densicomacin-2 86, 118, 119, 126, 127
4-Deoxyasimicin 86–88, 139, 180
4-Deoxycoriacin 86, 119, 120, 181
4-Deoxygigantecin 19, 86, 181, 186
Desacetyluvaricin 86, 88, 120, 180, 182, 184, 187, 229
4-Desoxycherimolin-2 86, 89, 171
4-Desoxyhowiicin 86, 88
4-Desoxyhowiicin-B 164, 182
Detoxinine 18
Deuteroacetone 55
Deuteroacetonitrile 19
Deuterochloroform 19
Deuteromethylene chloride 55
Diacetyl-corossolone 115
Dibutylboron triflate 22, 23, 27
α′-Dibutyl-2,5-tetrahydrofuran-dimethanol 213
Dichloromethane 40, 232
Dichloromethane-d_2 50, 52, 54
Didemnins 18, 45
17,18-Diepi-reticulatacin 159
15,16-Diepi-solamin 165
Diepomuricanin 120, 121, 182, 183, 228, 264, 265
Diepomuricanin-A 86, 120–124, 182, 186, 187, 228, 259, 260
Diepomuricanin-B 86, 121, 122, 186
Dieporeticanin-1 86, 121, 122, 124, 183, 186, 187
Dieporeticanin-2 86, 121, 122, 183, 186
Dieporeticenin 86, 122, 183, 228, 229, 259–261
Diepoxymontin 86, 122, 182
Diepoxyrollin 86, 123, 186
Diethylamine 141, 142, 188
Diethylcarbamazine 271
Diethyl carbonate 22
Diethyl D-tartrate 256
Diethyl ether 40
Diethyl isopropylidene D-tartrate 246
Diethyl phosphorocyanidate 13, 33, 35, 43
(+)-Diethyl-tartrate 246
2,2-Dimethoxypropane 252
Dihydroalmunequin 90
2,33-Dihydroannonacin 95
2,35-Dihydrobullatacin 163
Dihydrocherimolin-1 108
Dihydrocherimolin 86, 88, 156, 181
Dihydrocorossolone 115
2,33-Dihydrogoniothalamicin 131
Dihydromotrilin 145
2,33-Dihydrosolamin 165
Dihydrovenezenin 177
2,5-Di-(1-hydroxyheptyl)-tetrahydrofuran 216
(*E*)-1,9-Diiodo-1-nonene 243
Diisopropylethylamine 27
Dimesitoyl-α 213
Dimethylacetamide 256
4-Dimethylaminopyridine 47
Dimethyl sulfoxide 15, 19, 25, 39, 219
Dimethylsulfate 47
Dioxane 13
Dipeptide derivatives 33
2,5-Disubstituted tetrahydrofurans 231
Dodecyl bromide 238, 255
Dodecylmagnesium bromide 232, 233, 252, 255
Dodecylmagnesium
Dolabella andersoni 4
Dolabella auricularia 3–7, 11, 12, 18, 36, 48, 57, 64, 70
Dolabella californica 4
Dolabella ecaudata 4
Dolabellanin A 11
Dolabellanin P 11
Dolabella scapula 4
Dolabella sp. 4
Dolaisoleuine 18, 24, 25, 27, 30, 42
N-*Z*-(3*R*,4*S*,5*S*)-Dolaisoleuine 29
Dolaisoleuine *tert*-butyl ester hydrochloride 26, 33
Dolaphenine 14, 15, 17, 18, 35, 39, 64
R-Dolaphenine 36
S-Dolaphenine 36

Z-Dolaphenine 35
(6*R*)-*iso*-Dolaphenine 40
Dolaproine 18, 19, 22, 23, 27, 32, 39, 42
Dolaproinyl-dolaphenine 32–34
Dolapyrrolidone 56
(*S*)-Dolapyrrolidone 45
Dolastatin 1 5, 6
Dolastatin 2 5, 6
Dolastatin 3 6, 9–14, 55
(–)-Dolastatin 3 6, 9, 13, 14
Dolastatin 4 6
Dolastatin 5 6
Dolastatin 6 6
Dolastatin 7 6
Dolastatin 8 6
Dolastatin 9 6
Dolastatin 10 7–11, 14–15, 17–19, 21, 23, 25, 27, 32–39, 43–45, 48–51, 55, 57–70
Dolastatin 11 7, 8, 10, 11, 48, 51, 57, 60, 63, 67, 69, 70
Dolastatin 12 7, 8, 10, 11, 48, 51, 57, 60, 67,
Dolastatin 13 7–11, 48, 52–55, 57, 60, 67
Dolastatin 14 7, 8, 10, 11, 48, 54, 55, 57, 60, 63, 67
Dolastatin 15 7, 8, 10, 11, 45–48, 51, 55–57, 60, 61, 63, 67, 69, 70
Dolastatin C 11, 47
Dolastatin D 11, 48
Dolastatin H 11
Dolastatins 3, 4
Dolatrienoic acid 55
Dolatriol 5, 11
Dolatriol 6-acetate 5
Dolavaline 43, 55
Dolavalyl-valyl-dolaisoleuine 32–34
Dolavalyl-valyl-dolaisoleuine *tert*-butyl ester 36
Doliculide 11, 47, 48
Doliculol A 11
Doliculol B 11
Drosophila sp. 271
Dysidamide 45
Dysidea sp. 45
Dysidin 45

Epilachna varvestis 271
Epomuricenin 121, 264, 265
Epomuricenin-A 86, 123, 124, 182, 228, 259, 260
Epomuricenin-B 86, 123, 182
Epoxy-gigantetronenin 230
21/22-Epoxy-gigantetronenin 128
Epoxymurin-A 86, 123, 124
Epoxymurin-B 86, 123, 124, 182
Epoxyrollin-A 86, 121, 124, 187
Epoxyrollin-B 86, 120, 124, 187
Erythroleukemia 61
Escherichia sp. 271
Ethanol 4, 5, 7, 13
Ethyl acetate 5–9, 21, 25
N,*N*′-Ethylenebis(iodoacetamide) 68
(4*S*)-Ethyl-4-hydroxy-pentyl-2-oate 241
Evans' method 22

Filaricidal activity 270
Formal acetal 173
Formaldehyde 219
α-(ω-Formyldecyl)-α,β-Angelica lactone 155
α-(ω-Formyldecyl)-χ, β-α-Acetonyl-γ-lactones 231

Ghedoic acid 227
Giganenin 86, 124, 186
Giganin 86, 125, 186
Gigantecin 86, 114, 125, 181, 186, 208, 230, 261–263, 268
18/21-*cis*-Gigantecin 128
18/21-*trans*-Gigantecin 128
Gigantetrocin 126, 127, 186, 188, 192, 267
Gigantetrocin-A 86, 118, 119, 126, 127, 181, 182, 184, 185, 268, 269
Gigantetrocin-B 86, 118, 127, 182, 185, 268, 269
Gigantetrocinone 86, 87, 127, 185, 186
Gigantetronenin 86, 126, 128, 181, 182, 186, 188, 230, 261, 262
Gigantriocin 86, 128, 186
Gigantrionenin 86, 129, 186
Glaucanisin 86, 129, 182
Glutamic acid 232
D-Glutamic acid 232
L-Glutamic acid 216, 231–234
Gly-γ-carboxyl-γ-aldehyde 52
Glycerol 97, 102, 106, 117, 126, 128, 131, 133–137, 139, 178, 180
Goniocin 86, 116, 130, 180, 201, 262
Goniodenin 86, 116, 130, 186, 201, 262, 263
Gonionenin 86, 117, 131, 186, 262, 263

Goniothalamicin 86, 131, 181, 182, 185, 186, 268, 269, 271
Goniothalamicinone 86, 132, 185, 268, 269
Goniothalamus giganteus 92, 95, 101, 119, 124–126, 128–131, 186, 229, 262
Goniothalamus howii 114, 126–128, 131, 147, 152, 186
Grignard addition 248, 250
Grignard reaction 255
Grignard reagent 248, 250

Hantzsch condensation 15
Hantzsch method 17
Hantzsch reaction 18
Hexane 5–7, 10, 40
5-Heptyl-2-(1-hydroxyheptyl)-tetrahydrofuran 216
Howiicin-A 84, 86, 95, 182, 186
Howiicin-B 86, 88, 152, 182, 186
Howiicin-C 85, 86, 114, 186
Howiicin-D 86, 128, 186
Howiicin-E 86, 87, 147, 186
Howiicin-F 86, 118, 126, 182, 186
Howiicin-G 86, 126, 127, 182, 186
Human breast carcinoma 268
Human Burkitt lymphoma 69
Human cancer 2, 60, 65, 66
Human cervical carcinoma 273
Human chronic B-leukemia 61
Human colon adenocarcinoma 268
Human leukemia 60
Human lung carcinoma 268
Human lymphoma 60
Human melanoma LOX 63
Human ovarian carcinoma A2780 270
Human ovary cancer OVCAR-3 63
Human pancreatic carcinoma 61
Hydrobromic acid 18
Hydrogen bromide 35
Hydrogen chloride 25
Hydrogen peroxide 23
Hydrogen peroxide 29
4-Hydroxyacetogenins 188
8-Hydroxyannonacin 87, 132, 181
30-Hydroxybullatacin 87, 133, 180
30*R*-Hydroxybullatacin 133
30*S*-Hydroxybullatacin 133
31-Hydroxybullatacin 87, 133, 180, 267
32-Hydroxybullatacin 87, 134, 180, 267
10-Hydroxybullatacinone 87, 134, 180, 267
12-Hydroxybullatacinone 87, 134, 180, 267
28-Hydroxybullatacinone 87, 135, 181
29-Hydroxybullatacinone 87, 135, 181
30-Hydroxybullatacinone 87, 136, 181
31-Hydroxybullatacinone 87, 136, 181
32-Hydroxybullatacinone 87, 137, 181
Hydroxy-bullatacinones 206
Hydroxystannanes 252
4-Hydroxy-25-desoxyneorollinicin 87, 88, 161, 164, 187
4-Hydroxy-25-desoxyrollinicin 87
14-Hydroxy-25-desoxyrollinicin 262, 183
2-Hydroxyisovaleric acid 45, 56, 96
2-Hydroxy-3-methylpentanoic acid 51

Imidazole 233, 239, 243, 244, 248, 249, 253, 256
Immunosuppressive activity 82, 264, 271
Insecticidal activity 272
Iodoform 246
Iodomethane 23
Isoacetogenins 84, 109, 193, 197, 265, 270
Isoannonacin 87, 137, 181, 183, 185, 186, 265, 267, 269
Isoannonacin-A 84, 87, 96, 97
Isoannonacin-10-one 181, 183
Isoannonacinone 87, 88, 138, 141, 156, 267
Isoannoreticuin 87, 138, 183
Isobutane 206
Isobutyric acid 51
Isocherimolin-1 85, 87, 109, 139, 181, 196
Isodensicomacin-1 118, 119
Isodensicomacin-2 118, 119
Isodensicomacin 86, 87
Isodesacetyluvaricin 86–88, 139, *174*, 180, 187, 229, 261, 262
2′-(*S*)-Isodolaproine 21
Isodolastatin 10 19, 41, 64
Isodolastatins 63
(6*R*)-Isodolastatin 10 36–41, 43
(19*aR*)-Isodolastatin 10 63
(6*R*)-Isodolastatin 10[149] 36–41, 43, 63
cis-(6*R*)-Isodolastatin 10 42, 43
trans-(6*R*)-Isodolastatin 10 42, 43
L-Isoleucine 30
Isomolvizarin-1 87, 140, 139, 181
Isomolvizarin-2 87, 140, 181

Isomurisolin 87, 140, 152
Isoneoannonacinone 87, 141, 183
Isopropenyl chloroformate 47
Isorolliniastatin-1 87, 88, 141, 163, 270
Isorolliniastatin-2 85, 87, 107, 272
Isorollinicin 87, 141, 187
(3*S*, 4*S*, 5*S*)-Isostatine 18
Isosylvaticin 87, 142, 172
2,4-*cis*-Isosylvaticin 172
2,4-*trans*-Isosylvaticin 172
Itrabin 87, 142, 181, 195
Ivermectine 271

Jasplakinolide 11
Jetein 87, 143, 195, 281
Jones oxidation 166, 202, 248

Kedde reagent 188, 190
Krebs-2 ascites tumor 2

Leishmania donovani 270
Leishmania major 270
Lacceroic acid 227
(*S*)-Lactic acid 120, 177, 220
Laherradurin 87, 143, 181, 195
Larvicidal activity 271
L-Leucine 30
Leukemia L1210 67, 69, 269
Ligroin 5
Lindlar hydrogenation 238–240
Linoleic acid 227
1-Lithio-1-pentyne 243, 244
Lithium 190, 205
Lithium acetate 255
Lithium acetylide 258
Lithium amide 238
Lithium diisopropylamide 25
Lithium dinonyl cuprate 241
Lithium enolate 31
Lithium ethoxyacetylide 257
Lithium hydroxide 23, 29
Lithium trimethylsilyl-acetylide 243, 245
Loliolide 5, 11
Longicin 87, 144, 185, 268
2,4-*cis*-Longicinone 144
2,4-*trans*-Longicinone 144
Lu-103793 70
Lung cancer NCI-H522 63
Lung large cell carcinoma 60
Lymphocytic leukemia 3
Lymphocytic leukemia 3PS 264
Lymphocytic leukemia L1210 62
Lymphocytic leukemia P388 4, 5, 62, 63, 65, 66
Lymphoid (L1210) leukemia 3

Magnesium bromide 19
Magnesium chloride hexahydrate 256
Magnioliales 227
Malyngamide 45
Mammary cancer 269
Manganese dioxide 16, 17, 39, 250
Maytansine 68
Melanoma 60
Melanoma B16 62
Meldrum's ester 47
Membranacin 87, 144, 186
Methanesulfonamide 236
Methanol 5, 6, 14, 25, 35, 36, 47, 188, 194, 212
2-Methyl-3-aminopentanoic acid 51
N-Methyl-*N*-(benzyloxycarbonyl)-(*S*,*S*)-isoleucinal 25
Methyl 12-bromododecanoate 250
Methylene chloride 5, 6, 14, 27, 39
Methylene chloride-d_2 13
2-*O*-Methylglyceric acid 52
N-Methylisoleucine derivatives 25
(*S*)-Methyl lactaldehyde 248
(*R*)-γ-Methyl-γ-lactone 232
(*S*)-γ-Methyl-γ-lactone 232
γ-Methyl-γ-lactones 231
(*S*)-Methyl mandelate 19
N-Methyl-phenylalanine 52
(2*R*,4*S*)-4-Methyl-2-phenylthio-γ-butyrolactone 236
(3*S*,4*S*)-4-Methyl-2-phenylthio-γ-butyrolactone 236
2-Methylpropene 29
N-Methylvaline 55
Methyl undecenoate 234, 235
Mitosis 68, 69
Mitsunobu reaction 17, 245, 256
Molinema dessetae 270, 271
Molvizarin 87, 144, 180, 181, 183, 184, 185, 270
Monoacyl-2-glycerides 227
Monocytic leukemia 61
Montanacin 87, 145, 182
Morpholine 243, 244

Mosher method 90, 145, 166, 167, 221, 222
Motrilin 84, 87, 98, 145, 181, 184, 186
Mucocin 87, 146, 187, 202
Muricadienin 228
Muricatacin 87, 95, 96, 146, 183, 201, 233, 238, 255
(−)-Muricatacin 251, 254–258, 265
(−)-(4*S*,5*S*) Muricatacin 96
(+)-Muricatacin 232, 251–254, 256, 257, 265
(−)-(4*R*, 5*R*)-Muricatacin 96
(4*S*, 5*R*)-Muricatacin 96
(+)-(4*S*, 5*S*)-Muricatacin 96
Muricatacin-A 87, 149, 183
Muricatacin-B 87, 150, 183
Muricatacin-C 87, 150, 183
Muricatalin 87, 146, 183
Muricatetrocin-A 86, 87, 147, 148, 183, 185
Muricatetrocin-B 87, 147, 148, 183, 185, 187
Muricatin-A 87, 148, 183
Muricatin-B 87, 149, 183
Muricatin-C 87, 149, 183
Muricatocin-A 87, 149, 183
Muricatocin-B 87, 150, 183
Muricatocin-C 87, 150, 183
Murihexocin-A 88, 151, 183
Murihexocin-B 88, 151, 183
Murisolin 86, 88, 140, 152, 153, 183, 186, 196, 231, 234, 264, 271
16,19-*cis*-Murisolin 88, 153, 186
Murisolin-A 88, 152, 186
Myeloid leukemia 61

Narumicin-I 88, 153, 154, 187
Narumicin-II 88, 153, 154, 187
Neoannonacin-10-one 88
Neoannonacin-B 88
Neoannonacinone 154, 183
Neoannonin 85, 88, 154, 155, 168, 180, 184, 271
Neoannonin-B 154, 184
Neodesacetyluvaricin 86, 88, 120, 184
Neoisoannonacinone 88, 155, 183
Neoreticulatacin 88, 184
Neoreticulatacin-A 159
(1*S*,2*R*)-Norephedrine 22
(1*S*,3*R*)-Norephedrine 22
Neurospora crassa 272
Nitric oxide 206
Nitrobenzylalcohol 101, 105, 139, 145, 155, 167, 168
Nudibranchia 3

Oleic acid 227
Oligotetrahydrofuran 227
Optical activity 15
Otivarin 86, 88, 156, 181, 195, 196, 271
Ovarian adenocarcinoma 60
Ovarian teratocarcinoma 269
Ovicidal activity 271
Oxalyl chloride *232*
Oxazolidinones 27
Oxone™ 244
5-Oxoundecanoic acid 166
ent-4-Oxo-2,33-dihydrosolamin 231, 234

Palladium-on-carbon 25, 35
Palmitic acid 227
Panalicin 88, 156, 187
Parikh-Doering method 15, 19
Parvifloracin 88, 156, 185
Parviflorin 85, 88, 157, 167, 181, 185, 186
Patellamide A 14
Patellamide B 14
Patellamide C 14
Payne rearrangement 240
Pentadecanoic acid 118, 119
8-Pentadecanol 174
Pentafluorophenol 13
Pepsin 18
Pepstatin 18
Perchloric acid 178, 259–262
Pesticidal activity 82, 264, 271, 272
Phenethylamine 68
Phenylacetic acid 17
Phenylalanine 51, 56
Phenylmagnesium bromide 19
Phenylselenium chloride 250
Phomopsin A 68
Pivaloyl chloride 33
Plagionicin-A 88, 157, 186
Plasmodium falciparum 270
Polyalthia plagioneura 157, 186
Polyhydroxy-*bis*-tetrahydrofuran 165
Polyketides 81

Potassium bicarbonate 18
Potassium *tert*-butoxide 21
Potassium carbonate 19, 47, 245
Potassium hydroxide 188
Pre B-cell leukemia 61
(*S*)-Prolinal 19, 21
Proline 18
(*R*)-Proline 19
(*S*)-Proline 21
Promyelocytic leukemia HL-60 62
2-Propanol 6, 38
Propionyl chloride 22
(*R*)-Propylene oxide 243, 244
(*S*)-Propylene oxide 249–251
Proteus sp. 271
Pukeleimide 45
Purpureacin-1 85, 88, 110, 113, 183
Purpureacin-2 88, 158, 183
Pyridine-d_5 54
Pyridine 15, 19, 25, 232, 233, 236, 238, 239, 241, 244, 249, 252
4-Pyrrolidinopyridine 13

Ramberg-Backlund olefination 239, 240
Renal carcinoma 60
Renin 18
Reticulacinone 88, 158, 183
Reticulatacin 88, 89, 122, 158, 175, 176, 183, 223, 226, 237–239, 259, 260
Reticulatain-1 88, 159, 180, 183, 189
Reticulatain-2 88, 160, 183, 189
Reticulatamol 88, 160, 183, 201, 228, 250, 251
15-(*SR*)-Reticulatamol 160
Reticulatamone 88, 160, 183, 201, 228, 250, 251
(+)-Reticulatanon 248, 249, 251
Rhizoxin 68
Rioclarin 88, 161, 186
Rollinia membranacea 120–123, 144, 161, 163, 165, 172, 174, 186, 229
Rollinia mucosa 107, 146, 147, 161, 162, 172, 187
Rollinia papillionella 141, 161, 163, 165, 187
Rollinia sericea 161, 187
Rolliniastatin 84
Rolliniastatin-1 87, 88, 141, 161, 163, 164, 183, 186, 187, 189, 212, 250, 269, 270
Rolliniastatin-2 84, 85, 87, 88, 107, 113, 162, 163, 180–184, 187, 189, 225, 244–246, 269, 272
(+)-Rolliniastatin-2 163
(−)-Rolliniastatin-2 163
Rollinia sylvatica 172, 187
Rollinia ulei 120, 121, 124, 172, 174, 187, 270
Rollinicin 88, 187
Rollinone 88, 163, 186, 187
Rubrenolide 220

Sacoglossa 3
Sarcoma-180 2
Scandium perchlorate 233
L-Selectride 232, 233
Senegalene 88, 164, 184, 270
Sephadex LH-20 5–7, 9, 10
Sharpless asymmetric dihydroxylation 231
Sharpless asymmetric epoxidation 254, 255, 258
Sharpless epoxidation 231, 239, 242
Sharpless method 237
Sharpless procedure 236, 238
Silica gel 5–10, 13, 111, 188, 232
Sodium amalgam 248
Sodium bicarbonate 18
Sodium *tert*-butoxide 233
Sodium 1-hexanesulfonate 38
Sodium hydride 23
Sodium iodide 232, 234
Sodium methanolate 255
Sodium periodate 118, 119, 232
Sodium sulfite 29
Solamin 86, 88, 100, 121, 123, 164, 165, 182–184, 228, 231, 234, 236–240, 264
Solamin diacetate 259
Squamocin 84, 88, 141, 145, 165–167, 170, 180–182, 184, 186, 187, 212, 213, 220, 269–271
Squamocin-A 84, 88
Squamocin-B 88, 166, 184
Squamocin-C 87, 145, 184
Squamocin-D 85, 88, 102, 167, 184
Squamocin-E 85, 88, 157, 167, 184
Squamocin-F 88, 167, 184
Squamocin-G 88, 162, 184
Squamocin-H 85, 88, 100, 184
Squamocin-I 88, 155, 167, 168, 184
Squamocin-J 85, 88, 154, 155, 168, 184

Squamocin-K 85, 88, 105, 168, 184
Squamocin-L 86, 88, 120, 185
Squamocin-M 86, 87, 88, 139, 185
Squamocin-N 88, 168, 185
Squamocin-28-one 169
Squamocinone 89, 169, 187
Squamone 89, 99, 169, 184, 185
Squamosinin-A 89, 169, 185
Squamostanal-A 89, 166, 170, 185, 201
Squamostatin-A 84, 85, 89, 90, 170, 184, 185
Squamostatin-B 85, 89, 107, 113, 170, 185
Squamostatin-B′ 89
Squamostatin-C 85, 89, 110, 113, 171, 185
Squamostatin-D 89, 170, 185
Squamostatin-E 86, 89, 171, 182, 185
Squamosten-A 89, 171, 185
Squamous cell carcinoma *62*
Staphylococcus sp. 271
Statine 18
Streptomyces depsipeptides 18
Streptomyces sp. 45
Sulphomolybdic reagent 188
Sulfuric acid 29
Sulfur trioxide 15, 19, 25
Swern method 19
Swern oxidation 232, 234, 236, 248, 250
Sylvaticin 89, 142, 172, 183, 187, 271
12,15-*cis*-Sylvaticin 89, 172, 187

D-Tartaric acid 234–236
Tetraacetylannonacin 271
Tetrabutylammonium fluoride 234
Tetrahydrocoriacin 114
Tetrahydrofuran 15, 19, 25, 47, 83, 227
Tetrahydrofuran acetogenins 208
(*S*)-O-Tetrahydropyranyl lactal 234
Tetrahydropyrrolo[1,2-*c*]oxazole derivative 23
Tetrahydrosquamone 169
2,35,29,30-Tetrahydrosenegalene 164
2,35,29,30-Tetrahydrotetra-acetylsenegalene 164
Tetraodon stellatus 2
Tetrodotoxin 2
Thiazole-4-carboxylic acids 15
Thiazole-4-carboxylic esters 15
Thiazoline 15, 52
Toluene 47, 233–235, 238, 242, 247, 250, 251
Toluenesulfonic acid 116
Tosylic acid 130
Tosyl chloride 234
Toxic activity 6
Tri-*n*-butyltin hydride 232, 234
Tributyltin hydride 27, 232
Trichophyton sp. 271
Trieporeticanin 89, 122, 174, 184, 260, 261
Triethylamine 13, 15, 22, 33
Triethylphosphonoacetate 254
Trifluoroacetate 36
Trifluoroacetate derivatives 35
Trifluoroacetic acid 13, 19, 33, 35, 40, 43
Trifluoroacetic acid anhydride 18
Trilobacin 89, 173, 186
Trilobacin-4,24-diacetate 173
Trilobin 89, 173, 186
Trimethyloxonium tetrafluoroborate 19, 23, 25, 28
Trimethylsilyl chloride 233
Trimethylsilyl cyanide 233
Tripoxyrollin 89, 122, 174, 187, 228, 229, 231, 261, 262
Trypanosoma brucei 270
Tubulin 63, 68, 69
Tunichlorin 12

Uleicin-A 89, 187
Uleicin-B 89, 187
Uleicin-C 89, 172, 174, 187
Uleicin-D 89, 187
Uleicin-E 89, 187
Uleirollin 89, 174, 187
Ulicyclamide 14
Ulithiacyclamide 14
Undecanoic acid 171
6-Undecanol 114
Uvaria acuminata 81, 120, 177, 187, 269
Uvaria grandiflora 139
Uvaria grandiflora 187
Uvaria hookeri 139, 153, 154, 158, 165, 169, 175, 187
Uvariamicin-I 89, 158, 175, 176, 181, 187
Uvariamicin-II 88, 89, 158, 175, 176, 180, 181
Uvariamicin-III 89, 175, 176, 180, 181, 184, 187
Uvariamicin-IV 89, 158, 175, 176, 181

Uvaria narum 139, 153, 154, 156, 158, 165, 169, 175, 187
Uvaricin 81, 89, 120, 142, 143, 177, 187, 220, 226, 241, 264, 269
Uvaricinone 177

Valine 52, 55
Z-L-Valine 33
Venezenin 89, 92, 93, 177, 188, 267, 268, 269
Vicinal diols 206, 218
Vinblastine 68, 69
Vincristine 68
Vinylmagnesium bromide 232, 250
Voriacin 262

Walker 256 carcinoma 3
Weiler dianion alkylation 241
Wilkinson's catalyst 240, 243, 245
Wittig homologation 234, 236, 254
Wittig–Horner homologation 246
Wittig reaction 244
Wityak's procedure 257

Xylomatenin 84, 89, 91, 178, 188
Xylomaticin 89, 178, 185, 188
Xylopia aromatica 92, 95, 100, 126, 128, 177, 178, 179, 188
Xylopiacin 89, 179, 188
Xylopianin 89, 179, 188
Xylopien 89, 179, 188

MIX
Papier aus verantwortungsvollen Quellen
Paper from responsible sources
FSC® C105338

If you have any concerns about our products,
you can contact us on
ProductSafety@springernature.com

In case Publisher is established outside the EU,
the EU authorized representative is:
Springer Nature Customer Service Center GmbH
Europaplatz 3, 69115 Heidelberg, Germany

Printed by Libri Plureos GmbH
in Hamburg, Germany